Manufacturing Technology

Volume 1

Third Edition

R. L. Timings

Prentice
Hall

An imprint of **Pearson Education**

Harlow, England · London · New York · Reading, Massachusetts · San Francisco
Toronto · Don Mills, Ontario · Sydney · Tokyo · Singapore · Hong Kong · Seoul
Taipei · Cape Town · Madrid · Mexico City · Amsterdam · Munich · Paris · Milan

Pearson Education Limited
Edinburgh Gate
Harlow
Essex CM20 2JE
England

and Associated Companies throughout the world

Visit us on the World Wide Web at:
http://www.pearsoneduc.com

First published 1990
Second edition 1992
Third edition 1998

British Library Cataloguing in Publication Data
A catalogue entry for this title is available from the British Library

ISBN 0-582-35693-8

10 9 8 7 6 5
06 05 04 03 02

Set by 32 in Times $9\frac{1}{2}/12$ and Frutiger
Produced by Pearson Education Asia Pte Ltd
Printed in Singapore (MPM)

Manufacturing Technology

Volume 1

Contents

Preface

Manufacturing Technology, Volume 1 was originally written to provide a comprehensive text on the technology and organisation of manufacture to satisfy the requirements of the Business and Technician Education Council (BTEC) standard units for students studying for an engineering qualification at the OND/ONC (N level). In 1992 a second edition was published in response to the many letters and suggestions received, which requested coverage of many additional topic areas.

Whilst still satisfying the original aims, this new edition has been written to include additional material in order to give comprehensive coverage of the further topic areas that form part of the requirements for the Advanced GNVQ Engineering qualification. It reflects changes in technology and provides an introduction to BS 5750/ISO 9000. At the same time, the opportunity has been taken to:

- Make the language less formal and more reader friendly.
- Expand some of the explanations to improve their clarity.
- Correct some residual errors and omissions.
- Modernise the general format of the book.
- Introduce self-assessment exercises at key points in the text.
- Make more use of bulleted summaries.

Each chapter is now prefaced by a summary of the topic areas to be covered and finishes with a selection of practice exercises.

This book leads naturally into *Manufacturing Technology*, Volume 2, and is intended for students studying at the HNC/HND level and beyond. The broad coverage of Volumes 1 and 2 ensures they not only satisfy the requirements of technician engineers up to the highest level, but also provide an excellent technical background for undergraduates studying for a mechanical engineering, manufacturing engineering or combined engineering degree.

R. L. Timings
1998

Acknowledgements

The author and publishers are grateful to the following for permission to reproduce copyright material:

Kenrick Hardware Ltd for our Fig. 1.10; Davy-Loewy Ltd for our Fig. 1.31; Sir James Farmer-Norton & Co. (International) Ltd for our Fig. 1.33; T. Norton & Co. Ltd for our Fig. 1.40; Sweeney & Blocksidge (Power Presses) Ltd for our Fig. 1.42; A. A. Jones and Shipman Ltd for our Figs 1.55, 3.48 & 10.18; Sandvik Coromant UK Ltd for our Fig. 2.15; Cincinnati Milacron Ltd for our Fig. 2.27; Colchester Lathe Co. for our Figs 4.5 & 12.9(d); PMG Ball Screws Ltd for our Fig. 4.23; Rushworth and Co. (Sowerby Bridge) Ltd for our Figs 7.4 & 7.7; Tucker Fasteners Ltd for our Fig. 7.16; Moore and Wright Ltd (James Neill Group) for our Figs 10.2 & 10.3; Mitutoyo (UK) Ltd for our Fig. 10.9; Wyndley Bros Ltd for our Fig. 10.17; Rank Taylor Hobson Ltd for our Figs 10.36 & 10.37; Training Publications Ltd for our Figs 12.2, 12.3 & 12.7; Silvaflame Ltd for our Fig. 12.9(c).

How to use this book

There are many ways of using a text book. You can read through it from cover to cover and try to remember all that you have read. This is rarely successful unless you have a photographic memory, and few of us have. You may consider using it just as a reference book by looking up individual topics as and when you need that specific information. This can be useful as a reminder once you know the subject thoroughly but until then it can lead to misconceived ideas.

So, here are a few thoughts on how to maximise the benefit that you can get from this book.

This book consists of a number of chapters. Each chapter covers a major syllabus area. For example, **Chapter 1** covers alteration of shape. It is divided up into sections. If you turn back to the Contents page you will find that:

- **Section 1.1** deals with factors affecting the selection of manufacturing processes.

The remaining sections in Chapter 1 look at some important manufacturing processes in greater detail. These are divided between those in which alteration of shape occurs with *no loss of volume* (casting, moulding and flow forming) and those in which alteration of shape occurs with *loss of volume* (cutting). For example:

- **Section 1.2** deals with sand casting processes where there is little or no loss of volume.
- **Section 1.7** deals with the flow forming of metal in processes such as forging and extrusion. Again there is little or no loss of volume.
- **Section 1.8** deals with machining where material is cut away to provide the required shape. Thus there is loss of volume in this instance, and so on.

Sometimes it is necessary to divide these sections up further. For example, Section 1.8 subdivides into:

- **Section 1.8.1** which deals with drilling processes.
- **Section 1.8.2** which deals with turning processes (centre lathe work).
- **Section 1.8.3** which deals with milling processes, and so on.

I hope that the technical jargon has not put you off. To obtain the maximum benefit from this book (this may seem obvious!) start at the beginning. The first chapter lays the foundation for all that is to follow. As with all the chapters, this chapter is prefaced with a list of the main topic areas you are going to find in it.

Unlike the companion volume *Engineering Materials*, Volume 1 (2nd edition) – where after the first two introductory chapters you can immediately move to the sections on ferrous metals, non-ferrous metals or plastics – in this book there is a *progressive build up* of knowledge from the first chapter to the last. If you wish to work in some other order and be selective in your reading, then this should only be done in consultation with your tutor.

As you work through the chapters, you will come across **self-assessment tasks** at key points. If you have understood what you have read previously, you should be able to complete these exercises *without looking back* at the text. If you have to look back, then you are not yet sure of your ground and there is no point in moving on. Try again and, if you are still not sure, have a chat with your tutor to clear up your difficulty.

At the end of each chapter there is a selection of more extended **exercises**. Your tutor will guide you as to their use and check your responses to ensure that you have the background knowledge required to understand the next topic area to be studied.

No matter whether you work through the book from beginning to end systematically, or whether you work through the foundation chapter and then move on to the specialist areas you require, you must complete the self-assessment tasks and the end-of-chapter exercises as you come to them. This will ensure that you will understand the next stage of your journey through your study of manufacturing technology.

Although you may be tempted to take short cuts, it is recommended that you work through the book from beginning to end systematically. This is because all manufacturing engineers should have a general, background knowledge of the processes described before moving on to the more specialised processes described in *Manufacturing Technology*, Volume 2.

Finally, once you have qualified, you can still obtain benefit from this book. The numbering of the sections and subsections, the comprehensive list of contents and the extended index, will all help these texts (including the companion volumes: *Manufacturing Technology*, Volume 2 and *Engineering Materials*, Volumes 1 and 2) to act as quick reference books during your future career in engineering.

1 Alteration of shape

The topic areas covered in this chapter are:

- Factors affecting the selection of a manufacturing process.
- Casting: shaping with no loss of volume.
- Die casting: shaping with no loss of volume.
- Compression moulding of plastics: shaping with no loss of volume.
- Injection moulding and extrusion of plastics: shaping with no loss of volume.
- GRP, vacuum and blow moulding: shaping with no loss of volume.
- Flow forming of metals: shaping with no loss of volume.
- Machining: shaping with loss of volume.
- Shearing: shaping with loss of volume.

1.1 Factors affecting the selection of manufacturing processes

There are many factors which affect the selection of a suitable manufacturing process for a component or an assembly of components. Some of these factors are commercial, such as the unit cost of production and whether the existing plant is to be used, or whether there is to be investment in new and more economical plant. Other factors are of a more technical nature. Let's now take a look at these factors.

1.1.1 *Design specification*

This is fundamental to the choice of manufacturing processes to be used as it specifies:

- The shape of the component.
- The material from which the component is going to be made.
- The quality of the component. That is, the dimensional accuracy, the geometrical accuracy and the surface finish. *Its fitness for purpose.*
- Any heat treatment processes.
- Any corrosion-resistant, wear-resistant or decorative finishing processes.

1.1.2 *Geometry*

Cylindrical, conical and plain surfaces, where they form shaft ends, shoulders, etc., are most easily produced on a lathe. Screw threads may also be produced on a lathe. However, mutually parallel and perpendicular plane surfaces and slots are generally produced by shaping, planing, milling or surface broaching. More complex components may need to be cast, forged or moulded to shape before machining. Thus the shape of a component will dictate the processes which must be used to produce that component.

1.1.3 *Quality*

Quality control has become increasingly important in modern manufacturing industry. Until comparatively recently it was only necessary to specify the linear dimensional tolerances on a component and the appropriate machining process was expected to provide suitable geometrical accuracy and surface finish. This is no longer the case, and as well as dimensional tolerances, the component drawing may also specify geometrical tolerances and surface roughness (finish). The manufacturing process will need to be carefully chosen not only to enable the quality specification to be met quickly and cheaply, but to achieve a level of repeatability which is economical. That is, scrap must be kept to a minimum. Generally, it can be assumed that the closer the tolerances, the greater the unit cost of production.

Modern production techniques involving automation and the computer control of processing (including automatic gauging built into the machining process) have substantially reduced the production costs for high-quality components. The relationship between linear dimensional tolerance and typical manufacturing processes is considered in Section 6.6, and the relationship between surface finish and typical manufacturing processes is considered in Section 10.9.

At one time, quality was purely a function of the inspection department. Nowadays it is a total corporate philosophy with everyone in the company involved. A customer may specify that the quality of the goods and services that they are purchasing shall be under the control of a management regime complying with BS 5750/ISO 9000. Without this approval much business will be lost. Therefore, the sooner you are introduced to the concept of *total quality control* (TQC), the better. It will be considered in Chapter 10 of this book, and in even greater detail in *Manufacturing Technology*, Volume. 2.

1.1.4 *Quantity*

The quantity, or batch size, dictates the economics of process selection. Single components are generally produced (expensively) by hand, or by manually controlled machining processes. For example, the cost of producing a single component during the restoration of a veteran motor car is very much greater than the original cost when the same component was manufactured in batches. Generally, the cost falls as the batch size increases and more productive techniques can be employed. For example, it is quicker (and cheaper) to drill all the holes in a component using a drill jig than it is to mark out and drill each hole individually. However, when a drill jig is used, its cost has to be allowed for in the overall cost of the component. To determine whether the use of a jig is economical, a break-even diagram can be used.

Figure 1.1 shows an example of a simple break-even diagram which is based on the following data:

Fig. 1.1 *Break-even diagram*

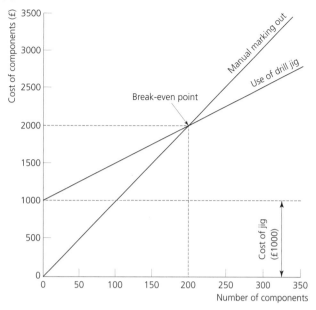

- When a batch of components are made by individually marking out the holes and then drilling them, the cost of material is £4.00 for each component, and the labour cost is £6.00 for each component. Therefore the total cost is £10.00 for each component.
- When a batch of the same components are made using a drill jig to position the drilled holes, the material cost is again £4.00 for each component, but the labour cost is reduced to £1.00 for each component. Therefore the total cost is now only £5.00 for each component.

However, in this second example, £1000.00 for the cost of the jig has to be allowed for and this is a fixed cost irrespective of the number of components made. Plotting these two sets of data on the same axes produces the results shown in Fig. 1.1.

Notice that the lines representing batch size plotted against cost cross each other for a batch of 200 components. This is the *break-even point* and the production cost is the same for both methods.

For smaller quantities (e.g. 50 components), the cost of production by the first method would be £500.00 for the batch, whilst production by the second method would cost £1250.00. Thus for quantities of less than 200 components the cost and use of a jig is not justified.

For batches larger than 200 components (e.g. 500 components) the cost of production by the first method would be £5000.00 for the batch, whereas the cost of production by the second method would be £3500.00. Thus for batches larger than 200 components the jig is justified on grounds of cost. However, the jig might be justified for smaller batches on grounds of accuracy and repeatability of quality.

This is a very simple example and ignores such factors as overhead expenses and depreciation of the jig. Further and more comprehensive examples of the use of break-even diagrams for cost analysis are considered in *Manufacturing Technology*, Volume 2.

1.1.5 *Material*

The material specified by the designer also influences the method of production. For example, a thermoplastic material would usually be injection moulded to shape, whereas a thermosetting plastic material would be pressure moulded in a press (Section 1.5).

A zinc-based alloy such as Mazak would be ideal for die casting but unsuitable for sand casting. However, cast iron would be ideal for sand casting but unsuitable for die casting because of its high melting temperature.

Even within a selected process the material can influence the tooling chosen. For example, turned components in free-cutting low-carbon steels can be economically manufactured using high-speed steel tools. The same components turned from a high-tensile alloy steel would require more costly carbide-tipped tools. However, modern practice favours the use of carbide and coated carbide-tipped tools because of the greater rates of material removal that can be achieved. Also, with identical disposable carbide inserts, greater consistency of performance can be achieved than when using tooling that needs to be resharpened from time to time by the setter or operator. To change an insert is quicker and cheaper than regrinding, and it reduces the downtime of the machine with associated loss of production.

1.1.6 *Machine or equipment process capacity*

Figure 1.2(a) shows a simple component which can be stamped out of strip metal using a power press. The second operation would punch the holes in the blanks. The production cost for each component can be substantially reduced by combining the two operations in a progression tool as shown in Fig. 1.2(b).

This tool produces the pierced blanks in a single operation. The cost of the more complex progression tool will be approximately the same or even slightly less than the cost of two separate tools. The decision as to which method of production is adopted will depend upon the capacity of the press available. If the press is capable of producing enough force to enable the progression tool to be used, then this is the obvious choice. However, there are two possibilities if the press is not capable of producing the force required by the progression tool: use the existing press and make the component in two operations, which would increase the production costs and reduce the profit margin; or purchase a new and more powerful press.

The purchase of plant such as a new press is extremely costly and can only be justified if the quantities of components being produced, and the savings in production costs, will ensure repeat orders to keep it continuously and profitably employed.

Sometimes a compromise can be agreed between the product designer and the production engineer so that the components are kept within the limitations of existing plant. However, market forces often dictate changes in product design which necessitate major changes in production methods, so the purchase of new plant becomes unavoidable. If the producer cannot afford, or cannot borrow the funds necessary to purchase the new plant, or does not consider that the profit margin justifies the investment, then the work will be lost to a competitor who has suitable plant capacity. Alternatively it may be more economical to subcontract specific operations or even the manufacture of whole components to specialist companies who have the necessary facilities.

Fig. 1.2 *Blanking and piercing: (a) separate operations; (b) combined operations (follow-on tool)*

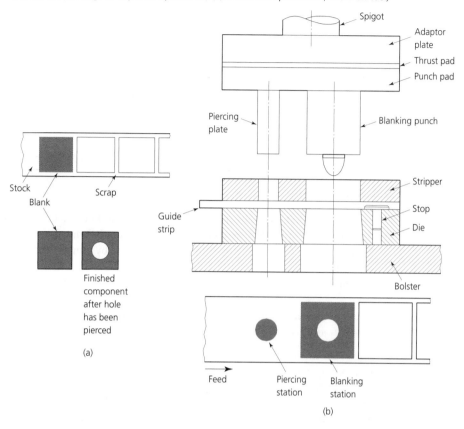

(a)

(b)

Wherever possible, standard components should be used that can be bought off the shelf. This makes for greater consistency of quality and facilitates after-sales service and maintenance.

1.1.7 *Cost*

So far the cost of production has been a recurring theme. The sole purpose of production is to:

- Produce goods to satisfy customer demand at a price the customer is prepared to pay.
- Produce the goods at a profit so that the producing company can flourish and invest for its future.
- Produce goods of a quality that ensures customer satisfaction.
- Deliver the goods on time as required by the customer.

These are the basic requirements that will ensure customer loyalty, leading to the repeat orders that will keep the producer in business. To achieve these ends, a manufacturing company must be constantly updating its production plant and techniques so as to reduce its production costs below that of its competitors, and ensure that it has the capacity to produce the goods its customers require, when they require them and to the quality specification they require.

1.2 Sand-casting processes: shaping with no loss of volume

Casting is the oldest method of shaping metal that is known. Cast brass and bronze artefacts from the earliest human civilisations are constantly being found by archaeologists.

Metals can be formed directly from the molten state by pouring them into moulds and allowing them to cool and solidify. Obviously, the mould must be made from a material with a higher melting point than that of the molten metal from which the casting is to be made.

There are several different casting processes and the selection of the most suitable processes for any given casting depends upon such factors as:

- Size of the casting.
- Complexity and section thickness of the casting.
- Quantity of castings required.
- Quality of the castings (e.g. dimensional accuracy and surface finish).
- Metal from which the casting is to be made (melting temperature).
- Mechanical properties of the casting.
- Unit cost of production.

One of the most versatile casting techniques is sand casting, which gets its name because the mould is made from a sand and clay mixture that is self-supporting when rammed into shape. Sand casting can be used for large, medium or small castings ranging from simple shapes to highly complex shapes. It is used for metals with medium and high melting points, such as copper alloys and cast irons.

The mould contains a cavity in the form of the finished product into which the molten metal is poured. In sand moulding, the form of this cavity is determined by ramming sand around a wooden pattern. The pattern is the same shape as the finished casting but slightly larger, dimensionally, to allow for volumetric shrinkage of the molten metal as it cools and solidifies. After ramming, the mould is opened so that the pattern can be removed from the cavity. The mould is then reassembled ready for pouring. Where small quantities are required, the moulds are made by hand; where large quantities are required, the moulds are made by semi-automated or fully automated processes.

Figure 1.3 shows a section through a typical two-part sand mould for a simple component. Poured into the runner using a ladle, the molten metal displaces air from the cavity and the air escapes through the risers. There must be a riser above each high point in

Fig. 1.3 *Typical sand mould*

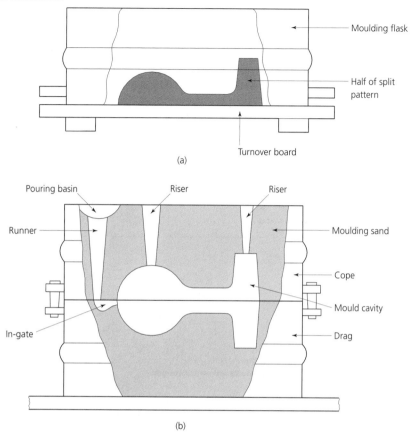

(a)

(b)

the cavity to prevent airlocks. Pouring continues until the molten metal appears at the top of each riser. This ensures that the mould cavity is full and also provides surplus metal which can be drawn back into the mould as shrinkage takes place during cooling. This avoids shrinkage cavities occurring in the casting.

The mould must also contain vents. These are fine holes made with a wire after the mould is complete. These vent holes stop just short of the mould cavity. The purpose of the vents is to release steam and other gases which are generated when the hot metal comes into contact with the moist moulding sand. (Moisture is required in moulding sand so that it will bind together and keep its shape). If the mould is not vented, the release of steam and gases causes bubbles to collect in the casting. Large bubbles are called *blowholes*; small bubbles are called *porosity*.

The mechanism of solidification and the defects which can occur in castings are fully covered in *Engineering Materials*, Volume 1. When the metal has solidified to form the casting, the mould is broken open and the casting is removed. The runners and risers are cut off and melted down again. The casting is now ready for machining.

The pattern has to be made oversize to allow for shrinkage as the metal cools. This is called the shrinkage allowance. Table 1.1 lists some common metals and the magnitude of the shrinkage allowance required. The pattern must also be made oversize wherever the casting is to be machined; that is, a machining allowance has to be superimposed on top of the shrinkage allowance. Not only does sufficient machining allowance have to be provided to ensure that the casting cleans up, but sufficient metal must be present to allow the tip of the cutting tool to operate well below the hard and abrasive skin of the casting. This is shown in Fig. 1.4.

Table 1.1 *Shrinkage allowance*	
Material	Shrinkage allowance
Aluminium	21.3 mm/m
Brass	16.0 mm/m
Cast iron	10.5 mm/m
Steel	16.0 mm/m

Fig. 1.4 *Need for shrinkage allowance*

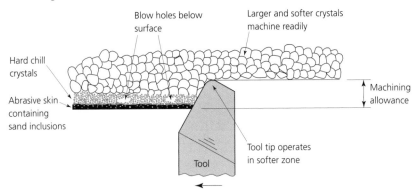

The mould for a hollow casting has an extra part called a core. The core is made from a mixture of moulding sand and a binder, usually a synthetic resin nowadays. Let's look at the component shown in Fig. 1.5. The inside shape is indicated by dotted lines. The core, which is the same as the inside shape of the component, is made in a core box. The mould is

Fig. 1.5 *Hollow component*

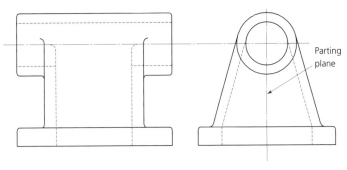

then made in two parts (drag and cope). The process is the same as for a non-hollow casting except the pattern is designed to leave locations for the core as shown in Fig. 1.6(a). These locations are called core prints. When the mould is opened to remove the pattern, the core is inserted and located in the core prints. Figure 1.6(b) shows the core in position in the lower half of the mould (the drag).

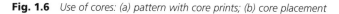

Fig. 1.6 *Use of cores: (a) pattern with core prints; (b) core placement*

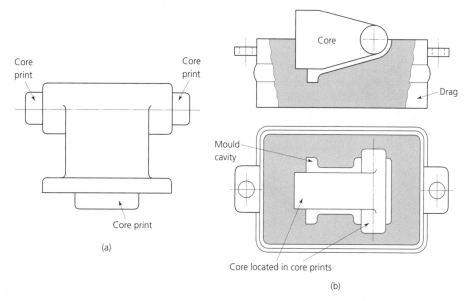

(a)

(b)

SELF-ASSESSMENT TASK 1.2

Explain why the cylinder-block casting for a heavy goods vehicle engine is more likely to be made by a sand-casting process than by a die-casting process.

1.3 Investment casting and die casting: shaping with no loss of volume

As well as making moulds from sand by ramming the sand around a pattern, moulds can also be made from other materials. Where large numbers of similar components are to be produced from materials such as cast iron, which has a relatively high melting temperature, shell moulding may be used for speed, repeatability and cost reduction. A sand–resin mixture is used, similar to the mixture for making cores. This process is called shell moulding and is fully described in *Manufacturing Technology*, Volume 2.

1.3.1 *Investment casting*

Another process that is increasingly being used is investment casting by means of the lost wax process. In this process the pattern is injection moulded from wax. The wax is coated with a ceramic slurry that is baked hard. During the baking process, the wax melts and runs out of the mould to be reused. The mould cavity that is left behind has a very smooth finish and a high dimensional accuracy. And there are no joins in the mould.

Patterns for sand casting have to be removed in the solid state, so their shape and the shape of the component is limited to allow easy removal of the pattern. But the wax pattern in the lost wax process is poured out of the mould in the molten, liquid state, so there are no similar constraints on its shape. The principle of this process is shown in Fig. 1.7, and the process is described in greater detail in *Manufacturing Technology*, Volume 2.

Fig. 1.7 *The lost wax process*

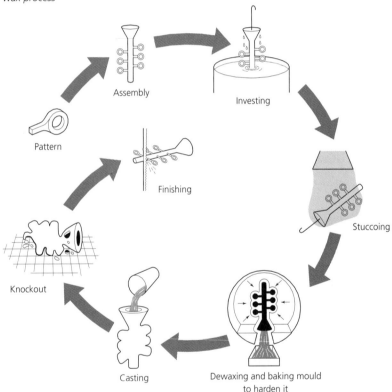

The advantages of this process can be summarised as follows:

- Since the moulds are made from ceramic materials that have very high melting temperatures, any metal that can be melted can be cast by this process.
- The surface finish and dimensional accuracy is excellent and compares with die casting but without the constraints imposed by the use of metal dies.

- The unit cost of investment castings is high compared with other casting processes, but the reduction in the need for extensive and expensive machining is removed. Thus the overall component cost can be lower.
- There are no design constraints on the shape of the component being cast.

1.3.2 *Gravity die casting*

The molten metal is poured into dies from a ladle in exactly the same way as for sand casting, except in this instance the mould is made of metal. In gravity die casting the parting line is usually vertical instead of horizontal; a simple mould is shown in Fig. 1.8. Unlike the refractory moulds previously described, metal dies are impervious, venting is difficult and adequate runners and risers must be provided. The faces of the die cavities can be coated with a refractory wash immediately before pouring to increase die life and to allow the casting of metals with a higher melting point, e.g. brasses and bronzes.

Fig. 1.8 *Gravity die casting: the dies are heated before pouring, so they do not chill the casting*

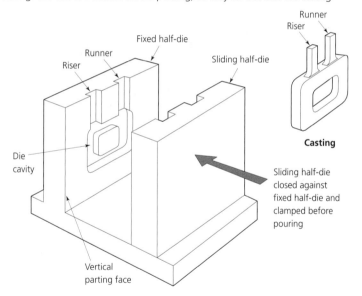

Dies for small components have to be preheated to prevent the casting being chilled, whereas larger dies have to be cooled after each pouring. Simple hollow components can be formed with metal cores, but more complex components use sand cores so that they can be broken out of the casting after solidification. The dies are very costly but the cost per component is low since up to 50,000 components can be made before the dies require replacement. The rate of production is higher than for sand casting and this reduces the unit labour cost. Only metals with low melting temperatures and which do not react chemically with the die materials can be cast in this way.

1.3.3 *Pressure die casting*

The principle of a typical pressure die-casting machine is shown in Fig. 1.9. Notice that the injection unit is submerged in the molten metal and coupled to the die blocks by a gooseneck. Since the injection chamber is immersed in the molten metal, this type of machine is called a *hot chamber* pressure die-casting machine. The die blocks are closed by a hydraulically operated toggle mechanism of great strength since the forces tending to separate the two halves of the die can be as high as 1.2 MN, in a machine capable of injecting 2.86 kg of zinc-based alloy per shot.

Fig. 1.9 *Hot chamber pressure die casting*

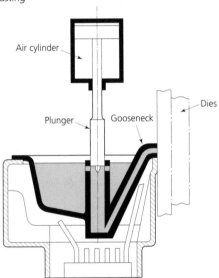

By current standards this is only a medium-sized machine. Automatic machines can achieve 1000 shots per hour for small, simple components and up to 200 shots per hour for larger, more complex components. When small components are being produced, multiple-impression dies are frequently employed so that the machine can be operated at its optimum capacity. Figure 1.10(a) shows components as ejected from a multiple-impression die. Figure 1.10(b) shows the clipped spray which can be melted down again. Figure 1.10(c) shows a set of components produced in one shot from a multiple-impression die.

Figure 1.11 shows the sequence of operations for a typical pressure die-casting machine cycle:

- The injection ram rises to allow molten metal to flow into the injection chamber and the dies are closed.
- The injection ram descends and forces the molten metal into the die and maintains it under pressure to make good the contraction losses on cooling. This also consolidates the metal and improves the mechanical properties of the casting.
- The dies open and, at the same time, separate themselves from the injection nozzle. A positive knock-out ejects the casting from the dies.
- The cycle is repeated.

Fig. 1.10 *Multiple impressio casting: (a) the spray of castings ready for clipping – note the peripheral web that improves metal flow during casting and gives the rigidity essential during flash clipping; (b) the flash after clipping – this is melted down and used again; (c) the eight components after clipping them from the spray*

(a)

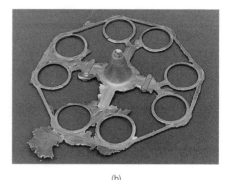

(b)

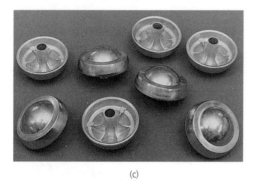

(c)

The die-casting machine and its dies represent a very costly capital investment which can only be warranted where large quantities of components are involved. Minimum economic quantities are usually considered to be about 20,000 components for pressure die casting, compared with 5000 components for gravity die casting. The advantages of this process include high accuracy and good surface finish, minimising machining costs; high production rates; low material consumption rates (clipped sprays and sprues can be recycled without loss and without contamination); and the faithful reproduction of complex shapes and decorative motifs.

The principle of cold chamber pressure die casting is shown in Fig. 1.12; the process gets its name because the injection chamber is not immersed in the molten metal. This allows the higher-melting aluminium-based alloys to be cast successfully. Apart from aluminium-based alloys, the cold chamber process can also be used with magnesium-based alloys and some brass alloys.

Fig. 1.11 *Pressure die casting: (a) dies close; (b) metal is injected; (c) dies open, component is ejected*

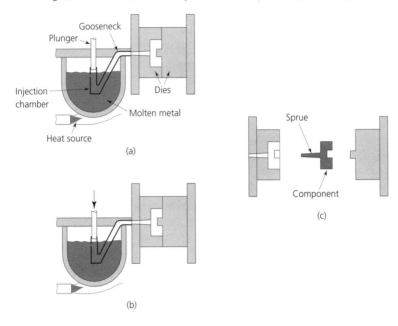

Fig. 1.12 *Cold chamber pressure die casting: the metal is only poured into the chamber at the moment of injection*

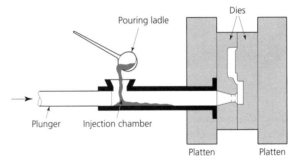

1.4 Compression moulding: shaping with no loss of volume

Thermosetting plastic materials, such as those based on phenol formaldehyde (Bakelite), urea formaldehyde, or melamine formaldehyde, are usually moulded by a compression moulding technique. The most widely used compression moulding techniques are positive moulding, flash moulding and transfer moulding. During moulding there occurs a chemical change in the moulding powder which makes it hard and strong and prevents it being softened by reheating; this is known as *curing*. Since curing is accompanied by the release of water vapour (steam), the moulds have to be designed to allow for this water vapour to escape and for volumetric shrinkage to take place.

The moulds are fitted into a hydraulic press (Fig. 1.13). The platens are provided with steam or electrical heating elements so that the moulds can be raised to the curing temperature of the material being processed. Modern presses are automatic in operation to ensure that mouldings of consistent quality are produced. The three factors which have to be preset are:

- The pressure.
- The time for which the mould is closed.
- The temperature of the mould.

Fig. 1.13 *Upstroke hydraulic press*

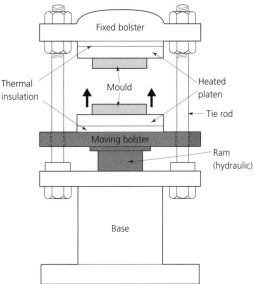

1.4.1 *Positive mould*

Figure 1.14 shows a section through a typical positive mould. A predetermined amount of thermosetting plastic moulding material in powder or granular form is placed in the heated mould cavity. The mould is closed by a hydraulic press. The moulding powder becomes plastic and is forced to flow into the shape of the mould. The mould remains closed whilst the moulding material cures. The mould is then opened, and the moulding is ejected.

Fig. 1.14 *Positive type compression mould*

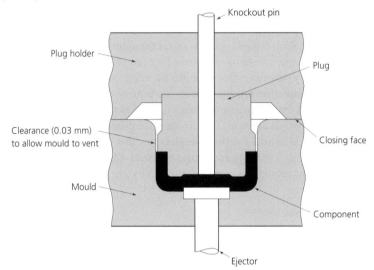

Positive moulding allows the two parts of the mould to close completely so that the thickness of the moulding is accurately controlled. The disadvantages of this technique are (1) the amount of moulding powder has to be very accurately controlled and (2) any gases liberated during curing tend to be trapped in the mould, causing porosity in the moulding. For this reason a venting clearance of 0.01–0.03 mm per side is left between the plug and the mould. This leaves a slight flash that has to be removed.

1.4.2 *Flash moulding*

This is the most common type of mould, and it is suitable for a wide range of components. A slight excess of moulding powder is placed in the mould cavity to ensure complete filling. The excess powder is forced out horizontally into the *flash gutter* as the moulds close. The *flash land* restricts the flow of the moulding powder and tends to hold it back in the mould cavity, ensuring complete filling and adequate moulding pressure. The flash produced by the excess moulding powder has to be removed from the finished moulding after it has been removed from the press. Figure 1.15 shows a section through a typical flash mould.

1.4.3 *Transfer mould*

This technique uses a more complex and costly three-part mould, as shown in Fig. 1.16, and is only resorted to where:

- The moulding is of complex shape with many changes of wall thickness which would make it difficult to ensure uniform filling of the cavity in a simple positive or flash mould.

Fig. 1.15 *Flash type compression mould*

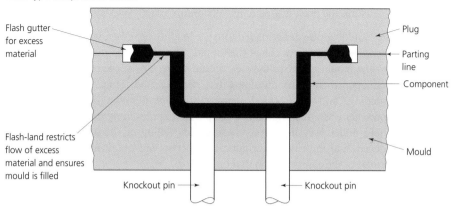

Flash gutter for excess material

Plug

Parting line

Component

Flash-land restricts flow of excess material and ensures mould is filled

Mould

Knockout pin → ← Knockout pin

Fig. 1.16 *Transfer type compression mould: (a) the die is open ready for removal of component and cull; (b) the floating plate closes on the mould and moulding powder is loaded into the chamber; (c) the transfer plunger descends and forces plasticised moulding powder through the sprue into the mould*

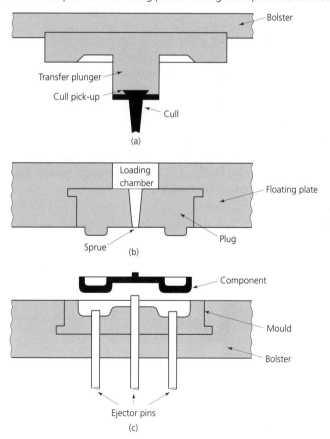

Bolster

Transfer plunger

Cull pick-up

Cull

(a)

Loading chamber

Floating plate

Sprue

Plug

(b)

Component

Mould

Bolster

Ejector pins

(c)

- A number of components are to be made at each stroke in a multicavity mould. This ensures the press is used to its maximum capacity and therefore economically.

The moulding powder is loaded into the loading chamber, where it becomes preheated and plasticised as the press closes the mould. The plasticised material is forced under pressure into the mould cavity or cavities via the sprue. The sprue is removed after the mould has been opened and the moulding has been ejected. Since the sprue is not part of the moulding, and since cured thermosetting plastic materials cannot be recycled, this represents waste material. Since the plasticised moulding powder is forced into the closed mould under very high pressure, complete filling of the mould is ensured, shrinkage is reduced, and improved mechanical properties are obtained from the moulding material because of the consolidation that takes place.

1.5 Injection moulding and extrusion: shaping with no loss of volume

Thermoplastic materials (polyethylene, polypropylene, nylon, etc.) are usually shaped by injection moulding or extrusion techniques. No curing takes place with these materials and any waste material, such as the flash and sprue, may be recycled. Since the moulds do not have to remain closed whilst curing takes place, they may be opened and the moulding ejected as soon as it has cooled sufficiently to retain its shape. Thus injection moulding has a faster cycle time than compression moulding.

1.5.1 Injection moulding

A measured amount of thermosetting plastic material is heated until it becomes a viscous fluid; it is then injected into the mould under high pressure. The principle of injection moulding is shown in Fig. 1.17. Like so many processes that are simple in principle, the practice is fraught with difficulties.

Fig. 1.17 *Injection moulding: the torpedo spreads the powder and ensures uniform heating*

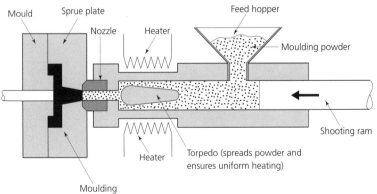

For instance, heating the moulding material until it is fluid is not easy. If care is not taken, the surface becomes degraded by the heat before the interior of the plastic mass has reached the moulding temperature. Uneven heating and cooling or poor mould design may cause the injected plastic material not to fill the mould cavity completely. Also there may be blisters, voids, sinks, shorts, and some cavities in a multicavity mould may not receive any plastic material at all. Ejection of the completed moulding is also difficult if distortion is to be avoided. Since plastic materials are adhesive, they will stick to the mould wherever an opportunity presents itself, even if a mould release agent is sprayed into the cavity immediately before moulding.

The principle variables which must be controlled are:

- The quantity of plastic material that is injected into the mould.
- The injection pressure.
- The injection speed.
- The temperature of the plastic.
- The temperature of the mould.
- The plunger-forward time (the time the plastic material in the mould is kept under pressure as it cools and becomes rigid).
- The mould-closed time whilst the moulding cools.
- The mould clamping force to ensure the pressure of the injected moulding material does not separate the moulds.
- The mould-open time whilst the moulding is ejected and the moulds are cleaned and sprayed with a release agent before commencing the next cycle.

1.5.2 *Extrusion*

The extrusion of thermoplastic materials is, in principle, a continuous injection-moulding process. Any thermoplastic material can be extruded to produce lengths of uniform cross-section such as rods, tubes, filaments and sections. The principle of this process is shown in Fig. 1.18, where in order to obtain a continuous flow of plastic material through the die, the plunger of the injection-moulding machine has been replaced by a feed screw. The plastic is still soft and lacks rigidity as it leaves the die and it must be carefully supported to prevent distortion. Cooling is provided by a water trough, mist spray or air blast. Except for rigid plastics such as unplasticised vinyl chloride, a conveyor draws the extruded material from the die ready for coiling or cutting to length.

Fig. 1.18 *Extrusion moulding*

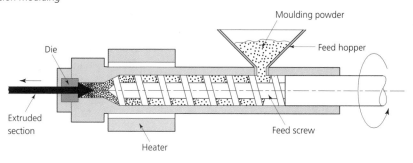

SELF-ASSESSMENT TASK 1.4

Explain briefly why thermosetting plastic materials are usually compression moulded whereas thermoplastic materials are usually injection moulded.

1.6 Miscellaneous moulding processes: shaping with no loss of volume

Let's now look at three quite different moulding processes that are extremely versatile. Processes that can be used for small or large batch production for small or large mouldings with a minimum of financial outlay in tooling.

1.6.1 *GRP moulding*

Glass fibre reinforced plastic (GRP) moulding is used to make things as diverse as motorcycle crash helmets, hulls for pleasure craft, the bodies for racing cars, and the wings for gliders. GRP is a composite material consisting of a polyester or an epoxy matrix reinforced with glass fibres. The glass used is quite different in its composition, properties and behaviour to window glass and imparts high strength and impact resistance to the moulding.

Since the resin used is also a powerful adhesive, the mould has to be polished and sprayed with a release agent before the process commences. The first layer of resin is the *gel coat* and contains the colorant (pigment). This is a thick, viscous resin that can be painted onto all the surfaces, horizontal and vertical, without running and dries to provide the highly polished surface finish associated with the sort of products named above. The *lay-up resin* is less viscous so that it can be worked into the fibres of the *chopped strand mat* or *woven cloth*.

Layers of glass fibre are built up until the required thickness is obtained. A catalyst is mixed with the resin immediately before use to accelerate the curing (setting) time and to ensure hardening occurs at room temperature. The surface in contact with the mould will have a high-gloss finish but the opposite surface will be rough unless a surfacing tissue is used for the final layer. Figure 1.19 shows a section through a GRP moulding. The mould can itself be built up quite simply from wood, plaster and GRP since there are no great forces involved.

1.6.2 *Vacuum moulding*

In this process a sheet of thermoplastic material is clamped over a former, heated until it softens and then drawn into the mould by creating a vacuum between the plastic sheet and the mould. The actual moulding pressure is provided by the atmosphere, so quite simple moulds are required and can be made from a variety of materials, depending upon the life required. For short runs, wood is quite adequate. The principle of this process is shown in Fig. 1.20. Products such as bathtubs, car bumpers, small boats and refrigerator liners can be made by this process. However, the process does have some limitations:

Fig. 1.19 *The composition of a GRP lay-up*

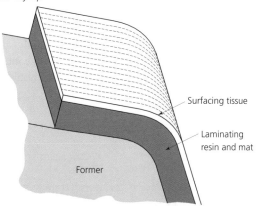

Surfacing tissue

Laminating
resin and mat

Former

Fig. 1.20 *Vacuum forming*

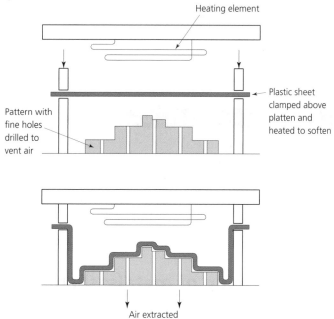

Heating element

Pattern with
fine holes
drilled to
vent air

Plastic sheet
clamped above
platten and
heated to soften

Air extracted

- It is very difficult to control the wall thickness, especially near bends.
- Compared with injection moulding and compression moulding processes, the dimensional accuracy is of a low order.
- Only components of simple shapes without sharp corners can be made.

1.6.3 *Blow moulding*

This process is used for making hollow containers such as plastic bottles. The basic stages are shown in Fig. 1.21:

- A plastic blank called a *parison* is moulded round a hollow mandrel.

- The mandrel, with the plastic blank still moulded round it, is transferred to the heated mould.
- Hot air is blown through the hollow mandrel to inflate the parison like a balloon. This continues until the inflated parison fills the mould.
- The mould, which is made in two halves, is then cooled and opened.
- Since there is no solid plug to be removed, the moulding can have re-entrant corners.

Fig. 1.21 *Blow forming*

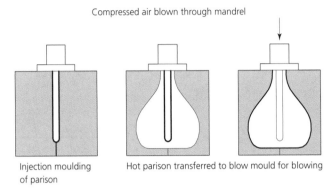

Compressed air blown through mandrel

Injection moulding of parison

Hot parison transferred to blow mould for blowing

SELF-ASSESSMENT TASK 1.5

List the most likely processes that would be used to manufacture the following articles, giving reasons for your choice:

(a) washing-up bowl
(b) glider fuselage
(c) mineral water bottle

1.7 Flow forming of metals: shaping with no loss of volume

The earliest known method of shaping metals was by melting them and casting them into moulds. The next oldest process for shaping metals was to work the metals in the solid state by hand forging them on an anvil. This is one of many processes for working metal to shape. Let's now look at some of these *flow-forming* processes.

Many metals and alloys, which are sufficiently malleable or ductile, may be formed whilst in the solid condition by squeezing, stretching or bending. That is, they may be *flow formed*. During flow-forming operations the grain structure of the metal becomes distorted and the metal becomes harder (work hardened). The grain structure of the formed metal may be restored to its original condition by heating the metal to above its temperature of recrystallisation (see *Engineering Materials*, Volume 1).

If the metal is flow formed (worked) above the temperature of recrystallisation, so that the grains can reform as fast as they are distorted, the metal is said to be *hot worked*.

If the metal is flow formed (worked) below the temperature of recrystallisation, so that the grains remain distorted, the metal is said to be *cold worked*.

When metals are flow formed to shape, their grain structure flows to the shape of the component and this greatly increases the strength of the component. This orientation of the grain is shown in Fig. 1.22 which compares a gear blank machined from a bar with one machined from a forged blank. Since metals break more easily along the lay of the grain than across the lay of the grain, the teeth of the gear cut from the forged blank will be the stronger. You can make a comparison with the behaviour of wood.

Fig. 1.22 *Grain orientation: (a) machining from a bar produces a plane of weakness where the tooth could break off under load; this is due to the grain lying parallel to the tooth; (b) machining from a forging produces a very much stronger tooth when the grain flows radially from the blank; the grain lies at right angles to the tooth*

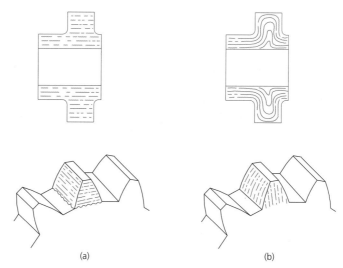

(a)   (b)

Fig. 1.23 *Bending wood: (a) wood bent at right angles to the grain is strong; (b) wood bent parallel to the grain is weak*

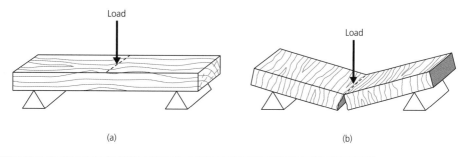

(a)   (b)

The advantages and limitations of hot working and cold working are compared in Tables 1.2 and 1.3. Let's now consider some examples of typical hot-forming and cold-forming processes.

Table 1.2 Hot-working processes

Advantages	Limitations
1. Low cost	1. Poor surface finish: rough and scaly
2. Grain refinement from cast structure	2. Due to shrinkage on cooling, the dimensional accuracy of hot-worked components is of a low order
3. Materials are left in the fully annealed condition and are suitable for cold working (heading, bending, etc.)	3. Due to distortion on cooling and to the processes involved, hot working generally leads to geometrical inaccuracy
4. Scale gives some protection against corrosion during storage	4. Fully annealed condition of the material coupled with a relatively coarse grain leads to a poor finish when machined, and low strength and rigidity for the metal considered
5. Available as sections (girders) and forgings as well as the more usual bars, rods, sheets, and strip and tube	5. Damage to tooling from abrasive scale on metal surface

Table 1.3 Cold-working processes

Advantages	Limitations
1. Good surface finish	1. Higher cost than for hot-worked materials. It is only a finishing process for material previously hot worked. Therefore, the processing cost is added to the hot-worked cost
2. Relatively high dimensional accuracy	
3. Relatively high geometrical accuracy	2. Materials lack ductility due to work hardening and are less suitable for bending, etc.
4. Work hardening caused during the cold-working processes increases strength and rigidity, and improves the machining characteristics of the metal so that a good finish is more easily achieved	3. Clean surface is easily corroded
	4. Availability limited to rods and bars, sheets and strip, and solid drawn tubes

1.7.1 Hot forging

Figure 1.24 shows basic forging operations such as drawing down, upsetting, piercing, drifting and swaging, as performed manually by the blacksmith on the anvil. These same operations can be applied to larger components by substituting pneumatic and steam hammers for the blacksmith's hand tools. The hot-working temperature ranges for a number of engineering materials are shown in Fig. 1.25.

Fig. 1.24 *Summary of forging processes*

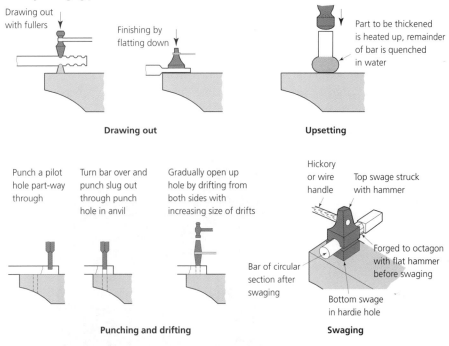

Drawing out with fullers

Finishing by flatting down

Part to be thickened is heated up, remainder of bar is quenched in water

Drawing out

Upsetting

Punch a pilot hole part-way through

Turn bar over and punch slug out through punch hole in anvil

Gradually open up hole by drifting from both sides with increasing size of drifts

Hickory or wire handle

Top swage struck with hammer

Bar of circular section after swaging

Forged to octagon with flat hammer before swaging

Bottom swage in hardie hole

Punching and drifting

Swaging

Fig. 1.25 *Forging temperatures*

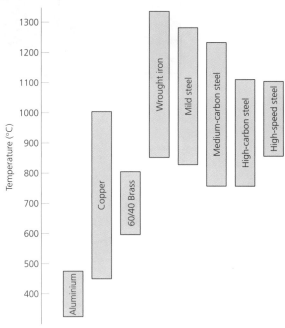

Temperature (°C)

Aluminium
Copper
60/40 Brass
Wrought iron
Mild steel
Medium-carbon steel
High-carbon steel
High-speed steel

Where large batches of components need to be forged, as in the car industry, closed-die forging is used. In place of the general smithying tools previously described, formed dies are used to produce the finished component. The dies may be closed by a hydraulic or mechanical press or, more usually for small components, by a drop stamp or drop hammer. Figure 1.26 shows a typical drop stamp being used for closed-die forging and Fig. 1.27 shows a section through forging dies in the closed position.

Fig. 1.26 *Drop stamp*

Fig. 1.27 *Closed forging dies*

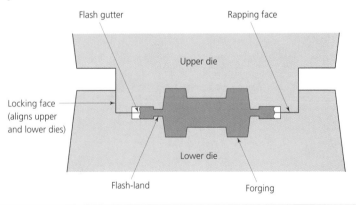

To ensure complete filling of the die cavity, the hot blank is made slightly larger in volume than the volume of the finished forging. As the dies close, the surplus metal is forced out into the flash gutter through the flash land. The flash land offers a restriction to the flow of surplus metal and holds it back in the die cavity to ensure complete filling. It also ensures that the flash is thinnest adjacent to the component being forged. This results in a neat, thin flash line being left after the flash has been trimmed off. The flash land should be kept as short as possible, otherwise the dies may fail to close.

The rapping faces are provided to ensure that the component is the correct thickness when the dies are closed. The sharp rap of the hardened surfaces coming into contact tells the operator that the operation is complete. The vertical walls of the die cavity are given a 7° taper, or draught, so that the forging can be easily removed. An example of a set of forging dies and the component they produce is shown in Fig. 1.28.

Fig. 1.28

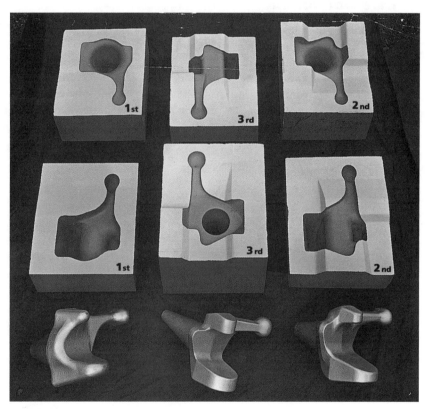

Modern manufacturing is increasingly using *near finished size* (NFS) techniques. That is, casting and flow-forming processes are used that can provide components of such accuracy and finish that only the minimum amount of machining is required to achieve the finished product. This saves on the amount of material required and the amount of energy used in manufacture. To this end, the more complex processes of cold forging and warm forging are being increasingly used together with sintered metal compacts. These processes are described fully in *Manufacturing Technology*, Volume 2.

Sintered powder metal compacts are now used in place of forgings for the connecting rods in the engines of many BMW cars. This process enables the connecting rods to be made to within a ±0.2 per cent tolerance of the target weight. This ensures a level of dynamic balance unattainable by conventional forging. A blend of metal powders that will give the required properties are compressed in dies under powerful hydraulic presses. These powder compacts are then *sintered* in a controlled-atmosphere furnace at a temperature of about 1100 °C, causing the particles to bond together. After sintering and finish machining of the big- and little-ends, the big-end of the connecting rod is cracked along a predetermined fault line by a purpose-built machine. When reassembled about the crankshaft, the result is a perfect fit and smooth, trouble-free running.

1.7.2 *Hot Rolling*

The effect of hot rolling on the grain structure of a metal ingot is shown in Fig. 1.29, and the hot rolling of an ingot into a slab is shown in Fig. 1.30. The white-hot slab is just leaving the rolls and is supported on a motorised roller bed (conveyor). The slab is passed backwards and forwards between the work rolls of the mill. The gap between the mill rolls is gradually reduced for each pass; the thickness of the slab is reduced and its length is increased. This process removes any discontinuities that may have been present in the ingot and also refines the grain structure and improves the mechanical properties of the metal.

Fig. 1.29 *Effect of hot rolling on the grain structure of a metal ingot*

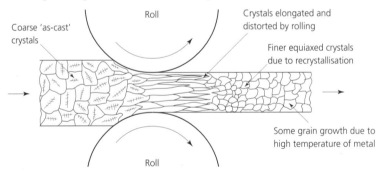

1.7.3 *Hot extrusion*

The principle of hot extrusion is shown in Fig. 1.31. A hydraulic ram squeezes a billet of metal, preheated above the temperature of recrystallisation, through a die rather like toothpaste being squeezed from a tube. The hole in the die is shaped to produce the required section and long lengths of material are produced by this process. Materials which are commonly extruded in this way are copper, brass alloys, aluminium and aluminium alloys. After hot extrusion the sections are often finished by cold drawing to improve the surface texture, dimensional accuracy and stiffness.

Fig. 1.30 *Hot rolling*

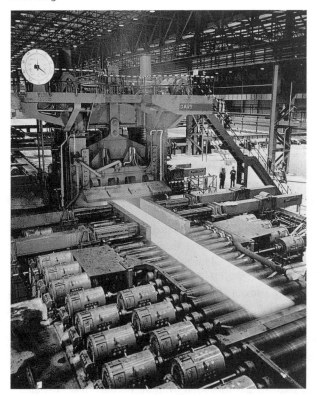

Fig. 1.31 *Hot extrusion: (a) commencement of stroke; (b) completion of stroke*

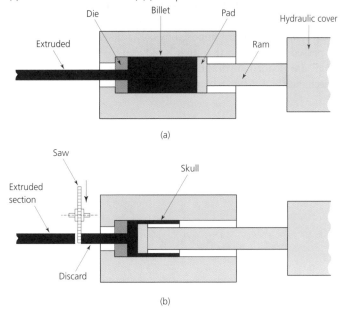

1.7.4 *Cold rolling*

Cold-working processes are essentially finishing processes. The forces required to produce quite modest reductions in cross-sectional area are very much higher than those for hot-working, so the amount of reduction is kept to a minimum. However, the finish and dimensional accuracy produced by cold working are much better than produced by hot working. Since cold working causes some work hardening of the metal, there is a corresponding improvement in its mechanical properties.

Because of the work hardening that occurs during rolling, the cold-rolled metal is supplied in various *condition grades*. If the metal has only received a light *pinch pass*, it will only be in the one-quarter hard condition. The hardness will increase through the half-hard condition and the three-quarters hard condition to the full-hard condition as the severity of the cold working is increased. Sometimes cold-rolled metals are required in the soft (annealed) condition ready for further processing. In order not to destroy the surface finish, great care has to be taken to avoid oxidation (scaling) of the metal surfaces. To achieve this, the metal is heat treated in atmosphere-controlled furnaces. Alternatively cold-rolled sheets are often annealed in packs. Only the outer sheets of the pack are oxidised; the inner sheets remain bright with some slight discolouration around the edges. Steel in this condition is known as cold-rolled close-annealed (CRCA).

The metal is usually broken down by hot working until there is only a finishing allowance left. The oxide film (scale) on the surface of the hot-worked metal is removed by pickling the metal in acid, after which it is passed through a neutralising bath and oiled to prevent corrosion before being passed to the cold-working process. Figure 1.32 shows the effect of cold rolling on the properties of the metal, and Fig. 1.33 shows a typical cold strip-rolling mill.

The pickled and oiled strip is unwound from the decoiler at the extreme left. This strip is straightened and flattened in the pinch rolls and leveller then passed to the mill rolls themselves for reduction in thickness. The mill in Fig. 1.33 is classified as a four-high single-stand reversing mill; that is, besides the reducing rolls there are a pair of backing rolls to help support the forces imposed by cold rolling, and there is only one mill stand. The metal is passed back and forth until it is reduced to the required thickness. The lower rolls in this mill are loaded hydraulically. (In earlier mills the top rolls were screwed down mechanically.) After passing through the mill, the strip (now reduced in thickness), is

Fig. 1.32 *Effect of cold rolling*

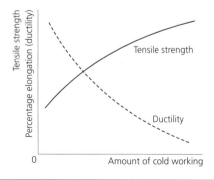

Fig. 1.33 *Cold rolling mill: the typical line shown in the diagram consists of coil storage, coil car, undriven decoiler with snubber roll, spade opener with debender, four-high reversing hydraulic mill, reversing coilers, coil car and storage station. Separate high- and low-pressure hydraulic packs provide the mill loading system and operate the ancillary equipment. A high-capacity soluble oil system supplies strip lubrication and roll cooling. The hydraulically operated mill has a loading rate of 0.1 in per second (6 in per minute)*

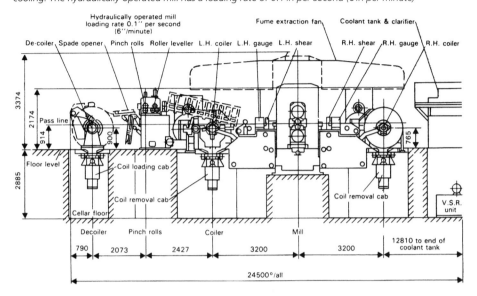

recoiled on the right-hand coiler. After the first pass, the mill is reversed and the strip is returned through the mill for further reduction; it is finally recoiled on the left-hand middle coiler.

1.7.5 *Tube drawing*

In this process a pickled and oiled hot-drawn tube is further reduced and finished by cold drawing it through a die on a draw bench as shown in Fig. 1.34(a). In order to control the wall thickness and internal finish of the tube, a plug mandrel is used as shown in Fig. 1.34(b). The floating mandrel, or plug, is drawn forward with the tube but cannot pass through the die. This technique is used for long thin-walled tube, but is of limited accuracy. Alternatively a fixed mandrel can be used as shown in Fig. 1.34(c). Obviously there are limitations to the length of tube which can be drawn with a fixed mandrel. However, it is widely used for thick-walled tubes and where greater accuracy is required.

1.7.6 *Wire drawing*

This process is shown in Fig. 1.35. It is similar in principle to tube drawing but, since a longer length of material is involved, the wire is pulled through the die by a rotating capstan or bull-block. The drawn wire may be coiled on the capstan or passed to a separate coiler after taking only one or two turns round the capstan. Fine wire, as used for electrical conductors, is produced on multiple-head machines as shown in Fig. 1.36. As the wire becomes thinner, it becomes progressively longer. Thus each successive capstan has to run

Fig. 1.34 *Tube drawing: (a) simple draw bench; (b) tube drawing using a plug; (c) tube drawing using a mandrel*

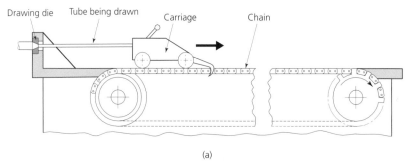

(a)

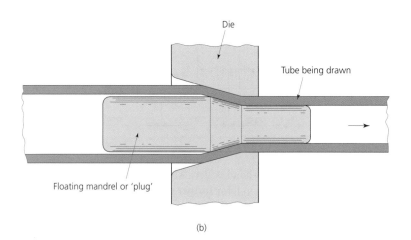

(b)

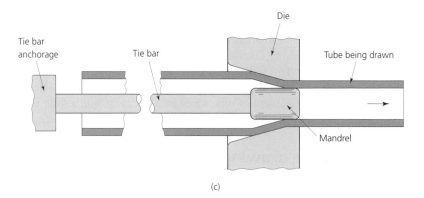

(c)

faster than the preceding one. The speed is controlled by the pull of the wire on the tension arm, which is coupled to the capstan motor speed control. If the tension on the arm increases, the capstan motor is slowed slightly; but if the tension on the arm decreases, the capstan motor is speeded up.

Fig. 1.35 *Single-die wire drawing*

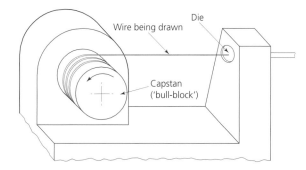

Fig. 1.36 *Multiple-die wire drawing*

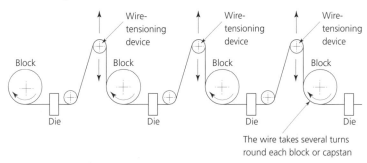

1.7.7 *Impact extrusion*

Impact extrusion differs fundamentally from hot extrusion. It is a cold-working process where a slug of metal is struck by a punch and made to flow up between the punch and the die (Fig. 1.37). Metal to be formed in this way is preferably soft and malleable, such as lead or pure aluminium. However, modern die materials and lubrication techniques enable alloy steels to be formed cold by impact extrusion as described in *Manufacturing Technology*, Volume 2.

Fig. 1.37 *Impact extrusion*

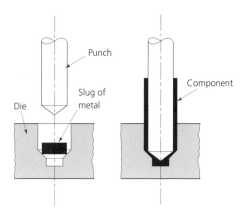

1.7.8 *Rivet heading*

Rivet heading is shown in Fig. 1.38. The preformed head of the rivet is supported by a hold-up or dolly, and the opposite end of the rivet is formed by a heading tool in a pneumatic hammer. This is a cold-forging process. Large rivets are frequently hot riveted. This allows the head to be formed more easily, and the contraction of the rivets on cooling draws the plates or sections being joined tightly together.

Fig. 1.38   *Cold heading a rivet*

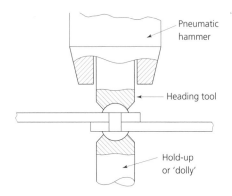

1.7.9 *V-Bending*

V-bending is the simplest of all bending process which can be carried out in a press using tools similar to those shown in Fig. 1.39 fitted into a fly press such as that shown in Fig. 1.40. The flat metal *blank* is laid across the die and the punch is lowered so as to press the metal into the V of the die. As the bending pressure is increased, the stress produced in the outermost grains of the metal (on both the tension and the compression sides) eventually

Fig. 1.39   *Simple V-bending tool*

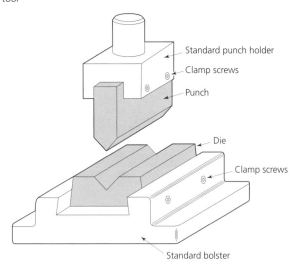

Fig. 1.40 *Fly press*

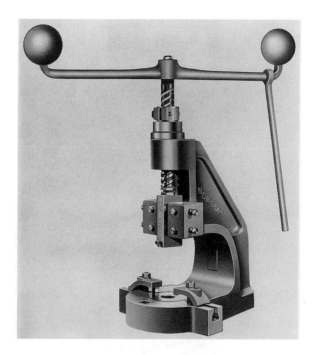

exceeds its yield strength. Once the yield strength has been exceeded, plastic deformation of the grains take place and the material takes a *permanent set*. That is, it remains bent even after it has been removed from the tools.

However, the grains adjacent to the *neutral plane* are only subject to elastic strain and try to spring back straight when the punch is raised and the bending force removed. The neutral plane is the layer of metal towards the centre of the section where there is no change of length (Fig. 1.41). This results in slight spring-back of the bent metal and has to be allowed for in the design of the tools by bending the metal beyond the angle required (overbend). A power press or a press brake would be used for larger components and thicker material. An example of a power press is shown in Fig. 1.42 and an example of a press brake is shown in Fig. 7.4.

1.7.10 *U-Bending*

U-bending is rather different to V-bending and an example is shown in Fig. 1.43. Notice that the blank is trapped between the punch and a spring-loaded pad, which also acts as the ejector. This pressure pad helps to prevent the blank from skidding and bowing as it is bent. The spring-loading of the pressure pad should be substantial if these objectives are to be achieved, and this has to be taken into account when determining the size of press required to close the tools and bend the blank.

To prevent skidding completely, the pressure pad may be fitted with pilot pegs, positioned to locate in holes in the blank. If no convenient holes are available, additional holes are sometimes pierced in the blank, especially for location purposes; these are called

Fig. 1.41 *V-bending*

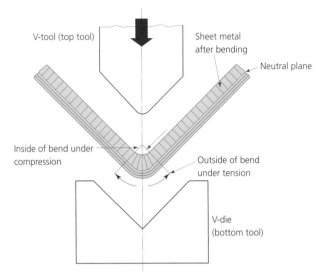

V-tool (top tool)

Sheet metal
after bending

Neutral plane

Inside of bend under
compression

Outside of bend
under tension

V-die
(bottom tool)

Fig. 1.42 *C-frame power press*

Fig. 1.43 *U-bending: (a) tools open; (b) tools closed*

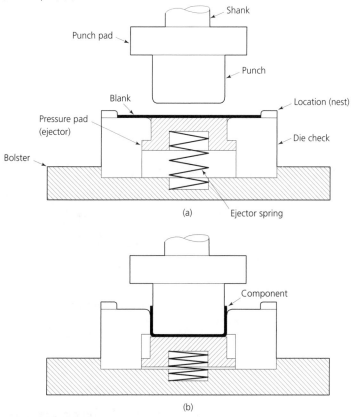

(a)

(b)

Fig. 1.44 *Grain orientation when bending*

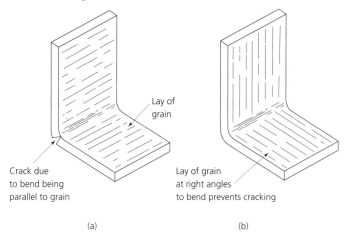

(a)

(b)

tooling holes. Bending tools should always be designed to balance the bending forces as far as possible.

When strip or sheet metal is bent, and especially when it is bent after cold rolling, pay particular attention to the orientation, or lay, of the grain (Fig. 1.44). The orientation of the grain will be parallel to the direction of rolling and there will be a tendency for the metal to tear and crack if the line of the bend is parallel to the orientation of the grain, as shown in Fig. 1.44(a). Therefore, metal should always be bent perpendicular to the orientation of the grain, as shown in Fig. 1.44(b).

SELF-ASSESSMENT TASK 1.6

List the most likely processes that would be used to manufacture the following articles, giving reasons for your choice:

(a) farm gate hinge
(b) light-aircraft propeller blade
(c) sheet metal for vehicle body pressings
(d) railway rails
(e) copper tube for domestic plumbing
(f) hexagon brass rod to be used for turned nuts and bolts
(g) steel shelf brackets

1.8 Machining: shaping with loss of volume

During the machining of metallic and non-metallic materials, surplus material is cut away from the blank until the finished shape of the required component is achieved. This is obviously wasteful and the amount of machining to produce any component should be kept to a minimum. The advantages of machining operations are their accuracy and the high surface finish which can be achieved when compared with casting or forging. Since tooling, workholding and the kinematics of manufacturing equipment will be considered in detail in subsequent chapters, we will only consider a brief outline of the more important machining operations in this chapter.

1.8.1 *Drilling*

Drilling and reaming processes are used for producing cylindrical holes, and for countersinking, counterboring and spot facing such holes. Figure 1.45 shows a typical heavy-duty column-type drilling machine suitable for the batch production of medium-sized components. Figure 1.46 shows typical drills and cutters for use with such a machine and the holes and surfaces they produce.

Fig. 1.45 *Column type drilling machine*

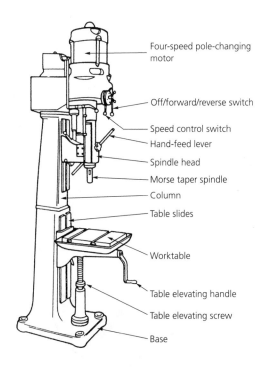

Four-speed pole-changing motor

Off/forward/reverse switch

Speed control switch
Hand-feed lever

Spindle head

Morse taper spindle

Column

Table slides

Worktable

Table elevating handle

Table elevating screw

Base

Fig. 1.46 *Drills, reamers and cutters: (a) twist drill; (b) long fluted machine reamer with Morse taper shank, right-hand cutting with left-hand helical flutes; (c) countersinking; (d) counterboring; (e) spot facing*

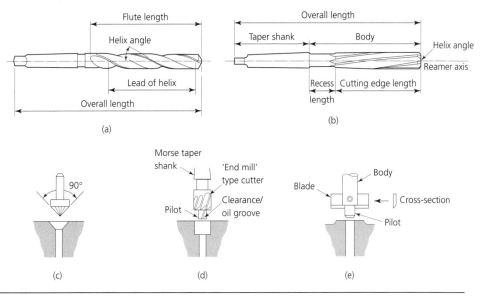

Flute length

Helix angle

Lead of helix

Overall length

(a)

Overall length

Taper shank

Body

Helix angle

Reamer axis

Recess length

Cutting edge length

(b)

90°

(c)

Morse taper shank

'End mill' type cutter

Clearance/oil groove

Pilot

(d)

Blade

Body

Cross-section

Pilot

(e)

1.8.2 *Turning*

Figure 1.47 shows a typical centre lathe. Turned components are produced on a centre and the more common cylindrical, conical and plain surfaces are summarised in Fig. 1.48. Notice that in turning operations the workpiece is rotated and the shape of the workpiece is determined by the path of the tool relative to the axis of the workpiece. Typical single-point turning tools and their applications are shown in Fig. 1.49.

Fig. 1.47 *Centre lathe*

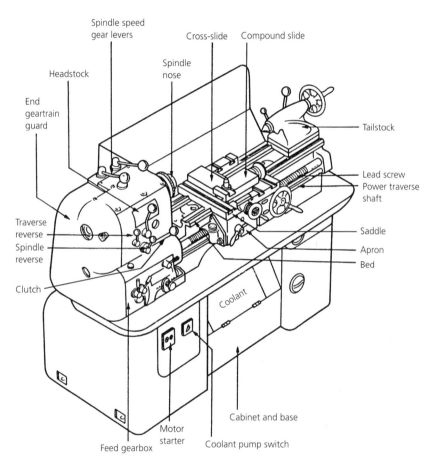

1.8.3 *Milling: horizontal machine*

Figure 1.50 shows how the horizontal milling machine gets its name from the horizontal axis of its spindle, hence the horizontal axis of its arbor that supports the cutter. Horizontal milling operations produce mutually parallel, perpendicular and inclined surfaces using multi-tooth cutters. Some examples of the cutters used on horizontal milling machines are

Fig. 1.48 *Surfaces produced on a lathe*

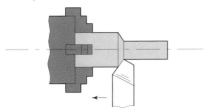

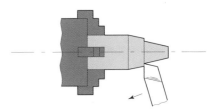

Cylindrical turning
The tool moves parallel to the axis of the workpiece

The feed/rev is kept small to improve the surface finish

Conical (taper) turning
The tool moves at an angle to the axis of the workpiece

The feed/rev is kept small to improve the surface finish

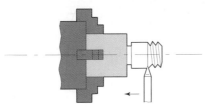

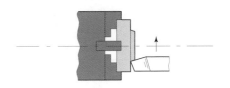

Screw cutting
The tool moves parallel to the axis of the workpiece

The feed/rev is coarse and equals the lead of the thread being cut: Lead = pitch × number of starts

Surfacing (facing)
The tool moves along a path perpendicular (90°) to the axis of the workpiece

The feed/rev is kept small to improve the finish

Fig. 1.49 *Lathe tool profiles: these tools are right-handed; left-handed tools cut towards the tailstock; the grey arrows indicate the direction of the rake angle of each tool*

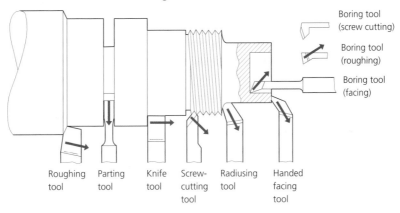

Boring tool (screw cutting)

Boring tool (roughing)

Boring tool (facing)

Roughing tool   Parting tool   Knife tool   Screw-cutting tool   Radiusing tool   Handed facing tool

Fig. 1.50 *Horizontal milling machine*

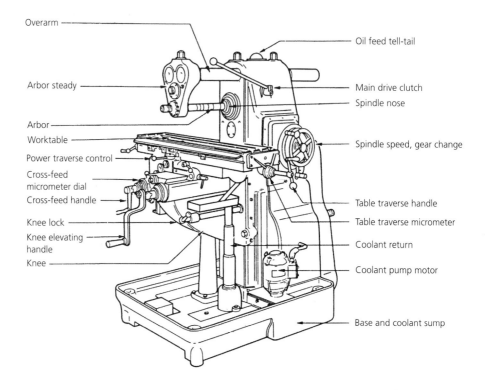

Overarm

Oil feed tell-tail

Arbor steady

Main drive clutch

Spindle nose

Arbor

Worktable

Power traverse control

Spindle speed, gear change

Cross-feed
micrometer dial

Cross-feed handle

Table traverse handle

Knee lock

Table traverse micrometer

Knee elevating
handle

Coolant return

Knee

Coolant pump motor

Base and coolant sump

shown in Fig. 1.51 together with some typical examples of the shapes and surfaces they produce.

1.8.4 *Milling: vertical machine*

Figure 1.52 shows how the vertical milling machine gets its name from the vertical axis of its spindle, hence the vertical axis of its cutter. Vertical milling operations also produce mutually parallel, perpendicular and inclined surfaces. They can also produce pockets and islands (Chapter 5). Some examples of vertical milling cutters are shown in Fig. 1.53 together with some typical examples of the shapes and surfaces they produce.

1.8.5 *Surface grinding*

Like milling machines, surface grinding machines may also be classified as horizontal spindle and vertical spindle machines. The grinding wheel rotates much more rapidly than

Fig. 1.51 *Horizontal milling machine cutters and the surfaces they produce: (a) slab milling cutter (cylinder mill); (b) side and face cutter; (c) single-angle cutter; (d) double equal angle cutter; (e) cutting a V-slot with a side and face mill; (f) double unequal angle cutter; (g) concave cutter; (h) convex cutter; (i) single and double corner rounding cutters; (j) involute gear tooth cutter*

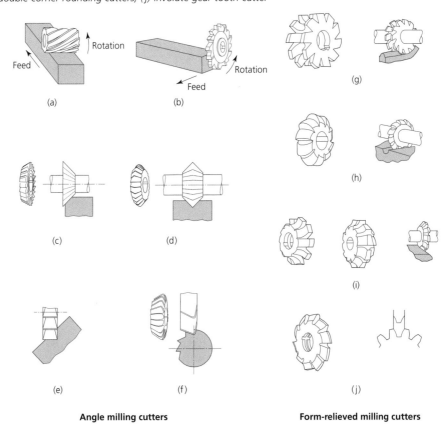

Angle milling cutters   **Form-relieved milling cutters**

a milling cutter and the table feed rate is also higher. Surface grinding machines produce the same surfaces as milling machines but to much higher standards of accuracy and surface finish. Apart from specialist and experimental production machines, the rate of metal removal by grinding is substantially less than achieved by milling. Grinding is essentially a finishing process. An example of a typical surface grinding machine is shown in Fig. 1.54.

1.8.6 *Cylindrical grinding*

In cylindrical grinding the slowly rotating workpiece is brought into contact with the edge of a rapidly rotating grinding wheel. The rate of metal removal is limited and the process is essentially a finishing process, producing work of high dimensional and geometrical accuracy with a high surface finish. A typical cylindrical grinding machine is

Fig. 1.52 *Vertical milling machine*

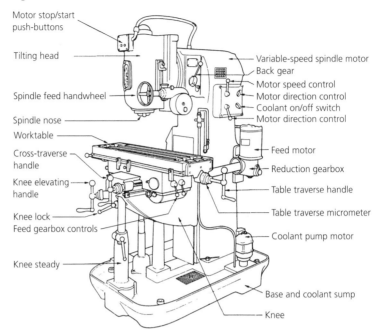

Motor stop/start push-buttons

Tilting head —————— Variable-speed spindle motor

Back gear

Motor speed control

Spindle feed handwheel ————— Motor direction control

Coolant on/off switch

Spindle nose ————— Motor direction control

Worktable —————

Cross-traverse handle

Knee elevating handle

Knee lock ————

Feed gearbox controls —

Feed motor

Reduction gearbox

Table traverse handle

Table traverse micrometer

Coolant pump motor

Knee steady —————

Base and coolant sump

Knee

Fig. 1.53 *Vertical milling machine cutters and the surfaces they produce: (a) end milling cutter; (b) face milling cutter; (c) slot drill; (d) recess A would need to be cut using a slot drill because it is the only cutter that will work from the centre of the solid; recess B could be cut using a slot drill or an end mill because it occurs at the edge of the solid; (e) this blind keyway would have to be sunk with a slot drill; (f) dovetail (angle) cutter*

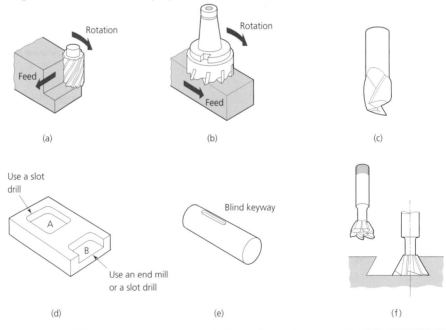

Rotation

Feed

(a)

Rotation

Feed

(b)

(c)

Use a slot drill

A

B

Use an end mill or a slot drill

(d)

Blind keyway

(e)

(f)

Fig. 1.54 *Surface grinding machine*

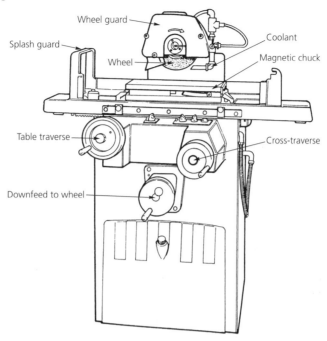

Fig. 1.55 *Cylindrical grinding machine*

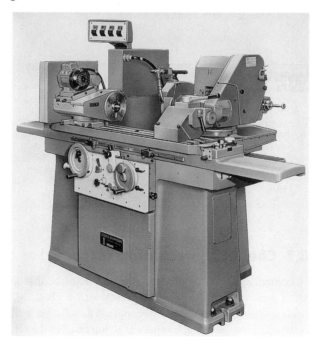

Fig. 1.56 *Some cylindrical grinding operations: (a) grinding tapered components, for parallel cylindrical grinding the subtable is set with $\theta = 0°$ so the work and wheel axes are parallel; (b) plunge cut grinding; (c) internal cylindrical grinding*

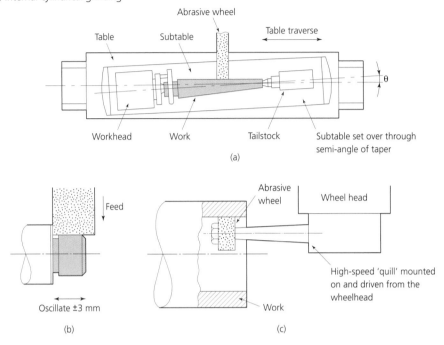

(a)

(b)

(c)

shown in Fig. 1.55 and some examples of cylindrical grinding operations are shown in Fig. 1.56.

SELF-ASSESSMENT TASK 1.7

With reference to Fig. 1.57, select a suitable machining process for producing the following features on the prototype component shown. Give reasons for your choice.

(a) key slot
(b) shank
(c) pockets
(d) holes

1.8.7 *Chemical machining (etching)*

This technique is used to produce printed circuit boards and similar components. The metal is removed by allowing it to react with a chemical reagent; for example, ferric chloride solution is used to remove copper. The solution is either sprayed onto the surface to be attacked, or the whole component is immersed in the solution.

- The circuit to be transcribed onto copper-faced Tufnol or fibreglass board is produced as a large-scale drawing. This is reduced photographically to an actual-size transparency. An example is shown in Fig. 1.57.
- The copper-faced board is pretreated with a photoresist coating which is sensitive to ultraviolet light. The transparency of the circuit is placed tightly in contact with the treated circuit board material and exposed to ultraviolet light.
- The exposed board is placed in a developer solution which removes the photoresist where it has been exposed to the ultraviolet light and hardens the photoresist wherever it was masked by the transparency. This makes such areas immune to the reagent.
- The board is now immersed in the reagent and the copper is stripped away wherever it was unprotected by the transparency of the circuit.
- The chemically machined (etched) circuit board is then washed to remove all traces of the reagent and any residual coating of photoresist.
- The printed circuit is then tinned ready for drilling, and for mounting and soldering the components into position.

Fig. 1.57 *Printed circuit*

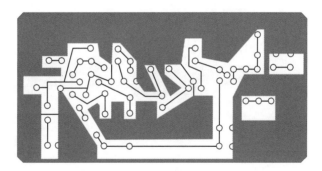

1.9 Shearing: shaping with loss of volume

Blanks cut from sheet or strip metal are produced by shearing processes. The principle of cutting a material by shearing is shown in Fig. 1.58.

- *Stage 1* As the top cutting blade is moved downwards and brought to bear on the material with continuing pressure, the top and bottom surfaces of the material are deformed.
- *Stage 2* As the pressure increases, the cutting blades close and there is plastic deformation of the material.
- *Stage 3* After a certain amount of plastic deformation has occurred, the cutting blades start to cut into the material. The remaining uncut material starts to work harden and become brittle in the zone indicated.
- *Stage 4* Fractures start to run into the work-hardened zone from the point of contact with the cutting blades. When the fractures meet, the material being cut fails in shear and the two portions separate.

Fig. 1.58 *Shearing action*

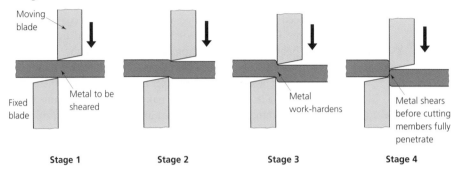

Moving blade

Fixed blade

Metal to be sheared

Metal work-hardens

Metal shears before cutting members fully penetrate

Stage 1   **Stage 2**   **Stage 3**   **Stage 4**

Fig. 1.59 *Clearance: (a) insufficient; (b) correct; (c) excessive*

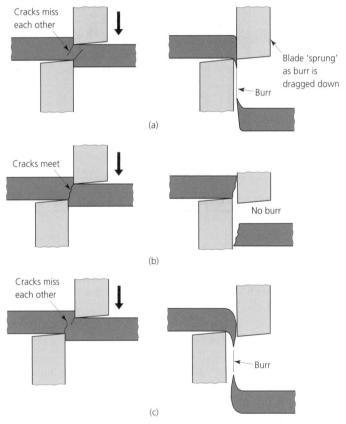

Cracks miss each other

Blade 'sprung' as burr is dragged down

Burr

(a)

Cracks meet

No burr

(b)

Cracks miss each other

Burr

(c)

It is a popular misconception that, in order to get a clean-cut blank free from burr, the edges of the cutting blades should be set close together. In fact, there should be a carefully controlled clearance between the cutting blades. The reason for this is shown in Fig. 1.59. If the blades are set too close together, the cracks, which run out from the points of contact with the cutting edges, miss each other and a clean shear does not take place. The excess metal tends to spring the blades apart and is dragged down between them to form a burr on the cut edge. This also causes excessive wear of the blades, which soon become blunt.

Where correct clearance is applied, as shown in Fig. 1.59(b), the cracks coincide and a clean shear occurs, leaving minimum burr. Wear on the shear blades is also reduced to a minimum.

Excessive clearance, as shown in Fig. 1.59(c), again causes the cracks to miss each other. There is also sufficient clearance through which the material may be dragged. This produces a radius on the outer edge of the material being cut, and a heavy burr on the inner edge of the material.

For correct shearing, the correct clearance must be carefully chosen to suit the type of material and its thickness. The cutting edges of the shear blades must be kept sharp. Correctly and incorrectly sheared edges are shown in Fig. 1.60.

Fig. 1.60 *Appearance of sheared edge when clearance is (a) correct and (b) incorrect*

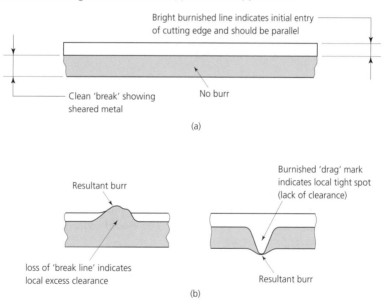

Where a large number of sheet metal blanks of complex shape are required it is usual to stamp them out in a press using a blanking tool. Figure 1.61(a) shows a simple blanking tool for use in a press, and Fig. 1.61(b) shows the strip and the blank stamped from it. Blanking is a very rapid process with up to several hundred components being produced each minute. In principle it is a shearing process and the force to cut the blank from the strip can be calculated as shown in Example 1.1.

Fig. 1.61 *Blanking: (a) blanking tool; (b) circular blank produced by tool in (a)*

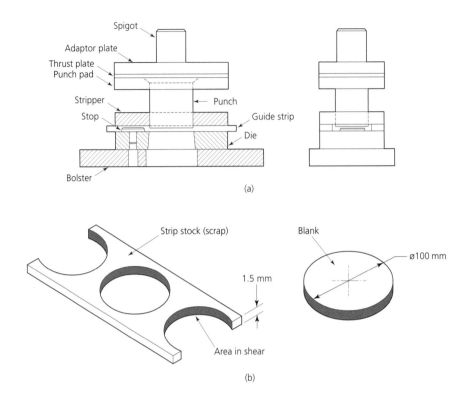

(a)

(b)

Spigot
Adaptor plate
Thrust plate
Punch pad
Stripper
Stop
Bolster
Punch
Guide strip
Die

Strip stock (scrap)
1.5 mm
Area in shear

Blank
ø100 mm

EXAMPLE 1.1

Figure 1.61(b) shows a circular component that is to be blanked out of strip metal whose ultimate shear stress is 450 N/mm². The shaded surface is the area in shear upon which the cutting force calculation is based ($\pi = 3.14$).

Area in shear = circumference of hole × thickness

$\qquad = 2\pi R \times$ thickness

$\qquad = 2 \times 3.14 \times 50 \times 1.5$

$\qquad = \mathbf{471\,mm^2}$

Blanking force = area in shear × ultimate shear stress of the metal

$\qquad = 471\,mm^2 \times 450\,N/mm^2$

$\qquad = \mathbf{212\,kN}$

Since the cutting action of a blanking tool is the same as for the shearing process, it is equally important that the correct clearance exists between the punch and the die. Some typical examples of clearances are given in Table 1.4, and the calculation of suitable punch and die diameters for cutting blanks from low-carbon steel strip is given in Example 1.2.

- For blanking: the die is made to the required size and the clearance is deducted from the punch.
- For piercing (hole punching): the punch is made to the required size and the clearance is added to the die.

Table 1.4 _Die clearances_

Material	Clearance per side*
Aluminium	$\frac{1}{60}$ material thickness
Brass	$\frac{1}{40}$ material thickness
Copper	$\frac{1}{50}$ material thickness
Steel	$\frac{1}{20}$ material thickness

*For diameters, double this value

EXAMPLE 1.2

Calculate the punch and die diameters for producing a circular blank 100 mm diameter from low-carbon mild steel 1.5 mm thick (Fig. 1.61(b)).

Die diameter = blank diameter = **100 mm**

Punch diameter = die diameter − clearance

$$= 100 \, \text{mm} - 2 \times [\text{metal thickness (mm)}/20] \quad \text{(see Table 1.4)}$$
$$= 100 \, \text{mm} - 2 \times [1.5 \, \text{mm}/20]$$
$$= 100 \, \text{mm} - 0.075 \, \text{mm}$$
$$= \mathbf{99.85 \, mm}$$

1.10 Blank layout

In order to avoid waste, the positioning of the blank in the strip material is very important. Figure 1.62 shows some examples of alternative blank layouts. For instance, the L-shaped bracket may be cut from the strip as shown in A or B. You can see that B is more economical with material. Alternatively, the strip may be passed through the tools twice, as shown in C and D.

Fig. 1.62 *Blank layouts*

A

B

C

D

SELF-ASSESSMENT TASK 1.8

1. Calculate the punch and die diameters for a piercing tool used for punching 30 mm diameter holes in 3 mm thick sheet brass.

2. Calculate the force required to pierce the hole described in Question 1 if the ultimate shear stress for the brass is 250 N/mm^2.

EXERCISES

1.1 Prepare a design specification for a simple component of your own choice and explain how this specification influences the manufacturing processes to be used in terms of geometry, quality, materials, machine or equipment process capability, and cost.

1.2 The component shown in Fig. 1.63 consists of a number of geometrical surfaces. Identify these surfaces and specify a suitable machining process for the production of each surface.

1.3 The unit manufacturing cost for producing a particular component by hand is £5.00. Using a machine reduces this cost to £3.00 but the machine costs £1500.00. Draw a break-even graph and determine the minimum number of components which have to be manufactured before the cost of a machine can be justified.

Fig. 1.63

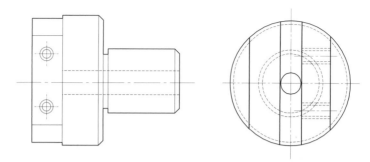

1.4 List the advantages and limitations of sand casting compared with pressure die casting.

1.5 List the advantages and limitations of forging and finish machining compared with machining from the solid for the component (a cluster gear blank) shown in Fig. 1.64.

Fig. 1.64

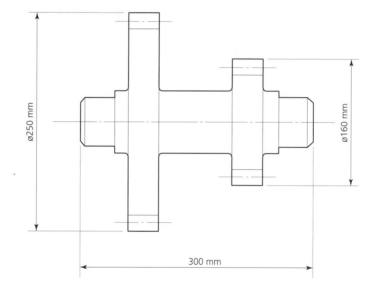

1.6 (a) Compare and contrast the processes of drawing solid rod and drawing tube on a draw bench. Use sketches to illustrate the principles in each case.
(b) With the aid of sketches, show how small-diameter wire is drawn on a multiple-die machine and explain how the machine compensates for the increase in length of the wire as its diameter is progressively reduced at each stage.

1.7 Explain with the aid of sketches how the brass door-bolt section shown in Fig. 1.65 is produced by a combination of hot extrusion and cold drawing.

Fig. 1.65

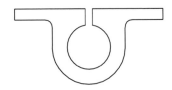

1.8 (a) Labelling the more important features, sketch a simple blanking tool for producing blanks of the bracket shown in Fig. 1.66. Take into account the grain orientation of the strip so as to avoid cracking during the bending process.

(b) Sketch a simple piercing tool for piercing the fixing holes in the blank for the bracket shown in Fig. 1.66. Pay particular attention to the method of locating the blank in the tool. Both holes are to be produced at the same time. Label the more important features of the tool.

(c) Labelling the more important features, sketch a simple press-bending tool for forming the bracket shown in Fig. 1.66. Pay particular attention to the method of locating the pierced blank in the tool.

Fig. 1.66

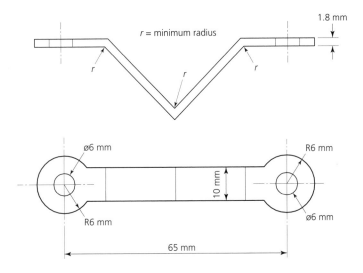

1.9 Discuss the types of component and plastic moulding materials for which the following moulding processes are most suitable:

(a) flash moulding
(b) transfer moulding
(c) injection moulding;
(d) extrusion moulding

1.10 State **two** applications of chemical machining and explain why chemical machining is the most appropriate process in each instance.

2 Tooling

The topic areas covered in this chapter are:

- Factors affecting the choice of tooling.
- Basic principles of metal-cutting tools.
- Cutting-tool materials.
- Cutting-tool geometry: single- and multi-point tools.
- Abrasive wheels: cutting action, construction and safe use.
- Cutting fluids.

2.1 Factors affecting the choice of tooling

Many factors affect the choice of tooling for the manufacture of a given component, and some of the more important are listed below:

- Machining process.
- Workpiece material.
- Cutting-tool materials.
- Cutting-tool geometry (single-point tools).
- Cutting-tool geometry (multi-point tools).
- Influence on chip formation.
- Process variables: cutting speed, feed, depth of cut, power available.
- Application of a cutting fluid.

Before we consider these factors in detail, it is necessary to understand the principles of cutting. For simplicity they will be applied to single-point cutting tools as used on lathes, shaping and planing machines. The same principles can be generalised to multi-point cutting tools such as milling cutters.

2.2 Basic principles of cutting tools

Perhaps one of the first controlled cutting operations you performed was the sharpening of a pencil with a penknife. It is unlikely you will have received any formal instruction before your first attempt but, most likely, you soon found out (by trial and error) that the knife blade had to be presented to the wood at a definite angle if success was to be achieved. This is shown in Fig. 2.1(a).

Fig. 2.1 *Clearance angle: (a) β = 0, the blade skids along the pencil without cutting; (b) β > 0, the blade bites into the pencil and cuts*

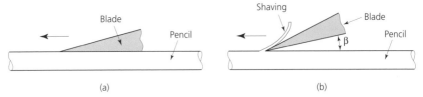

If the blade is laid flat on the wood, as shown in Fig. 2.1(a), it just slides along without cutting. Tilted at a slight angle, as shown in Fig. 2.1(b), it will bite into the wood and start to cut. Tilted at too steep an angle, it will bite into the wood too deeply and it will not cut properly. The best angle will vary between a knife which is sharp and a knife which is blunt. A sharp knife will penetrate the wood more easily, at a shallower angle, and will be easier to control.

Let's look at that knife blade. It is the shape of a wedge. In fact, all cutting tools are wedge shaped (more or less), so let's now look at the angles of a typical metal-cutting tool.

2.2.1 *Clearance angle*

We have just seen that for a knife to penetrate the wood, we need to incline it to the surface being cut, and that we have to control this angle carefully for effective cutting. This angle, is called the *clearance angle* and we give it the Greek letter beta (β). All cutting tools have to have this angle. It has to be kept as small as possible to prevent the tool digging in or to prevent the tool from chattering. On the other hand, it has to be large enough to allow the tool to penetrate the workpiece. The clearance will vary slightly depending upon the cutting operation and the material being cut. It is usually between 5° and 7°.

2.2.2 *Wedge angle*

If, in place of a pencil, we tried to sharpen a point on a piece of soft metal such as copper, we would find that the knife would quickly become blunt. Examined under a magnifying glass, the blunt edge would show signs of crumbling away. To cut metal successfully, the cutting edge must be ground to a less acute angle to give greater strength. This is shown in Fig. 2.2.

Fig. 2.2 *Wedge angles of sharpened blades: (a) γ_w for cutting wood; (b) γ_m for cutting steel; notice that $\gamma_m > \gamma_w$*

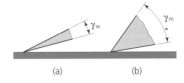

The angle to which the tool is ground is called the *wedge angle* or the *tool angle* and it is given the Greek letter gamma (γ). The greater the wedge angle, the stronger the tool. And the greater the wedge angle, the quicker the heat of cutting will be conducted away from the cutting edge. This will prevent the tool overheating and softening, and help to prolong the life of the tool. Unfortunately, the greater the wedge angle, the greater the force to make the tool penetrate the workpiece material. The choice of the wedge angle becomes a compromise between all these factors.

2.2.3 Rake angle

The *rake angle* is given the Greek letter alpha (α). It is very important, for it alone controls the geometry of the chip formation for any given material, hence it controls the cutting action of the tool. The relationship of the rake angle (α) to the angles β and γ is shown in Fig. 2.3.

Fig. 2.3 *Rake angle α is given by the formula $\alpha = 90° - (\beta + \gamma)$ where β is the clearance angle and γ is the wedge angle*

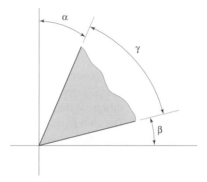

Let's now look at the geometry of metal-cutting tools in greater detail and how this geometry controls chip formation. Figure 2.4(a) shows a cutting tool ground so that it has a zero rake angle and a zero clearance angle. The workpiece, which is slightly narrower than the width of the cutting tool, is a low-strength ductile material such as low-carbon (mild) steel. Figure 2.4(b) shows what happens when the tool starts to cut. The metal just ahead of the tool is compressed until it starts to shear away from the workpiece and piles up ahead of the cutting tool. This mainly plastic deformation of the metal ahead of the cutting tool and mainly elastic deformation of the metal below the cutting tool sets up reaction forces to the movement of the cutting tool. The most important of these, as shown in Fig. 2.4(b), are:

- F_c which is the cutting reaction force.
- F_t which is the thrust reaction force.

Assuming the machine tool in which cutting is taking place is strong enough to keep cutting without mechanical failure, the thrust force (F_t) would gradually push the tool off the workpiece, by springing the tool and workpiece apart, as shown in Fig. 2.4(c). The deformation of the workpiece material below the tool will be mainly elastic and it is the spring-back of this zone that produces the reaction force F_t. The underside of the tool would be heavily worn and scored, the cutting edge would be destroyed, and the newly cut surface of the workpiece would be very rough and uneven.

Fig. 2.4 *The effect of trying to cut without clearance: (a) a cutting tool ground with $\alpha = \gamma = 0$, $\beta = 90°$; (b) what happens when the tool starts to cut; (c) F_t gradually pushes the tool off the workpiece*

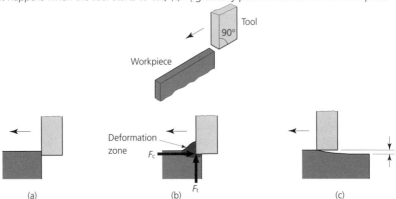

(a) (b) (c)

The situation can be greatly improved by grinding a clearance angle of about $7°$ on the tool, as shown in Fig. 2.5(a). This allows the tool to bite into the workpiece and also prevents the underside of the tool from rubbing on the newly cut surface, thus reducing the thrust force (F_t). Any slight spring-back of the deformed zone will be behind the cutting point of the tool, so the tool can now cut relatively freely without lifting off the workpiece. This agrees with our earlier pencil-sharpening analogy.

The chip produced by workpiece material shearing ahead of the cutting tool has the shape shown in Fig. 2.5(b). The chip parts from the workpiece along a path called a shear plane. In Fig. 2.5(c) the chip has been lifted away from the workpiece so that you can see the area in shear. Notice that the length of the shear plane (L) multiplied by the width of the cut (W) gives the area (A) in shear for the material being cut.

For any given material, the smaller the area in shear, the smaller the cutting force (F_c) and the greater the cutting efficiency. Since any reduction in the width of cut would cause a corresponding reduction in the rate of metal removal, the most effective way of reducing the shear area is to reduce the length of the shear plane (L). It has been shown by

Fig. 2.5 *Improved cutting: (a) a gliding clearance of about $7°$; (b) the chip parts from the workpiece along the shear plane; (c) the area in shear is given by $A = WL$ where W is the width of the cut and L is the length of the shear plane*

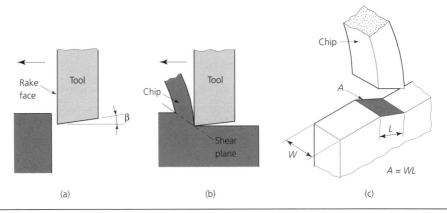

(a) (b) (c)

Fig. 2.6 *How the rake angle affects the shear plane: (a) L_1 is a shear plane where $\alpha = 0$; (b) L_2 is a shear plane where $\alpha > 0$; notice how L_2 is shorter than L_1*

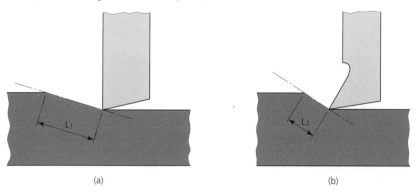

(a)   (b)

experimentation that if the rake face of the tool is inclined away from the perpendicular by giving the tool a *positive rake angle*, the length of the shear plane is reduced (Fig. 2.6).

Furthermore, a rake face with a positive rake angle enables the chip to peel away from the parent material without having to turn through such an acute angle. Thus, a high positive rake angle reduces the cutting force (F_c) by reducing the cutting area, and it also reduces the pressure of the chip on the rake face of the tool. Both these factors lead to increased cutting efficiency and reduced tool wear.

Unfortunately, there is a limit to how much the rake angle can be increased. Figure 2.7 shows the cutting-tool angles and how they are interrelated. Notice that three angles are involved:

- Clearance angle.
- Rake angle.
- Tool or wedge angle.

The clearance angle is generally fixed at the following angles by the surface being cut:

- External cylindrical surfaces, 5–7°.
- Flat surfaces, 6–8°.
- Internal cylindrical surfaces, 8–10° plus secondary clearance.

Clearance angles less than these can lead to rubbing, whereas angles greater than these can lead to chatter and a tendency for the tool to dig in. Any increase in the clearance angle leads to a corresponding reduction in the wedge angle and reduction in the strength of the tool at the cutting edge.

Thus, with the clearance angle fixed within narrow limits, the rake angle and the wedge angle have to be balanced to form a compromise between cutting efficiency (large rake angle) and tool strength (large wedge angle). A small wedge angle not only reduces the strength of the tool at the cutting edge, it also reduces the mass of metal behind the cutting edge which is available to conduct away the heat generated by the cutting process. This can lead to overheating of the tool, resulting in softening and early failure of the cutting edge. Notice how the theory fits in with the pencil-sharpening analogy at the start of the chapter.

Fig. 2.7 *Rake angles for high-speed tool steels*

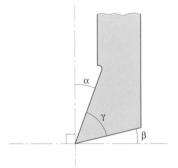

Material being cut	α
Cast iron	0°
Free-cutting brass	0°
Ductile brass	14°
Tin bronze	8°
Aluminium alloy	30°
Mild steel	25°
Medium-carbon steel	20°
High-carbon steel	12°
Tufnol plastic	0°

SELF-ASSESSMENT TASK 2.1

Summarise the basic cutting angles and explain briefly, in your own words, how they influence the cutting action of the tool.

2.3 Chip formation

The chips formed when machining metals and non-metals fall into one of three categories, depending upon the characteristics of the workpiece material and the geometry of the cutting tool.

2.3.1 *Discontinuous chip*

The shearing of the chip from the workpiece material has already been discussed in Section 2.2. In forming the chip, as shown in Fig. 2.8(a), the workpiece material is severely strained and, if it is a brittle, non-ductile material, it may completely fracture along the shear planes in the primary deformation zone to produce a discontinuous or granular chip as shown in Fig. 2.8(b). Discontinuous chips are associated with non-ductile materials such as grey cast iron, free-cutting brass and thermosetting plastics.

Fig. 2.8 *Discontinuous chips: (a) general chip formation; (b) discontinuous chip formation*

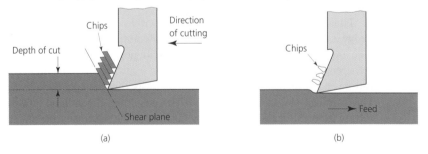

2.3.2 *Continuous chip*

This is the long ribbon-like chip which is produced by tools with a high positive rake angle when machining ductile materials such as low-carbon (mild) steels, copper and aluminium. The workpiece material behaves as a rigid plastic but, although the chip shears from the workpiece material along shear planes, it does not separate into completely separate plates as shown in Fig. 2.8(a). Instead, only partial separation occurs and a continuous chip is produced as shown in Fig. 2.9(a). The inside of the curvature of the chip is usually rough and the slip planes are visible even to the unaided eye in the large chips produced by heavy machining. The outer surface of the chip is burnished smooth as it flows over the rake face of the cutting tool.

Fig. 2.9 *Continuous chips: (a) general continuous formation; (b) formation for soft, ductile, low-strength metals (tear type)*

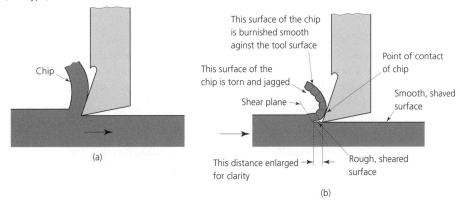

Chips produced from some very soft and ductile materials, with a low strength, tend to tear away from the parent workpiece material just ahead of the cutting tool instead of shearing cleanly. This creates a rough surface which has to be cleaned up by a cutting tool with a very keen edge. If the cutting edge becomes dull, or if the torn surface is below the cutting plane of the tool, the surface will not clean up and a rough finish to the workpiece will result. This can be partly overcome by reducing the rate of feed and increasing the cutting speed. Some materials such as aluminium and copper have to be finish-machined using diamond-tipped tools and very high cutting speeds. Under these conditions the material behaves as though it were very much stiffer and harder, and very good surface finishes can be obtained.

2.3.3 *Continuous chip with built-up edge*

Under some conditions the friction between the chip and the rake face of the tool becomes very great. This causes particles of metal from the chip to become pressure welded to the rake face of the tool, making it rough, masking the cutting edge and changing its geometry. This leads to increased friction and heat at the cutting edge, and layer after layer of chip material is built up (Fig. 2.10).

Fig. 2.10 *Chip welding (built-up edge): (a) layers of chip material form on the rake face of the tool; (b) excessive chip welding produces an unstable built-up edge; particles of built-up edge material flake away and adhere to the workpiece, making the machined surface rough; they also adhere to the chip, making it jagged and dangerous; the result is a poor surface finish*

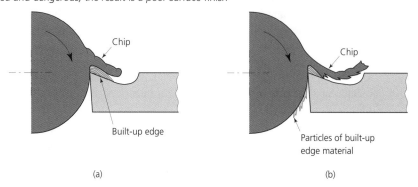

(a) (b)

The result is called a *built-up edge* and the mechanism by which it is produced is known as *chip welding*. Eventually the amount of built-up material becomes unstable and it breaks down. The particles of built-up material that flake away weld themselves to the chip and to the surface of the workpiece. This leaves a rough surface on the workpiece and a dangerously jagged chip.

Since chip welding has a considerable adverse influence on tool life, power consumption and surface finish, every attempt must be made to prevent it. This is achieved by reversing one or more of the causes of chip welding as follows:

- *Reducing friction* This can be achieved by increasing the rake angle, polishing the rake face, and introducing a lubricant between the chip and the tool.
- *Reducing the temperature* The temperature at the cutting zone can be lowered by reducing the friction between the chip and the tool and by using a coolant.
- *Reducing the pressure* The pressure between the chip and the tool can be reduced by increasing the rake angle, reducing the feed rate, and/or using oblique instead of orthogonal cutting.
- *Preventing metal-to-metal contact* This can be achieved by introducing a high-pressure lubricant between the chip and tool interface. Such lubricants contain chlorine or sulphur additives which build up a thin film of high lubricity on the tool rake face. This non-metallic film reduces the friction between the chip and the tool, and it prevents metal-to-metal contact. Alternatively non-metallic cutting-tool materials such as metallic carbides and ceramics may be used.

Continuous chips have razor-sharp and ragged edges which are extremely dangerous. These long ribbon-like chips tend to tangle with the cutter and workpiece, and they are difficult to dispose of. Swarf removal is particularly important in high-production automatic and computer-controlled machine tools. To remove the dangers and difficulties associated with continuous chips, the cutting tools should be fitted with a chip breaker as shown in Fig. 2.11(a). The action of the chip breaker is shown in Fig. 2.11(b). Notice that the chip breaker forces the chip to curl up into a tight spiral. This work-hardens the chip, making it brittle, so it breaks up into short lengths which are safely and easily disposable.

Fig. 2.11 *Chip breaker: (a) inserted-tip tool with chip breaker; (b) action of chip breaker*

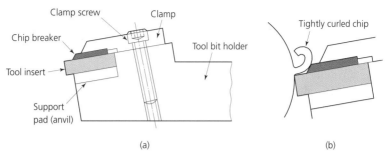

(a)   (b)

SELF-ASSESSMENT TASK 2.2

Suggest brief explanations for the following. Give an example for part (a) and an example for (b).

(a) some materials produce a continuous chip when being cut
(b) some materials produce a discontinuous chip when being cut
(c) why it is an advantage for a turning tool to have a chip breaker when cutting ductile materials

2.4 Workpiece materials

The material from which the workpiece is made has a considerable influence upon the choice of cutting-tool geometry and cutting-tool material. Generally, when using high-speed steel cutting tools, low-strength ductile materials are cut with tools possessing a high positive rake angle to take advantage of the increased cutting efficiency and to leave a good surface finish. High-strength materials are cut with low-rake-angle tools to give the cutting edge a large wedge angle that provides adequate strength and heat dissipation. Figure 2.7 lists typical rake angles for single-point cutting tools in high-speed steel. The values given are for roughing cuts; it usually helps to increase them slightly for finishing cuts.

So far, only ductile metals have been considered and the only factor influencing the choice of rake angle has been the shear strength of the workpiece. However, other materials often have to be machined. For example, sand castings and thermoset plastic mouldings are highly abrasive, so carbide tooling is used for these materials to give an adequate tool life, even though the shear strengths of sand castings and thermosets are relatively low. Carbide tooling is also required when machining high-strength alloys. Such carbide tooling is also used where the economics of manufacture demands high rates of metal removal and where machine tools are available with sufficient power to exploit the tooling.

Non-ductile materials such as grey cast iron and free-cutting brass do not form the continuous, ribbon-like chip associated with the ductile metals, but form a granular, discontinuous chip (Section 2.7). For such non-ductile materials the rake angle of the cutting tool can be reduced to very low values (Fig. 2.7) and advantage can be taken of the corresponding increase in tool strength.

2.5 Cutting-tool materials

Cutting-tool materials must have the following properties:

- Sufficient strength to resist the cutting forces.
- Sufficient hardness to resist wear and give an adequate life between regrinds.
- The ability to retain its hardness at the high temperatures generated at the tool point when cutting.

Figure 2.12 shows the relationship between hardness and temperature for some typical cutting-tool materials.

Fig. 2.12 *Hardness–temperature curves for cutting-tool materials*

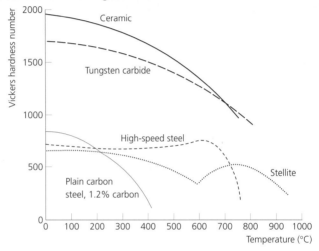

2.5.1 *High-carbon steels*

High-carbon steels are plain carbon steels with a carbon content of 0.87–1.2 per cent. As can be seen from Fig. 2.12, although they can have an initial hardness greater than high-speed steel, their hardness is rapidly reduced when the temperature of the material rises. Therefore they are unsuitable for production machining and are mainly used for hand tools.

2.5.2 *High-speed steels*

These are amongst the most widely used cutting-tool materials for such applications as drills, reamers, milling cutters, turning tools, thread-cutting tools, planing and shaping tools. In the annealed (soft) condition, high-speed steels can be forged and machined with relative ease; yet when they are hardened, they retain sufficient strength to work unsupported with high positive rake angles and maintain their hardness almost up to red-heat. Table 2.1 lists the composition of some typical high-speed steels. The more costly super high-speed steels contain an appreciable amount of metal cobalt. This has the effect of substantially raising the temperature at which the steel can retain its hardness. It also increases the strength and toughness of the steel so that it can resist the cutting forces present when machining modern high-strength alloys.

Table 2.1 Typical high-speed steels (HSS)

Type of steel	Composition (%)*						Hardness (VNP)	Uses
	C	Cr	W	V	Mo	Co		
18% tungsten	0.68	4.0	19.0	1.5			800–850	Low-quality alloy, not much used
30% tungsten	0.75	4.7	22.0	1.4			850–950	General-purpose cutting tools for jobbing workshops
6% cobalt	0.8	5.0	19.0	1.5	0.5	6.0	800–900	Heavy-duty cutting tools
Super HSS 12% cobalt	0.8	5.0	21.0	1.5	0.5	11.5	850–950	Heavy-duty cutting tools for machining high-tensile materials

*C = carbon, Cr = chromium, W = tungsten, V = vanadium, Mo = molybdenum, Co = cobalt; the remaining percentage is iron.

2.5.3 *Stellite*

Stellite is a cobalt-based alloy containing little or no iron. It can only be cast to shape or deposited as a hard facing onto a medium-carbon steel shank using an oxyacetylene torch to melt the Stellite rod. It requires no heat treatment and is so hard that it can only be machined by grinding. It is much more expensive than high-speed steel, and although slightly softer, it retains its hardness even when the cutting edge is glowing red-hot. It is sufficiently strong and tough to be used in standard tool holders at high values of positive rake angle. A typical composition is cobalt 50 per cent, tungsten 33 per cent, carbon 3 per cent, various 14 per cent.

2.5.4 *Cemented carbides*

Preformed tool tips made from metallic carbides are harder than Stellite, cheaper than Stellite, and capable of operating at the same temperatures. Carbides can only be machined by grinding using silicon carbide (green grit) abrasive wheels. Carbide cutting tools fall into three categories.

Tungsten carbide is very hard and brittle and is used to machine such materials as grey cast iron and cast bronzes. These metals have a relatively low tensile strength and form a discontinuous chip. However, they have a hard and abrasive skin as a result of the casting process. Tool tips made from tungsten carbide tend to be porous, so particles of metal from the workpiece material tend to become embedded in the tool tip. Although the metal being cut will not chip-weld directly to the carbide, it will adhere to the embedded particles to form a built-up edge (Section 2.8). Therefore, owing to its brittleness and low strength, coupled with the tendency for a built-up edge to form on the tool tip, care has to be taken when using straight tungsten carbide inserts for machining ductile materials.

The tungsten carbide particles are mixed with metallic cobalt particles to form a *composite* cutting material after sintering. The sintered material shows tungsten carbide particles are dispersed through a matrix of metallic cobalt (Fig. 2.13). The size of the carbide particles and the percentage by volume of carbide in the composite will greatly affect the properties and performance of the material.

Fig. 2.13 *Sintered carbide cutting-tool material*

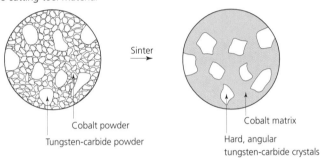

Let's compare this material with the surface of a road. Too much tar and not enough stone chippings will create a road that will quickly wear away. Conversely, too many chippings and not enough tar will give a very wear-resistant surface, but without the toughness for the job; it would crack up at the first glimpse of a juggernaut.

Similarly, more carbide and less cobalt in the matrix gives a very wear-resistant cutting-tool material but it would lack toughness. It would be very suitable for taking finishing cuts and for machining low-strength and abrasive materials (such as grey iron castings and laminated plastics). Less carbide and more cobalt in the matrix will produce a less wear-resistant cutting-tool material, but one that is very much tougher and suitable for rough machining high-tensile materials.

The grain size of the carbide particles is also important. A tool tip with very fine grains of carbide will hold a sharp-cutting keen edge and is suitable for finish machining. It is particularly useful when machining non-ferrous metals and alloys. Tool tips with larger grains will not sharpen to such a keen edge, and often a reinforced edge has to be employed. Coarse-grained tips are used for heavy cutting operations on materials which produce long, continuous chips that tend to lead to heavy wear and cratering of the rake face of the tip.

Mixed carbides are mixtures of tungsten and titanium carbides. They are less hard and abrasion resistant than straight tungsten carbide, but they are very much stronger and tougher and are used for cutting high-strength materials. They are also less porous than straight tungsten carbide and are therefore less prone to form a built-up edge. The comments concerning particle size and quantity apply equally to mixed carbides.

Coated carbides are more expensive than tungsten or mixed carbides but can be run at cutting speeds up to 30 per cent greater than those recommended for tungsten or mixed carbides without any reduction in tool life. Tungsten or mixed-carbide tool tips are coated with a very hard and abrasion-resistant film of titanium nitride (TiN). They cannot be reground as this would destroy the coating, hence coated carbides are only available as *disposable inserts*. Table 2.2 lists the standard grades of carbide together with some typical applications.

Table 2.2 Carbide grades for metal-cutting tools

ISO code	ISO grades	General applications
P (blue)	P01–P50	Ductile materials such as plain carbon and low-alloy steels, stainless steel, long-chipping malleable cast iron, ductile non-ferrous metals and alloys
M (yellow)	M10–M40	Tough and difficult materials such as high-carbon steels and high-duty alloy steels, manganese steels, cast steels, alloy cast irons, austenitic stainless steel castings, malleable cast iron, heat-resistant alloys
K (red)	K01–K30	Materials lacking in ductility and components which cause intermittent cutting. Cast iron, chill-cast iron, short-chipping malleable cast iron, hardened steel, non-ferrous free-cutting alloys, free-cutting steels, plastics, wood, titanium alloys

Cutting properties

P01 ◄── K30
Increasing hardness and ability to withstand wear
High cutting speeds and fine feeds

P01 ──► K30
Increasing toughness and ability to withstand
interrupted cutting with coarse feeds
Rough machining high-strength materials

Since carbides are very brittle compared with metallic cutting-tool materials they have to be securely supported by the tool shank, to which they may be brazed or clamped (Fig. 2.14). Modern practice favours the use of disposable, clamped, indexable tips so they can be simply turned round to expose a new cutting edge when blunt, and finally thrown away when all the cutting edges of the insert have been used.

The use of throwaway inserts has a number of advantages:

- The inserts are mass-produced and, relatively speaking, they are very cheap, much cheaper than the cost of regrinding.
- The tolerances on the inserts are such as to ensure consistent cutting performance from one insert to the next.
- The tolerances on the inserts also ensure there is no need to reset the tooling after changing an insert. This reduces the downtime and lost production through tool replacement. Some examples of tool holders and inserts are shown in Fig. 2.15.

Fig. 2.14 *Carbide-tipped single-point tools fixed by (a) brazing and (b) clamping*

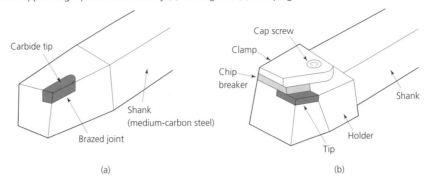

(a)

(b)

Fig. 2.15 *Tool shanks and the different shapes of disposable inserts*

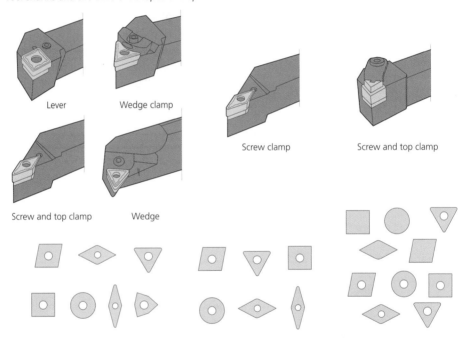

2.5.5 *Ceramics*

Ceramic tips are even harder than those made from carbides and are even more brittle. The ceramic material most commonly used is aluminium oxide, either commercially pure or mixed with other metallic oxides such as chromic oxide. Ceramic tips cannot be brazed to their shanks and can only be used with clamped-tip holders. Ceramic tips are weak in tension and susceptible to edge chipping. They are used for high-speed finishing cuts with fine feed rates where high standards of surface finish are required.

Machines used with ceramic tooling must be very powerful and rigid to exploit this material's special properties. Cutting speeds of 150–300 m/min are common when using ceramic tooling, but any vibration or chatter will cause immediate failure of the cutting edge. Because of its low transverse strength, ceramic tooling is normally used with a *negative rake* angle of $-5°$ to $-7°$ and the chips frequently appear red-hot as they leave the tool (further proof of the high power required with ceramic tooling).

SELF-ASSESSMENT TASK 2.3

Give examples of where the following cutting-tool materials can be used with the greatest advantage. Give reasons for your choice.

(a) high-carbon steel
(b) high-speed steel
(c) tungsten carbide

2.6 Cutting-tool geometry: single-point tools

So far we have only been concerned with establishing the principle of the metal-cutting wedge, together with the corresponding rake and clearance angles in a single plane. The plan profile of the tool is also important as it has a significant effect upon the tool geometry and cutting efficiency. The lathe tool shown in Fig. 2.16(a) is performing a parallel (cylindrical) turning operation. Since the cutting edge is perpendicular to the direction of feed, the tool is cutting *orthogonally*. The shaded area represents the cross-sectional area (shear area) of the chip. The area is calculated by multiplying the feed per revolution by the depth of cut ($d \times f$).

Fig. 2.16 *Feed and depth of cut for parallel cylindrical turning: (a) orthogonal cutting; (b) oblique cutting; (c) comparing the two cuts shows that oblique cutting reduces W without reducing A; the zigzag line is the ISO symbol for feed direction*

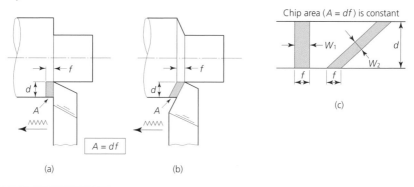

Figure 2.16(b) shows the same turning operation using a tool in which the cutting edge is inclined to the direction of feed; this tool is cutting *obliquely*. The cross-sectional area of the chip is the same as when cutting orthogonally since it is again equal to $d \times f$. However, when cutting obliquely, the chip thickness (W) is reduced; Fig. 2.16(c) shows that:

- The depth of cut d is constant for each example.
- The feed/rev f is constant for each example.
- The chip area ($d \times f$) is constant for each example (parallelogram theory).
- The chip thickness W varies so that $W_1 > W_2$.

Therefore, when cutting obliquely, the rate of metal removal is the same as when cutting orthogonally but the chip is thinner. This thinner chip is more easily deflected over the rake face of the tool, and the tangential cutting force on the tool is reduced as is the frictional wear on the rake face of the tool. Figure 2.17 shows the main forces acting upon orthogonal and oblique turning tools.

Notice that when cutting obliquely there is also a radial component of the feed force acting on the tool. Since this radial force keeps the flanks of the cross-slide traverse screw and nut in contact, any backlash that may be present is taken up, thus preventing the tool from being drawn into the work when taking a heavy cut. Care must be taken when cutting obliquely that the plan approach angle of the tool is not made excessive, else chatter will occur. Figure 2.18 shows how the principles of orthogonal and oblique cutting are applied during a perpendicular (surfacing) operation on a lathe. The area of cut is again $d \times f$.

Fig. 2.17 *Forces acting on a turning tool: (a) orthogonal cutting (no radial force on tool); (b) oblique cutting*

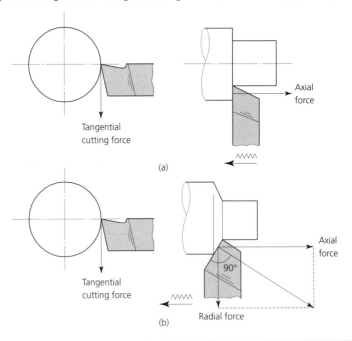

Fig. 2.18 *Area, feed and depth of cut for perpendicular turning (A = df); (a) orthogonal cutting (grooving and parting off); (b) oblique cutting (surfacing)*

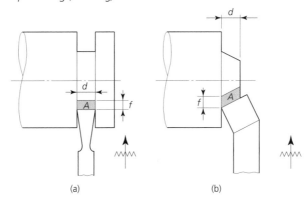

(a)   (b)

So far only tools with positive rake angles have been considered. Traditionally, these tools had to be made from materials with a high transverse strength such as high-speed steel or Stellite. Carbide tools are more brittle and were only used with small positive rake angles or, more usually, with negative rake angles. Negative rake geometry provides greater support for the brittle carbide tool tip (Fig. 2.19). Figure 2.19(a) shows that, with a positive rake angle, the tip of the tool is in shear; whereas in Fig. 2.19(b) the tip of the tool is in compression.

Fig. 2.19 *Rake: (a) positive; (b) negative*

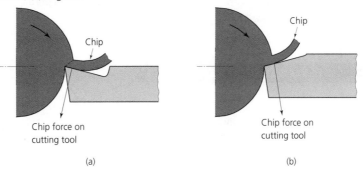

(a)   (b)

A positive rake angle was earlier claimed to be necessary for efficient cutting and to ease the path of the chip over the rake face of the tool, so it may appear strange that a tool with negative geometry can cut effectively. However, negative rake geometry transfers the greater part of the work done in cutting to the chip. This, together with the high cutting speeds possible with carbide-tipped tools, causes considerable heating at the cutting zone and in the chip. This in turn produces a local reduction of the shear strength of the workpiece material and produces an increase in ductility, both of which make it easier to cut high-strength and relatively hard materials. At the same time, the forces acting on the tool and abrasive wear of the rake face are both reduced.

Near finished size (NFS) strategies in modern manufacturing (Chapter 1) have created a move away from powerful machines removing large amounts of metal at very high rates; chips no longer come off the machine red-hot, once a spectacular sight at machine tool exhibition. With improvements in the manufacture of carbide and ceramic inserts, and the need only to finish-machine, there is an increasing use of carbide tooling with *positive* rake angles. This is achieved as shown in Fig. 2.20.

Fig. 2.20 *An insert with a positive rake angle*

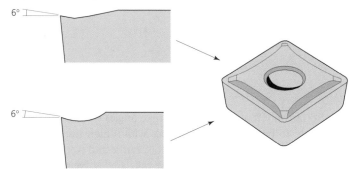

Machines are becoming smaller, faster and more versatile to cope with the changes that are taking place. This all leads to reduced material and energy consumption, together with lower capital and operating costs. This is good for the environment and for the consumer.

Even more sophisticated cutting-tool materials such as cubic boron nitride and industrial diamonds are discussed in Volume 2 of *Manufacturing Technology* and Volume 2 of *Engineering materials*.

SELF-ASSESSMENT TASK 2.4

1. Sketch examples of single-point tools:

    (a) cutting orthogonally
    (b) cutting obliquely

2. Sketch the difference between cutting with a positive rake and cutting with a negative rake, and compare the advantages and limitations of these two techniques.

2.7 Cutting-tool geometry (multi-point tools)

The fundamental metal-cutting wedge is equally applicable to multi-point cutting tools but is sometimes less obvious to see. The cutting geometry of a number of commonly used multi-point tools will now be considered.

2.7.1 *Twist Drill*

Figure 2.21 applies the basic cutting angles (including the inevitable metal-cutting wedge) to a twist drill. Notice that the helix angle of the drill flutes provides the rake angle at the outer edge of the lip of the drill. This angle is not constant and becomes reduced towards the centre of the drill, with a corresponding reduction in cutting efficiency. It is not possible to vary the rake angle of a twist drill to any great extent during regrinding; this is because the helix angle of the flutes is fixed at the time of manufacture. However, drills of various helix and point angles can be purchased for drilling different materials (Fig. 2.22).

Fig. 2.21 *Twist drill: (a) cutting angles; (b) variation in rake along the lip of the drill; the rake angle of the periphery is equal to the helix angle of the flute*

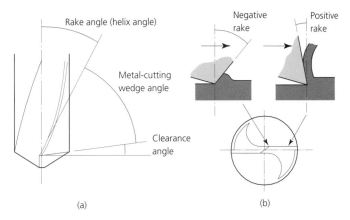

(a) (b)

The d/f ratio and area of cut for a drill is determined as shown in Fig. 2.23. The depth of cut is fixed and equal to the radius of the drill, whereas the feed is variable and equal to the axial movement of the drill per revolution. The cutting speed and feed rate depend upon the workpiece material properties. For any given workpiece material, the feed rate also depends upon the drill diameter, as this influences the strength of the drill.

2.7.2 *Milling cutters*

Figure 2.24(a) shows a single-point cutting tool, as previously described, cutting with linear motion relative to the workpiece. It is not necessary for the tool to move in a straight line and Fig. 2.24(b) shows a single-point cutting tool mounted in a rotating cutter block. The cutter now generates a curved surface (arc of a circle) and it can be seen that, when cutting with rotary motion, the heel of the tool has to be backed off to prevent it rubbing on the workpiece. The angle produced by backing off the heel of the cutting tool is called the *secondary clearance angle.*

If the axis of the rotating cutter is moved parallel to the workpiece, as shown in

Fig. 2.22 *Twist drill point and helix angles*

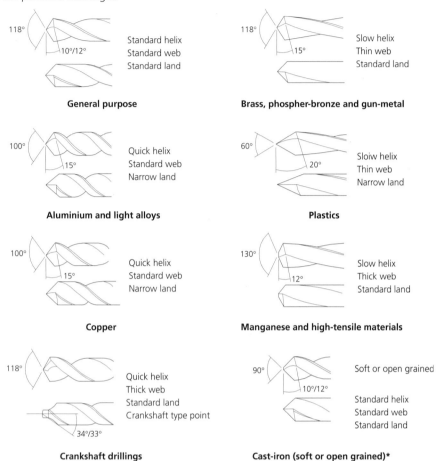

118°
10°/12°

Standard helix
Standard web
Standard land

General purpose

118°
15°

Slow helix
Thin web
Standard land

Brass, phospher-bronze and gun-metal

100°
15°

Quick helix
Standard web
Narrow land

Aluminium and light alloys

60°
20°

Sloiw helix
Thin web
Narrow land

Plastics

100°
15°

Quick helix
Standard web
Narrow land

Copper

130°
12°

Slow helix
Thick web
Standard land

Manganese and high-tensile materials

118°
34°/33°

Quick helix
Thick web
Standard land
Crankshaft type point

Crankshaft drillings

90°
10°/12°

Soft or open grained

Standard helix
Standard web
Standard land

Cast-iron (soft or open grained)*

* For medium or close grain use a standard drill
For harder grades of alloy or cast iron it may be necessary to use a manganese drill

Fig. 2.23 *Twist drill cutting*

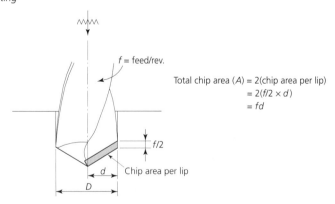

f = feed/rev.

Total chip area (A) = 2(chip area per lip)
= 2(f/2 × d)
= fd

f/2

Chip area per lip

d

D

Fig. 2.24 *Rotary cutting action: (a) linear movement of a single-point tool; (b) rotary movement of a single-point tool; (c) traversing work under a rotating cutter*

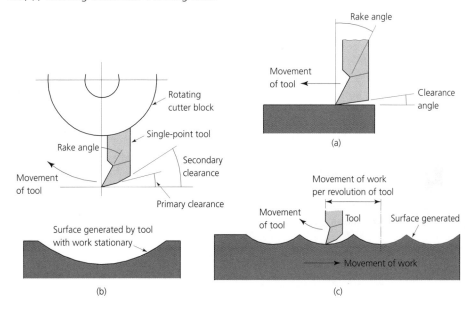

Fig. 2.25 *Multiple cutting edges: (a) increasing the number of tool inserts in the cutter block; (b) surface generated by one tool point; (c) surface generated by two tool points; (d) surface generated by four tool points*

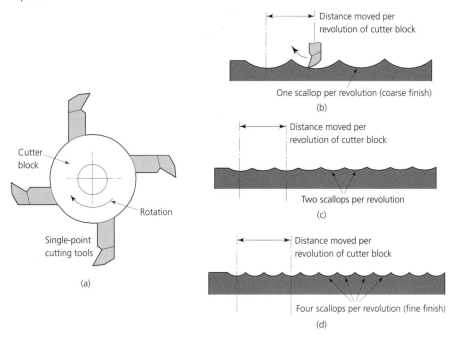

Fig. 2.24(c), a surface approaching a plane surface will be generated. The finer the feed per revolution of the cutting tool, the closer the surface generated will approach a plane surface since successive cuts will overlap. Unfortunately, reducing the feed per revolution also reduces the rate of material removal.

The performance of the rotating cutter can be greatly improved by inserting a number of single-point tools in the cutter block, as shown in Fig. 2.25(a). The feed per revolution of the cutter block is shared between the individual cutting tools, thus reducing the feed per tool without reducing the rate of material removal. This is shown in Fig. 2.25(b) to (d). The forces acting on each tool tip are also correspondingly reduced. The basic cutting geometry of a milling cutter tooth is compared with that of a single-point tool in Fig. 2.26.

Like single-point cutting tools, milling cutters can also be selected to provide orthogonal or oblique cutting actions. Figure 2.27(a) shows a straight tooth cutter cutting orthogonally, and Fig. 2.27(b) shows a helical tooth cutter cutting obliquely. The helical tooth cutter not only reduces the chip thickness but ensures that each successive tooth starts to cut before the preceding tooth ceases to cut. This maintains uniform torque loads on the machine drive and reduces the amount of chatter, giving an improved surface finish to the workpiece.

Fig. 2.26 *Basic tooth geometry: (a) single-point cutting tool; (b) milling cutter tooth*

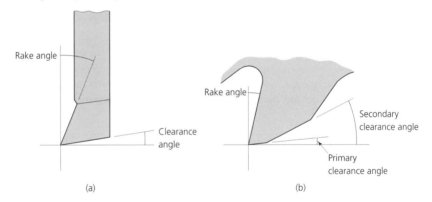

Fig. 2.27 *Tooth forms: (a) orthogonal (straight tooth cutter); (b) oblique (30° helical tooth cutter)*

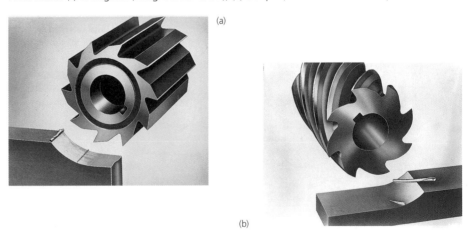

1. Some non-ferrous metals are difficult to drill with standard flute twist drills.
   (a) Find out and explain why this is so.
   (b) Select a suitable combination of rake angle and point angle for drilling brass.
2. (a) Show with the aid of sketches why a milling cutter tooth requires both a secondary clearance angle and a gullet.
   (b) Show with the aid of sketches how:
       (i) milling cutters can have orthogonal cutting action
       (ii) milling cutters can have an oblique cutting action
   (c) Discuss the advantages and limitations of such cutter tooth formations.

Unlike the drilling machine and the lathe, the longitudinal traverse rate of a milling machine table is not related to the rotational speed of the cutter but is stated independently as the distance travelled by the worktable per minute. This is limited by the optimum feed per tooth that will give a satisfactory cutter life between regrinds for any given application. Example 2.1 gives a typical calculation for the rate of material removal when milling.

EXAMPLE 2.1

Calculate the time taken to complete a cut using a slab mill under the following conditions:

Diameter of cutter	125 mm
Number of teeth	6
Feed per tooth	0.05 mm
Cutting speed	45 m/min
Length of cut	270 mm

(Take π as 3)

$$N = 1000S/\pi D$$
$$= \frac{1000 \times 45}{3 \times 125}$$
$$= \mathbf{125\,rev/min}$$

where $N =$ spindle speed
$S = 45$ m/min
$D = 125$ mm
$\pi = 3$

Feed per rev = feed per tooth $\times$ number of teeth
$$= 0.05\,\text{mm} \times 6$$
$$= \mathbf{0.3\,mm/rev}$$

Table feed per min = feed per rev $\times$ rev per min
$$= 0.3 \times 120$$
$$= \mathbf{36\,mm/min}$$

Time to complete $= \dfrac{\text{length of cut}}{\text{table feed per min}} = 270/36 = \mathbf{7.5\,min}$

SELF-ASSESSMENT TASK 2.6

Calculate the time taken to complete a cut using a slab mill under the following conditions:

Cutter diameter	100 mm	Cutting speed	30 m/min
Number of teeth	8	Length of cut	300 mm
Feed per tooth	0.05 mm	(Take π as 3)	

2.7.3 *Abrasive wheels*

The grinding process removes material by the use of rapidly rotating abrasive wheels. Abrasive wheels consist of a large number of abrasive particles, called grains, held together by a bond to form a multi-tooth cutter which cuts in a similar manner to a milling cutter (Fig. 2.28). The grains at the surface of the wheel are called *active grains* because they actually perform the cutting operation. In peripheral grinding, each active grain removes a short chip of workpiece material; the chips can be seen in Fig. 2.29 which shows highly magnified dross from a grinding wheel. Notice that the dross consists of particles of abrasive material stripped from the grinding wheel together with metallic chips which are remarkably similar to the chips produced by the milling process.

Fig. 2.28 *Cutting action of abrasive wheel grains*

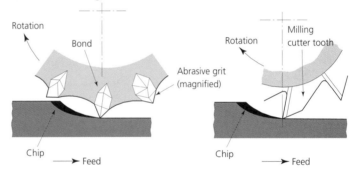

Fig. 2.29 *Dross from a grinding wheel*

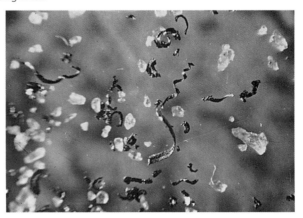

Since it is convenient to think of each grit particle in a grinding wheel as very small milling cutter teeth, the abrasive wheels used in grinding processes have many more teeth than a milling cutter. Because this reduces the chip clearance, the abrasive wheel produces a superior surface finish at the expense of a lower metal removal rate. The fact that the cutting points are irregularly shaped and randomly distributed over the active face of the wheel further enhances the quality of the surface finish, although *vibration patterns* can occur due to worn bearings and the wheel being out of balance. As grinding proceeds, the cutting edges of the grains become dulled and the cutting forces acting on the grains increase until either the blunt grains are fractured or they are ripped from the bond, exposing new active cutting points. A correctly selected grinding wheel should therefore have self-sharpening characteristics.

2.8 Abrasive wheel selection

The correct selection of a grinding wheel depends upon many factors and the following comments should only be considered as general guidelines. Manufacturers' literature should be consulted for more precise information.

2.8.1 *Material to be ground*

- Aluminium oxide abrasives should only be used on materials with relatively high tensile strengths.
- Silicon carbide abrasives should only be used on hard, brittle materials with relatively low tensile strengths.
- A fine-grained wheel can be used on hard, brittle materials.
- A coarser-grained wheel should be used on soft, ductile materials.
- A general guide is to use a soft grade of wheel with a hard workpiece, and a hard grade of wheel with a soft workpiece.
- It is permissible to use a close structure on hard, brittle materials, but a more open structure should be used on soft, ductile materials.
- The bond is chosen for a particular application and is rarely influenced by the material being ground.

2.8.2 *Rate of stock removal*

- A coarse-grained wheel should be used for rapid stock removal, but it will give a comparatively rough finish. A fine-grained wheel should be used for finishing operations that require low rates of stock removal.
- The structure of the wheel has a major effect on the rate of stock removal; an open structure with a wide grain spacing is used for maximum stock removal and cool cutting conditions.
- Note that the performance of a grinding wheel can be appreciably modified by the method of dressing, the operating speed and the workpiece traverse rate.

2.8.3 *Arc of contact*

Figure 2.30 explains the arc of contact. For efficient grinding, the arc of contact should generally be kept as small as possible. That is, the diameter of the abrasive wheel should be large compared with the workpiece diameter.

Fig. 2.30 *Arcs of contact: (a) large; (b) small*

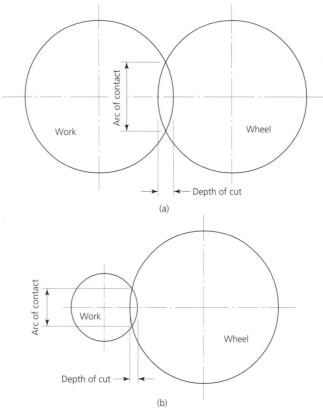

- For a small arc of contact, a fine-grained wheel should be used, whereas for a large arc of contact a coarser-grained wheel should be used to prevent overheating.
- For a small arc of contact, a 'hard' wheel may be used, whereas for a large arc of contact a 'soft' wheel should be used as the cutting edges will become dulled more quickly.
- For a small arc of contact a closed structure wheel may be used; this has the advantages of improved surface finish and closer dimensional control. For a large arc of contact an open-structured wheel should be used to maintain the free cutting conditions.

2.8.4 *Bond*

The bond is selected for its mechanical properties. It must achieve a balance between having enough strength to resist the rotational, bursting forces and the applied cutting forces, and satisfying the requirements of cool cutting together with the controlled release of dulled grains and the exposure of fresh cutting edges.

2.8.5 *Type of grinding machine*

A heavy-duty, rigidly constructed machine can produce accurate work using softer grades of wheel. This reduces the possibility of overheating the workpiece and drawing its temper (i.e. reducing its hardness) or, in extreme cases, causing surface cracking of the workpiece. Furthermore, broader wheels can be used, and this increases the rate of metal removal without loss of accuracy.

2.8.6 *Wheel speed*

Variation in the surface speed of a grinding wheel has a profound effect upon its performance. Increasing the speed of the wheel causes it to behave as though it is harder than marked. Conversely, reducing the surface speed of a grinding wheel causes it to behave as though it is softer than marked. Care must be taken to ensure the bond has sufficient strength to resist the bursting effect of the rotational forces. The safe working speed of an abrasive wheel must never be exceeded.

2.8.7 *Traverse rate*

The effect of the workpiece traverse rate on grinding wheel wear is shown in Fig. 2.31.

Fig. 2.31 *Wear conditions when work traverses (a) two-thirds of the wheel width per revolution and (b) one-third of the wheel width per revolution of work*

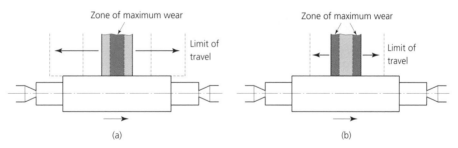

2.9 Abrasive wheel specification

Having selected the type of grinding wheel that is required for a particular job, we have to be able to specify the wheel required. Manufacturers code their wheels and we will now look at how this is done. Any grinding wheel consists of two constituents:

- The *abrasive grit* that does the cutting.
- The *bond* that holds the grit together.

The specification of a grinding wheel indicates its construction and its suitability for a particular operation. For example, let's consider a wheel with the marking 38A60-J5V:

38A is the abrasive type	Table 2.3
60 is the grain size	Table 2.4
J is the grade	Table 2.5
5 is the structure	Table 2.6
V is the bond	Table 2.7

Reference to the tables shows that such a wheel has an aluminium oxide abrasive, the grain size of the abrasive grit is medium to fine, the structure has a medium spacing, and a vitrified bond has been used. We will now consider this code in greater detail.

2.9.1 *Abrasive*

The abrasive is chosen to suit the material being cut. As a general classification:

- Brown aluminium oxide is used for grinding tough materials.
- White aluminium oxide is used for grinding hard die steels.
- Green silicon carbide is used for grinding hard, brittle materials such as cemented carbides.

Table 2.3 shows how the abrasive type may be coded and is based upon *Norton* abrasives. The British Standard marking system only calls for A = aluminium oxide abrasives or C = silicon carbide abrasives. It does permit the use of manufacturers' prefixes so that specific abrasives can be identified within each broad classification.

Table 2.3 *Abrasive types*

Manufacturer's type code*	BS code	Abrasive	Application
–	A	Aluminium oxide	A high-strength abrasive for hard, tough materials
32	A	Aluminium oxide	Cool, fast cutting for rapid stock removal
38	A	Aluminium oxide	Light grinding of very hard steels
19	A	Aluminium oxide	Milder than 38A, it is used for cylindrical grinding
37	C	Silicon carbide	For hard, brittle materials of high density such as cast iron
39	C (green)	Silicon carbide	For very hard, brittle materials such as tungsten carbide

*These type codes are based upon Norton abrasives.

2.9.2 *Grain size*

The number indicating the grain size represents the number of openings per linear 25 mm in the sieve to size the grains. The larger the number, the finer the grains. Table 2.4 gives a general classification. The sizes listed as very fine are called *flours* and are only used for polishing and for superfinishing processes.

Table 2.4 *Grit size*

Coarse	Medium	Fine	Very fine
10	30	70	220
12	36	80	240
14	40	90	280
16	46	100	320
20	54	120	400
24	60	150	500
		180	600

2.9.3 *Grade*

Grade indicates the strength of the bond, hence the hardness of the wheel. In a *hard* wheel the bond securely anchors the grit in place, thus reducing the rate of wear. In a *soft* wheel the bond is weak and the grit is easily detached, resulting in a high rate of wear. If the bond is too hard, then blunt grit will not break away and the wheel will become *glazed*. If the wheel keeps glazing and having to be dressed, a softer grade should be used so that new, sharp grains are always present at the cutting face of the wheel. If the wheel is too soft, it will wear away too quickly, causing a loss of accuracy. Table 2.5 shows a general classification for wheel grade.

Table 2.5 *Grade*

Category	Designations
Very soft	E F G
Soft	H I J K
Medium	L M N O
Hard	P Q R S
Very hard	T U W Z

2.9.4 *Structure*

Structure indicates the amount of bonding material between the grains and the closeness of adjacent grains; in terms of a milling cutter, it represents the chip clearance. An open structure cuts freely and tends to generate less heat in the cutting zone. That is, an open structure is free cutting and allows rapid material removal. However, it will not produce as good a finish as a close structure. Table 2.6 show a general classification of structure.

Table 2.6 **Structure**	
Category	Designations
Close spacing	0 1 2 3
Medium spacing	4 5 6
Wide spacing	7 8 9 10 11 12

2.9.5 *Bond*

A wide range of bonds are available, so care must be taken to ensure the bond is suitable for the application. The safe use of the wheel mainly depends upon the selection of the correct bond.

- *Vitrified bond* This is the most widely used bond and is similar to glass in composition. It has high porosity and strength, resulting in a wheel suitable for high rates of material removal. It is not adversely affected by water or cutting fluid at normal temperatures.
- *Rubber bond* This is used where a small amount of flexibility is required in the wheel, such as in thin cutting-off wheels and centreless-grinding control wheels.
- *Resinoid (Bakelite) bond* This is for high-speed wheels where the bursting forces are great. Such wheels are used on portable electric grinding machines used in foundries for dressing castings and in fabrication shops for dressing welds. They are strong enough to withstand more abuse than the other bonds.
- *Shellac bond* This is used for large, heavy-duty wheels where a fine finish and cool cutting is required. Such wheels are used for finishing rolling-mill rolls. Table 2.7 shows the code for the general classification of bonding materials.

Table 2.7 **Bond**	
Type of bond	BS code
Vitrified bond	V
Resinoid bond	B
Rubber bond	R
Shellac bond	E

2.9.6 *Wheel speed*

It is now necessary by law for the maximum safe operating speed to be marked on all abrasive wheels. And the maximum operating speed of the spindle must be marked clearly on all grinding machines. The speed of the machine spindle must never exceed the speed marked on the wheel. It is an offence under the Health and Safety at Work Act to cause a wheel to overspeed by using it on a machine for which it not suitable.

2.10 Abrasive wheel faults

2.10.1 *Loading*

When a soft material, such as a non-ferrous metal, is ground with a general-purpose wheel, the spaces between the grains become clogged with metal particles. Under these conditions the particles of metal can be seen embedded in the wheel. This condition is called *loading* and is detrimental to the cutting action of the wheel. Loading destroys the clearance between the grains, causing the wheel to rub rather than to cut freely. Excessive force will have to be applied to the work in an attempt to get the wheel to cut. This in itself may be sufficient to fracture the wheel. In addition, considerable heat will be generated by the wheel rubbing instead of cutting. This heat will not only soften the material being ground (draw the temper) but it will weaken the bond and may cause the wheel to burst.

2.10.2 *Glazing*

If a wheel is used with too strong a bond, so the blunt grains cannot break free, the surface of the wheel will become shiny. This is called *glazing*. The grains will rub rather than cut, so excessive force will have to be applied to the workpiece. As for loading, the work and the wheel will be overheated, resulting in possible softening of the work and possible failure of the grinding wheel.

2.10.3 *Dressing*

Both loading and glazing are overcome by dressing the wheel to remove the embedded metal particles or to remove the blunt abrasive grains.

The *Huntington* wheel dresser (Fig. 2.32) has star wheels that can dig into the abrasive wheel, removing the particles of embedded metal and any blunt grains. This device is widely used with pedestal-type double-ended off-hand grinding machines but lacks the accuracy for precision grinding machines.

Fig. 2.32 *Huntington wheel dresser*

Lugs hook over workrest

The diamond wheel dresser is shown in Fig. 2.33. This is used for dressing the wheels of precision surface grinding and cylindrical grinding machines. It not only dresses the wheel but trues it as well. Brown Burt industrial diamonds from South Africa are used, as they are very much cheaper and more plentiful than gemstones.

Fig. 2.33 *Dressing: (a) incorrect – the tip of the diamond will wear flat; this will blunt the new abrasive grains as they are exposed; (b) correct – the diamond leads the wheel centre and trails the direction of rotation; the diamond will keep sharp and dress cleanly*

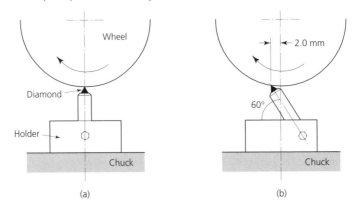

(a)   (b)

SELF-ASSESSMENT TASK 2.7

1. Describe an abrasive wheel with the specification 32A46-M7V and suggest a purpose for which it could be used.

2. (a) Describe the tell-tale signs that would lead you to believe that an abrasive wheel is:
   (i) loaded
   (ii) glazed
   (b) Describe the corrective action you would take before using a wheel in either of the conditions (i) and (ii).

2.11 Process variables

Process variables are the factors immediately under the control of the machine operator and are largely concerned with achieving a compromise between maximum rates of metal removal and maximum tool life.

2.11.1 *Cutting speed*

Cutting speed influences the power required for the cutting process and the tool life. It does not influence the forces acting on the cutting tool. The forces acting on the cutting tool depend solely upon the properties of the workpiece material, the cutting-tool geometry, and the depth of cut and rate of feed. Example 2.2 shows how increasing the cutting speed affects the power required for a given process. The upper limit is set by the rigidity of the machine tool being used and the power available at its spindle.

EXAMPLE 2.2

Compare the power required to turn a 100 mm diameter bar at 80 rev/min with the power required at 160 rev/min. The tangential cutting force is constant at 350 N in both cases.

Power required at 80 rev/min

$$\text{Power (W)} = \frac{\text{force (N)} \times \text{radius (m)} \times 2\pi \times \text{speed (rev/min)}}{60}$$

$$= \frac{350 \times 0.05 \times 2 \times 3.142 \times 80}{60}$$

$$= \mathbf{146.6\ W}$$

Power required at 160 rev/min

$$\text{Power (W)} = \frac{\text{force (N)} \times \text{radius (m)} \times 2\pi \times \text{speed (rev/min)}}{60}$$

$$= \frac{350 \times 0.05 \times 2 \times 3.142 \times 160}{60}$$

$$= \mathbf{293.2\ W}$$

Notice that doubling the spindle speed doubles the power required. Thus the power consumed in cutting is proportional to the spindle speed.

Expressed algebraically

$$P = \frac{2\pi R F_t N}{60}$$

where $2\pi R F_t / 60$ is constant for any given set of cutting conditions

thus $P \propto N$

The cutting speed also influences the tool life. It has just been shown that increasing the cutting speed increases the power required. Since power can be defined as the rate of doing work or using energy, increasing the cutting speed increases the rate at which energy is dissipated at the cutting zone. Since energy cannot be created or destroyed (law of conservation of energy), the mechanical energy of cutting is converted into heat energy in the cutting zone. This increases the temperature at the tool tip and reduces the tool life. The relationship between cutting speed and tool life is logarithmic; an empirical relationship has been derived experimentally:

$$(Vt)^n = C$$

where $V =$ cutting speed in m/min
$t =$ tool life in minutes
$n =$ tool life index (see Table 2.8)
$C =$ a constant for a given set of cutting conditions

Table 2.8 *Tool life index*

Material and conditions	Tool material	n
3.5% nickel steel	Cemented carbide	0.2
3.5% nickel steel (roughing)	High-speed steel	0.14
3.5% nickel steel (finishing)	High-speed steel	0.125
High-carbon, high-chromium die steel	Cemented carbide	0.15
High-carbon steel	High-speed steel	0.2
Medium-carbon steel	High-speed steel	0.15
Mild steel	High-speed steel	0.125
Cast iron	Cemented carbide	0.1

EXAMPLE 2.3

The life of a lathe tool is 4 hours when operating at a cutting speed of 40 m/min. Given that $Vt^n = C$, find the highest cutting speed that will give a tool life of 8 hours.

The value of n is 0.125.

Determine the value of $\log_{10} C$ for the initial conditions

$$C = V(t_1)^n \qquad V = 40\,\text{m/min}, \ t_1 = 240\,\text{min}, \ n = 0.125$$

$$
\begin{aligned}
\log_{10} C &= \log_{10} V + n \log_{10}(t_1) \\
&= \log_{10} 40 + (0.125 \log_{10} 240) \\
&= 1.6021 + (0.125 \times 2.3802) \\
&= 1.6021 + 0.2975 \\
&= \mathbf{1.8996} \ (C \text{ is a constant so there are no units})
\end{aligned}
$$

Using the value of C calculated above, determine V_{max} for the revised conditions

$$V_{\text{max}} = C/t_2^n \qquad t_2 = 480\,\text{min}$$

$$
\begin{aligned}
\log_{10} V_{\text{max}} &= \log_{10} C - n \log_{10} t_2 \\
&= 1.8996 - (0.125 \times 2.6812) \\
&= 1.8996 - 0.3352 \\
&= 1.5644
\end{aligned}
$$

$$
\begin{aligned}
V_{\text{max}} &= 10^{1.5644} \\
&= \mathbf{36.68\ m/min}
\end{aligned}
$$

You can see from this example that reducing the cutting speed by only 8.3 per cent doubles the life of the tool. Similarly, increasing the cutting speed by only 8.3 per cent would almost halve the life of the tool. Hence the need always to select a lower rather than higher cutting speed if the machine controls do not give the optimum value.

2.11.2 *Depth of cut and rate of feed*

The depth of cut, feed rate and area of cut have already been described for a number of machining operations. The same area of cut, hence the same rate of metal removal, can be achieved in two ways:

- Using a shallow depth of cut and a coarse feed.
- Using a deep cut and a fine feed.

Figure 2.34(a) shows the effect of using a shallow cut and a coarse feed. Notice that the chip is bent across its thickest section and, since the bending force increases as the cube of the thickness of the chip, doubling the rate of feed increases by eight times ($2^3 = 8$) the load on the tool resulting from the deflection of the chip.

Figure 2.34(b) shows the alternative effect of using a deep cut and a fine rate of feed. The chip is now bent across its thinnest section and, for the reasons set out above, the load on the tool is substantially reduced. Doubling the depth of cut only doubles the load on the tool, so increasing the depth of cut has far less effect on the cutting force than increasing the rate of feed. And a deep cut with a fine rate of feed will produce a better surface finish without reducing the rate of metal removal. Unfortunately, an excessively deep cut tends to promote vibration (chatter), and the ratio of depth of cut to rate of feed has to be a compromise between the load on the tool, the surface finish and the point at which chatter commences.

Fig. 2.34 *Effects of feed rates and cut depths: (a) coarse feed + shallow cut, and effect on chip; (b) fine feed + deep cut, and effect on chip*

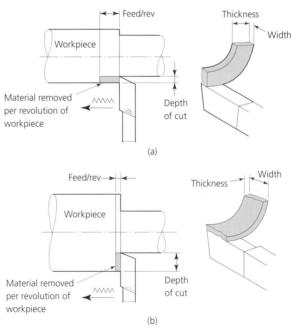

2.12 Cutting fluids

The correct selection and use of cutting fluids is one of the most important, and often one of the cheapest, factors in enhancing the performance of cutting tools. To obtain optimum rates of metal removal and to maintain optimum tool service life, it is necessary to lubricate and cool the chip-tool interface. Cutting fluids are designed to fulfil one or more of the following functions:

- To cool the tool and workpiece.
- To lubricate the chip-tool interface and reduce tool wear due to friction and abrasion.
- To prevent chip welding (formation of a built-up edge).
- To improve the finish of the machined surface.
- To flush away the chips from the cutting zone.
- To prevent corrosion of the work and the machine.

Experiments have shown that, for the majority of machining operations, the cooling and flushing action of the cutting fluid is most important as the rate of tool wear is extremely sensitive to small temperature changes at the chip-tool interface. For this reason, emulsified cutting fluids are used in most workshop applications since the high water content improves the cooling action and reduces the cost.

However, the lubricating action of the cutting fluid becomes of prime importance in reducing the wear rate of the tool in such operations as tapping, broaching and gear cutting, where expensive form tools are used. In these operations the cutting speeds are relatively low and the chip force on the rake face of the tool is very high, thus straight (undiluted) cutting lubricants are used. These lubricants often contain an extreme-pressure additive.

Mineral lubricating oils are unsuitable as cutting fluids. Their viscosity is too high and their specific heat capacity is too low to make them effective coolants; their lubricity is inadequate to withstand the high contact pressure between the chip and the tool, and they give off noxious fumes when raised to the cutting temperature. They also represent a fire hazard.

2.12.1 *Soluble oils*

High cutting temperatures cause tool softening and chip welding, and they can promote corrosive chemical reactions between the chip and the tool. In some instances a reduction in temperature of only 14 °C at the cutting zone can increase the tool life by 150 per cent. Control of the cutting temperature, leading to a reduction of thermal expansion, can also help to maintain the cutting accuracy and to prevent distortion of the workpiece.

Water and oil will not mix without an emulsifier. An emulsifier, e.g. a detergent, breaks up the oil into droplets which spread throughout the water to form an emulsion. This is what happens when the so-called soluble oils are added to water. The milky appearance of these emulsions is due to the light being refracted by the oil droplets. It is from this milky appearance that the emulsion gets the popular name of 'suds'.

Dilution with water reduces the lubricating properties of the oil, and soluble oils are unsuitable for very severe machining conditions. The high water content tends to cause corrosion of the workpiece and the machine, therefore soluble oils must contain a rust inhibitor.

2.12.2 *Compounded or blended oils*

These are mixtures of mineral and fatty oils. The film strength of the fatty oils is retained, even when diluted with 75 per cent mineral oil. As a result, they are much cheaper, more fluid and more chemically stable than neat fatty oils, whilst retaining the superior high-pressure lubricating properties of fatty oils. Blended oils are very versatile and are suitable for a wide range of machining operations.

2.12.3 *Synthetic cutting fluids*

In this group of cutting fluids, the mineral and fatty oil base of conventional cutting oils is replaced by aqueous solutions of inorganic chemicals together with corrosion inhibitors and extreme-pressure lubricating additives. These solutions are transparent, but colouring agents are added to differentiate them from water and from soda-ash solutions. Having a high water content, synthetic chemical cutting fluids possess excellent cooling properties. Other benefits include a high level of cleanliness in the cooling system and on the machine slideways with lack of sludging, long-term stability, easily removable residual films, and no tendency to foaming. Being transparent, they give improved visual control of intricate machining operations. Other advantages are:

- An absence of fire risk; some oil-based cutting fluids are highly flammable.
- Improved long-term stability over soluble oils, particularly in hard water areas.
- Reduced health hazards for operators.
- Spent fluids may be disposed of as normal trade effluent without first having to render them safe environmentally by complex and expensive chemical treatment.

2.12.4 *Extreme-pressure additives*

The very high pressures that exist at the chip-tool interface during machining do not allow conventional fluid film and boundary layer lubrication to be achieved to any significant extent under severe cutting conditions. Where extreme-pressure lubrication properties are required, chlorine and sulphur compounds are added to the cutting oil. These compounds react chemically with the tool and chip material at the chip-tool interface to produce a coating of high lubricity even when subjected to extreme pressures.

Sulphurised oils are probably the most useful and widely used group of extreme-pressure cutting fluids available either as straight oils or as soluble oils. They are compounded to avoid any free sulphur, which would attack and stain copper and high-nickel alloys. Sulphurised oils are used for processes such as gear cutting, broaching, thread grinding, thread cutting and automatic lathe work.

Sulphured oils contain free or elemental sulphur that is completely dissolved in a mineral or compounded oil. The sulphur is in a very active state, and although the oils exhibit the ultimate in extreme-pressure characteristics, they will attack and stain copper and alloys with a high nickel content.

1. Explain why lubricating oils are unsuitable as cutting fluids.

2. (a) Find out what precautions should be taken when mixing emulsified cutting oils (suds) to ensure a stable emulsion.
   (b) Explain under what conditions the emulsion will deteriorate and should be replaced.

EXERCISES

2.1 With the aid of sketches show how the fundamental metal-cutting wedge is applied to the following metal-cutting tools so as to provide rake and clearance angles:
(a) lathe parting-off tool
(b) twist drill
(c) thread-cutting tap
(d) slot milling cutter for a horizontal milling machine
(e) power hacksaw blade and cold chisel

2.2 Describe the influence of the workpiece material on the selection of cutting angles for a given cutting tool and on the type of chip produced.

2.3 (a) With reference to manufacturers' manuals, describe the essential differences between a P15 carbide tool insert and a K30 carbide tool insert. Suggest one appropriate use for each insert and give reasons for your suggestions.
(b) With the aid of sketches, show the difference between positive rake cutting and negative rake cutting; discuss the advantages and limitations of each technique.

2.4 Show with the aid of sketches what is meant by the terms **orthogonal cutting** and **oblique cutting**. Describe the effect of these tool geometries on the cutting action of a single-point lathe tool in terms of the d/f ratio and the chip thickness.

2.5 Discuss the factors which influence the choice of cutting speed, rate of feed and depth of cut when setting up a machine for a given job.

2.6 To the nearest second, calculate the time taken to complete a cut using a slab milling cutter under the following conditions:

Diameter of cutter	100 mm
Number of teeth	8
Feed per tooth	0.05 mm
Cutting speed	30 m/min
length of cut	250 mm

2.7 Describe with the aid of sketches:
(a) the difference between the formation of continuous and discontinuous chips
(b) what is meant by a built-up edge on a tool and how it may be avoided
(c) the function of a chip breaker

2.8 The life of a lathe tool is 3 hours when it is cutting at 80 m/min. Given that $Vt^n = C$, calculate the highest cutting speed which will give a tool life of 8 hours. Take $n = 0.15$.

2.9 (a) Discuss the factors which have to be considered when selecting an abrasive wheel (grinding wheel) for a specific application.

(b) Explain what is meant by **arc of contact** and how this affects the grade of wheel chosen.

2.10 List the main functions of a cutting fluid and discuss the factors which have to be considered when selecting a cutting fluid for a particular application.

3 Toolholding and workholding

The topic areas covered in this chapter are:

- Principles of location and restraint.
- Practical location and clamping.
- Drilling: toolholding and workholding.
- Milling: toolholding and workholding.
- Turning: toolholding and workholding.
- Grinding: toolholding and workholding.

3.1 Principles of location and restraint

Figure 3.1 shows a body in space. It is free of all restraints, and it has *six degrees of freedom*. That is, it is able to:

- Move back and forth along the X-axis.
- Move from side to side along the Y-axis.
- Move up and down along the Z-axis.
- Rotate in either direction about the X-axis.
- Rotate in either direction about the Y-axis.
- Rotate in either direction about the Z-axis.

In order that the metal block shown in Fig. 3.1 can be worked upon by hand or by machine tools it must be located in a given position by restraining its freedom of movement.

Figure 3.2 shows how a metal block similar to that shown in Fig. 3.1 can be located in a given position by the application of suitable restraints. The base plate supports the block and locates it in the vertical plane by restraining its downward movement. At the same time, it restrains rotation about the X and Y axes of the block. The addition of three location pegs adds restraint along the X and Y axes and positions the block on the plate. Finally screw clamps are provided to complete the restraint of the block by ensuring its contact with the plate and the location pegs at all times.

The use of screw clamps also provides for variation in size of the block due to manufacturing tolerances. In this example the block is restrained in every direction by contact with solid metal abutments, so it is said to be under *positive restraint*. Thus *locations position* the tool or workpiece relative to the machine, whereas *restraints prevent unwanted movement* of the tool or workpiece and ensure its contact with the locations.

Fig. 3.1 *Six degrees of freedom*

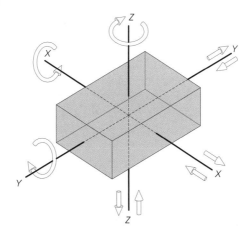

Fig. 3.2 *Location and restraint*

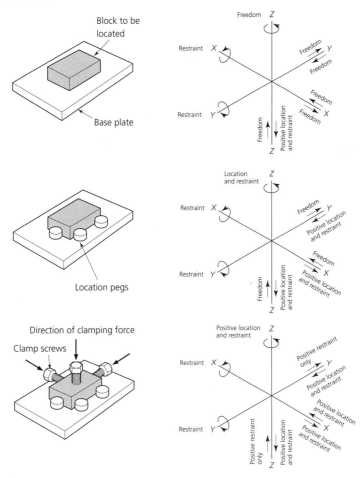

Restraints may be positive or frictional (Fig. 3.3). Wherever possible, cutting forces should be resisted by positive restraints (solid abutments), not by frictional restraint alone. Figure 3.3(a) shows the workpiece restrained by clamps alone (frictional restraint). This is not good practice and should be avoided if possible. The use of multiple clamps can lead to distortion of the workpiece and wastes time during set-up. Figure 3.3(b) shows better workholding practice. The main cutting force is resisted by a solid abutment (positive restraint), which also provides location of the workpiece in a longitudinal direction. This enables the frictional restraint to be reduced to only two clamps.

Fig. 3.3 *Positive and frictional restraints: (a) excessive clamping wastes time and is bad practice; (b) correctly placed clamps and abutment (stop)*

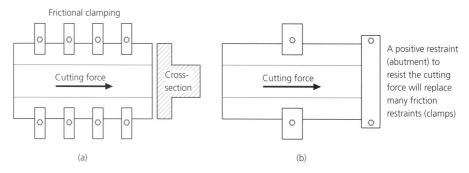

3.2 Practical locations

During the continuous or batch production of large numbers of components by repetitive machining, it is essential that each successive component is located in exactly the same position on the machine. The choice of location will depend upon whether the locating or datum surface of the component is a flat or curved external surface, or a hole. Simple, cylindrical location pegs are shown in Fig. 3.2; some additional locating devices are shown in Fig. 3.4. Where positive locations are used, they should be kept to a minimum to prevent distortion and interaction.

An example of the care needed when using multiple locations is shown in Fig. 3.5. The flat plate link is to be located from the two previously machined holes. However, the centre distance between the holes can vary because of the manufacturing tolerance allowed by the designer. It would be impossible for all the components to be located over two cylindrical location spigots fixed in position on the base plate. Only a component whose hole centre distance was exactly the same as the centre distance of the location spigots would fit. Thus only one cylindrical location spigot should be used; the second hole is located on a location spigot which has been form relieved to allow for the variation in hole centre distance.

Fig. 3.4 *Locations for jigs and fixtures: (a) V-location for external curved surfaces; (b) spigot location for datum hole; (c) plain button location for external flat surface; (d) screw jack adjustable location; (e) spring-loaded adjustable location; the adjustable locations are for locating irregular external flat surfaces*

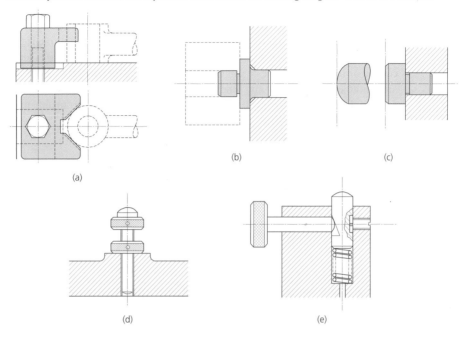

(a)

(b)

(c)

(d)

(e)

Fig. 3.5 *Use of locations: fixed locations must allow for component tolerances*

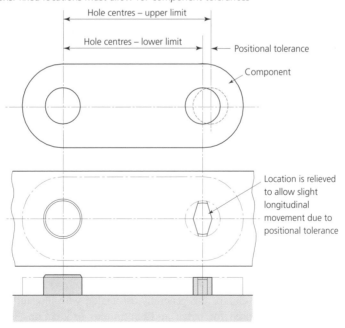

Hole centres – upper limit

Hole centres – lower limit

Positional tolerance

Component

Location is relieved to allow slight longitudinal movement due to positional tolerance

3.3 Practical clamping

Whilst offering adequate restraint, clamping devices should be designed to prevent damage to previously machined surfaces and to avoid distortion or damage to the workpiece. Figure 3.6 shows the difference between clamps and locations suitable for unfinished and finished surfaces. The arrangement shown in Fig. 3.6(a) would be satisfactory for a rough casting or forging. The line contact of the clamp, and the point contact of the location pad are suitable for a rough and uneven surface, but would mark a previously machined surface. For holding on previously machined surfaces, the arrangement shown in Fig. 3.6(b) would be preferable.

Fig. 3.6 *Restraint and location for (a) unfinished surfaces and (b) finished surfaces*

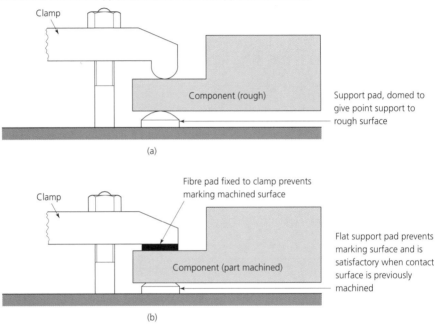

Clamping on thin, unsupported surfaces can cause distortion and even breakage. Figure 3.7 shows the incorrect and correct clamping of a cast-iron workpiece. Had the workpiece been made from a more ductile material, clamping on a thin, unsupported section would not have caused breakage; it would have caused the workpiece to bend and distort, leading to loss of accuracy during machining and even permanent distortion.

The main cutting force should be supported by a solid abutment. This abutment must be proportioned to support the workpiece as close to the point of cutting as possible, otherwise distortion or breakage can occur (Fig. 3.8).

Clamps should be quick and easy to apply in order to minimise the set-up time. Some typical clamping devices are shown in Fig. 3.9 (screw) and Fig. 3.10 (toggle and cam). Toggle and cam clamps are generally associated with jigs and fixtures (Sections 3.4 and 3.5), which are used for batch and continuous production. The clamping force on the

Fig. 3.7 *Clamping: (a) incorrect – breakage caused by clamping on thin, unsupported section; (b) correct – the clamping force is adequately supported*

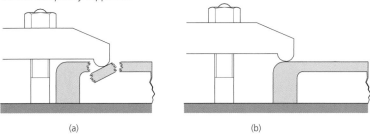

(a)   (b)

Fig. 3.8 *Resistance: (a) incorrect – although the location is adequate to resist the cutting force, the component is distorted due to inadequate support; (b) correct – the location backs up the component in resisting the cutting force and prevents the component from being distorted*

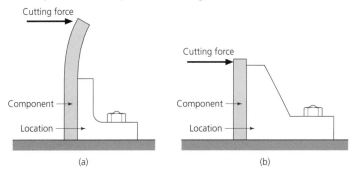

(a)   (b)

Fig. 3.9 *Screw type clamping devices: (a) simple bridge type clamps; (b) a two-way clamp; (c) edge clamps*

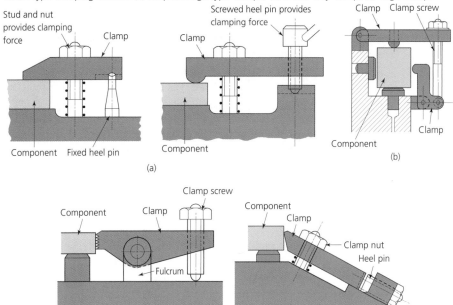

Fig. 3.10 *Toggle and cam-operated clamping devices: (a) simple cam-operated clamp; (b) hook cam clamp, for light operations where the clamp must swing clear for loading; (c) toggle clamp; (d) toggle open; (e) toggle closed; as the links come into a straight line, considerable clamping force is exerted for a small effort at the handle; no reaction in direction A will cause the toggle to open*

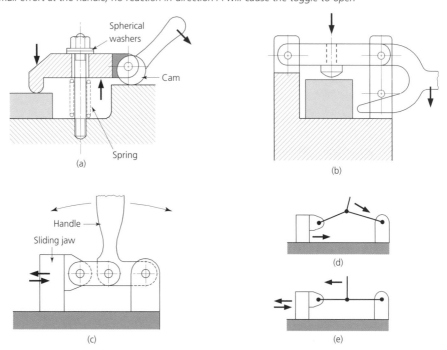

component should be maximised by ensuring that the bolt and nut securing the clamp are nearer to the workpiece than to the heel of the clamp or packing (moments of forces).

Let's now consider clamping by compressed air operating through pistons and cylinders. Figure 3.11(a) shows a section through a typical double-acting piston and cylinder, and Fig. 3.11(b) shows a suitable pneumatic circuit for use with such a piston and cylinder. The incoming air is first passed through a filter to remove dust and moisture which would cause rapid wear and early failure of the valves and cylinder bores. It is then passed through a pressure regulator; this reduces the pressure from the workshop air line to a suitable value for holding the workpiece securely but without damage. The air then passes through a lubricator and then to the control valve before finally passing to the cylinder. As shown, compressed air would pass into the valve at A and then to the cylinder via port B. This would move the piston to the right of the figure. Air already in the cylinder would be exhausted to the atmosphere through ports C and D. Reversing the valve, as shown inset, feeds the compressed air to the cylinder via ports A and D, thus moving the piston in the reverse direction – to the left of the figure. Air already in the cylinder would be exhausted to the atmosphere via ports B and C.

The piston and cylinder may be used for direct clamping as shown in Fig. 3.12(a) or indirect clamping as shown in Fig. 3.12(b). The arrangement in Fig. 3.12(a) allows variations in component size to be accommodated and maintains a constant clamping

Fig. 3.11 *Pneumatic equipment: (a) typical double-acting piston and cylinder; (b) simple air circuit*

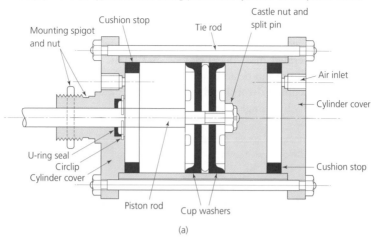

(a)

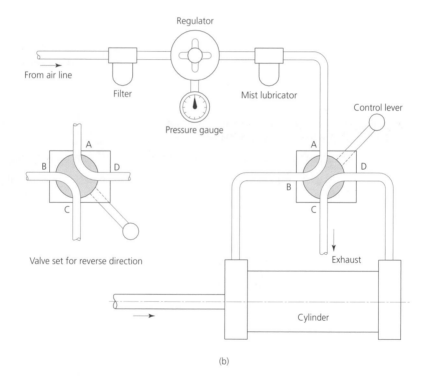

(b)

pressure for a given air pressure. The clamping pressure is easily adjustable by altering the air pressure at the regulator (Fig. 3.11).

But with direct clamping, if the air pressure fails, then the clamping pressure is lost; loss of pressure during machining could cause a serious accident. To overcome this danger, indirect clamping can be used, as shown in Fig. 3.12(b). Here the clamping pressure is provided by the toggle clamp and the air cylinder is only used to open and close the toggle. During a loss of air pressure the toggle would remain closed and the clamping force on the workpiece would not be lost. However, as with all toggle clamps, variation in component size cannot be tolerated. Although costly compared with manually operated mechanical clamps, pneumatic clamps have the advantage that the clamping pressure can be accurately controlled, they are quick to operate and, in the case of large and complex workholding devices, many clamps can be operated simultaneously or sequentially by a single lever.

Fig. 3.12 *Pneumatic clamping: (a) direct; (b) indirect*

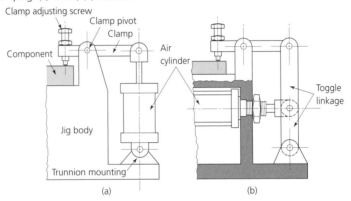

(a)   (b)

SELF-ASSESSMENT TASK 3.1

1. Explain the essential difference between a location and a restraint.

2. For each of the following items, find an example besides those in the text then make a sketch of it:

   (a) a screw-action clamp
   (b) a toggle-action clamp
   (c) a cam-action clamp
   (d) a pneumatically operated clamp

3. Explain how a pneumatically operated clamp can be made fail-safe and why this is advisable.

3.4 Drilling machine: toolholding and workholding

To drill a hole in a drilling machine so that it is correctly sized and positioned, four basic conditions must be satisfied:

- The drill must be located in the drilling machine spindle so that the axis of the drill is coincident with the axis of the machine spindle.

- The drill and spindle must rotate together without slippage. There must be total restraint between the drill shank and the spindle.
- The workpiece must be located so that the centre lines of the hole are in alignment with the spindle axis (Fig. 3.13).
- The workpiece must be restrained so that it resists the cutting forces and is not dragged round by the drill.

Fig. 3.13 *Basic drilling alignments*

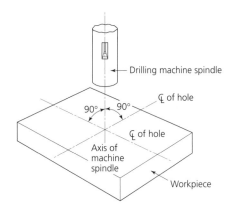

Fig. 3.14 *Taper location of drill shank: (a) spindle and shank maintain axial alignment under maximum metal conditions; (b) spindle and shank maintain axial alignment under minimum metal conditions; (c) misalignment due to dirt between drill and spindle*

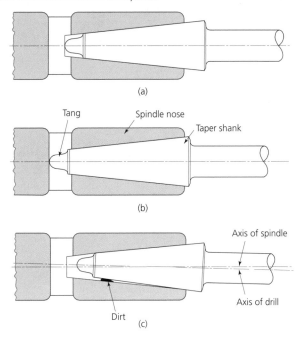

Large twist drills are normally held in the spindle nose directly by means of a taper shank (Fig. 3.14). Axial alignment of the drill in the spindle is assured and maintained despite variations in the taper size due to manufacturing tolerances and wear. However, it is essential that the internal and external tapers are carefully cleaned before the drill shank is inserted into the machine spindle; this prevents damage to the tapers, ensures correct alignment and ensures the drill does not slip.

The narrow angle of taper provides a wedging action that ensures the drill does not drop out of the spindle and there is adequate frictional restraint to drive the drill under normal cutting conditions. But should the drill be allowed to dig in or seize in the hole, it will slip and its taper will be damaged. The tang on the drill is provided solely for its removal (Fig. 3.15); it is not intended to provide additional drive.

Small-diameter twist drills normally have a parallel shank and are held in a drill chuck (Fig. 3.16). The drill chuck and its shank rely upon a system of concentric tapers to ensure axial alignment. The jaws are closed on the drill shank by rotation of the chuck sleeve. The chuck may be mounted directly onto a taper on the end of the drill spindle or it may be used with a taper arbor so it can be fitted to a spindle with a taper bore. Direct mounting is used on small bench drilling machines, whereas a taper arbor is used on pillar and column drilling machines, where taper shank drills in the larger sizes may need to be used. The Morse system of tapers is used on drills and drill spindles as it is *self-locking*.

Fig. 3.15 *Removing a taper shank drill using a tapered drift*

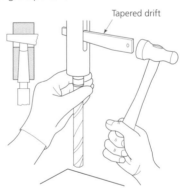

Tapered drift

Having dealt with the location and restraint of the drill, let's now consider the location and restraint of the workpiece. The location and restraint of the workpiece must follow the principles previously discussed. For one-off and jobbing work, small components are usually held in a machine vice bolted to the worktable of the machine, whereas large components are bolted directly to the worktable itself. Figure 3.17 shows the restraints and locations offered by a machine vice. To ensure the spindle axis is perpendicular to the workpiece, the following alignments should be checked:

- A matched pair of parallels are used to support the work and to ensure that its undersurface is parallel to the vice slideways (*a, a*).
- The working surfaces of the vice slideways are parallel to the machine worktable (*b, b*).
- The fixed jaw of the vice is perpendicular to the machine worktable.

Fig. 3.16 *Drill chuck: (a) typical chuck and accessories; (b) a series of concentric tapers are used to maintain axial alignment between the arbor, the chuck and the drill; for clarity, the jaws are shown at 180° and the mechanism for moving them has been omitted*

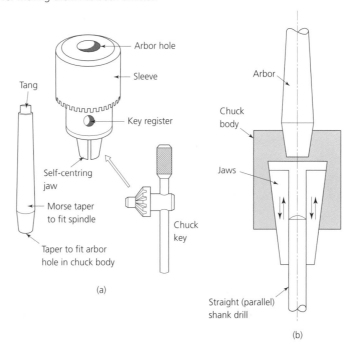

(a)

(b)

The parallels provide positive restraint by resisting the axial feed force of the drill in a downward direction. The vice jaws provide positive restraint by resisting the tendency of the rotary cutting forces to rotate the workpiece.

Cylindrical work is more difficult to hold. Figure 3.18 shows how a cylindrical component is held vertically in a machine vice using a V-block. To ensure the spindle axis is parallel to the workpiece axis, the following alignments must be checked:

- The V-block must be seated on the vice slideways so that the V is perpendicular; that is, its end face is parallel to the vice slides in each direction (a, a).
- As previously, the vice slideways must be parallel to the machine worktable (b, b), and the fixed jaw of the vice must be perpendicular to the worktable.

Figure 3.19 shows how a cylindrical component can be held horizontally on the drilling machine worktable. To ensure the axis of the component is parallel to the worktable (a, a), it is essential to use a matched pair of V-blocks as shown. It is difficult to start a drill on the curved surface of a cylindrical component and some guidance is required.

Figure 3.20 shows a drill jig suitable for drilling a cross-hole through a shaft end. A drill jig not only locates the workpiece in the correct place relative to the drill axis, it also guides the drill close to its point so that it cannot wander out of position. Since the bush is a close-running fit on the drill, the hole cannot be reamed without first removing the bush. Figure 3.21 shows a removable bush and liner sleeve. The drill bush is placed in the liner sleeve as shown to guide the drill during the drilling operation. The bush can be removed

Fig. 3.17 *Holding work in a vice: (a) restraints, a positive and b frictional; (b) locations*

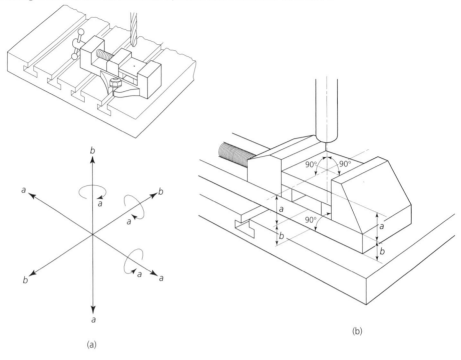

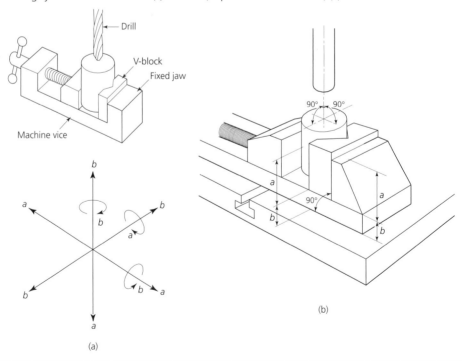

Fig. 3.18 *Holding cylindrical work in a vice: (a) restraints, a positive and b frictional; (b) locations*

Fig. 3.19 *Holding cylindrical work on a drilling machine table: (a) restraints, a positive and b frictional; (b) locations*

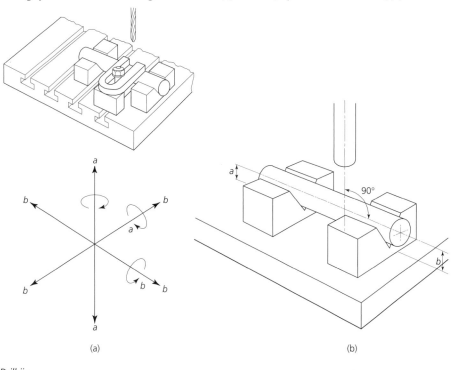

(a)

(b)

Fig. 3.20 *Drill jig*

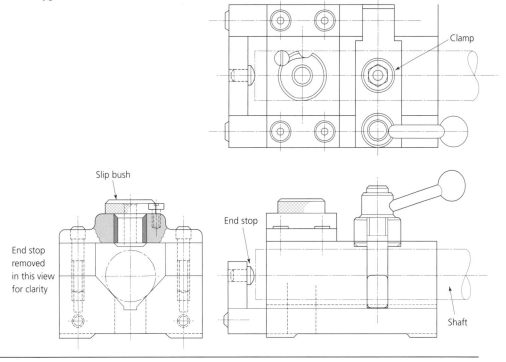

Slip bush

Clamp

End stop

End stop
removed
in this view
for clarity

Shaft

Fig. 3.21 *Removable bush and liner sleeve*

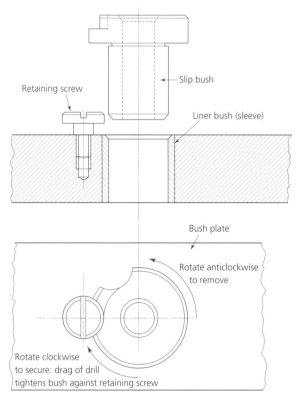

without disturbing the jig, hence without disturbing the axial alignment of the drill spindle and the workpiece. The hole can then be reamed. The reamer is a clearance fit in the liner sleeve and is guided solely by the drilled hole.

SELF-ASSESSMENT TASK 3.2

1. Sketch a simple drilling jig for a component of your choice.

2. A pillar drilling machine has a table that can be rotated and is supported on an arm that can be swung from side to side about the machine pillar. Explain with sketches how these movements allow a component that is clamped to the machine table to be positioned under the drill.

3. Suggest why small-diameter drills are unlikely to have a taper shank.

3.5 Milling machine: toolholding and workholding

There are two main types of milling machine and they are classified by the plane in which their spindle axes lie. A typical horizontal-spindle milling machine is shown in Fig. 1.50 and a typical vertical-spindle milling machine is shown in Fig. 1.52. The method of mounting the cutters depends upon which machine is being used.

3.5.1 *Toolholding in the horizontal-spindle milling machine*

In the horizontal milling machine the milling cutters are usually mounted on an arbor which locates and drives the rotating cutter. Unlike the Morse taper of the drilling machine spindle, the milling machine taper is a self-releasing taper that only locates the arbor. The arbor has a *positive drive* through dogs fixed to the spindle nose. The arbor is held into the spindle nose by a draw-bolt passing through the spindle of the machine. Details of the spindle nose and the corresponding end of the arbor are shown in Fig. 3.22.

Fig. 3.22 *Milling machine: (a) spindle nose; (b) taper register of arbor to fit spindle nose*

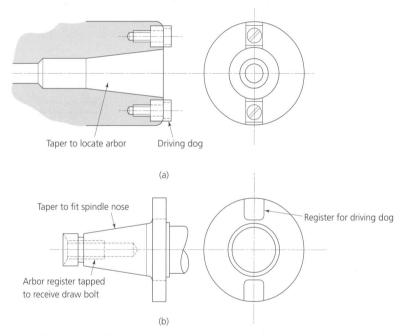

Figure 3.23(a) shows the restraints and locations acting upon a cutter mounted on a long arbor. In the example shown, the arbor is driven positively by the dogs on the spindle nose, but the cutter is driven by friction alone. This is adequate for the majority of jobbing applications, but for production milling where heavier cuts are likely to be taken the cutters should be keyed to the arbor to provide a positive drive as shown in Fig. 3.23(b).

The long arbor should always be supported as a beam rather than as a cantilever to keep the deflection of the cutter to a minimum (Fig. 3.25). Stub arbors are also used, supported as a cantilever in the spindle nose alone. Figure 3.24 shows a typical stub arbor used for mounting shell-end mills. The restraints and locations for this type of arbor are also shown. In addition, large face mills may be bolted directly to the spindle nose.

Fig. 3.23 *Mounting milling cutters on a long arbor: (a) restraints acting on the cutter when mounted on an arbor; (b) keying the cutter to the arbor; the length of the key is greater than the width of the cutter; any portion of the key that extends beyond the cutter is 'lost' in the spacing collars, which also have keyways cut in them*

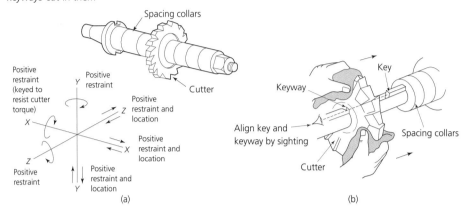

Fig. 3.24 *Stub arbor, and its locations and restraints*

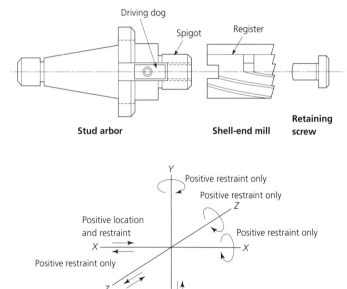

The forces acting on a milling cutter, when removing metal rapidly, are very great. Therefore the cutter arbor must be adequately supported and the cutter correctly mounted to avoid inaccuracies and chatter. Figure 3.25(a) shows a cutter incorrectly mounted as excessive overhang from the points of support allows the arbor to flex. This leads to inaccuracy, chatter and poor surface finish. In extreme cases the cutter teeth may be damaged.

Fig. 3.25 *How to use an overarm: (a) bad mounting; (b) good mounting; (c) good mounting*

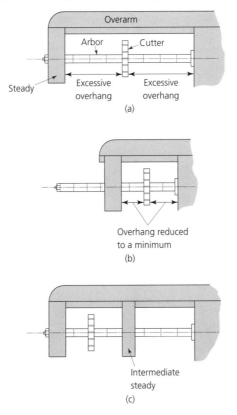

Figure 3.25(b) shows a cutter correctly mounted with the minimum of overhang and maximum rigidity. The cutter is kept as close to the spindle nose as possible, and the overarm and steady are adjusted to provide support as close to the bearing as possible. Sometimes the shape and size of the work itself prevents the cutter being mounted close to the spindle nose. Figure 3.25(c) shows how an additional, intermediate steady can be placed on the overarm to support the arbor close behind the cutter, minimising the effective overhang. A selection of cutter types and their applications is shown in Fig. 1.51.

3.5.2 *Toolholding in the vertical-spindle milling machine*

Large, face-milling cutters may be bolted directly to the spindle nose or mounted on a stub arbor and driven by the spindle-nose dogs. In either case the location and drive is positive.

Most of the cutters used on vertical-spindle machines are either end mills or slot drills; they are held in collet chucks. The collet chuck itself has a taper shank and is located in the spindle-nose taper by means of a draw-bar passing through the machine spindle. The cutters have a parallel shank with a screwed end. By screwing the cutter into the collet, there is no chance of the cutter being drawn out of the collet by the action of the helical

Fig. 3.26 *Mounting a screwed shank on a solid end mill: the main body (A) has an integral taper shank to fit a standard taper nose spindle; the locking sleeve (B) gives a precision fit, positions the collet (C) and mates with the taper nose of the collet; the collet has a split construction and is internally threaded at its rear end; hardened and ground, the male centre (D) centres the cutter and anchors the extreme end to ensure rigidity and true running; with the screwed shank cutter (E) any tendency for it to turn in the chuck during operation increases the grip of the collet on the shank; this ensures maximum feed rates and cutting speeds; the cutter cannot push up or pull down during operation*

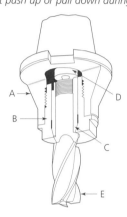

teeth. Figure 3.26 shows a section through a collet chuck. Some typical vertical milling cutters and their uses are shown in Fig. 1.53.

3.5.3 *Workholding*

Various methods of workholding are available on the milling machine. Figure 3.27(a) shows a small component held in a machine vice, and Fig. 3.27(b) shows a large component clamped directly to the machine table. The appropriate restraints and locations are shown in both instances. These examples show work held on a horizontal-spindle milling machine. The forces acting on the workpiece are very different when work is held on a vertical-spindle milling machine. In the vertical-spindle machine, the cutter will tend to rotate the work as well as push it along the table, so side abutments should be provided to prevent the work from rotating. Cylindrical components may be clamped and located using V-blocks (Fig. 3.28).

To reduce set-up time when large quantities of the same component are being produced, a fixture is usually provided. This locates the component in the correct position relative to the milling machine table and cutter, and it provides the necessary restraints to resist the cutting forces. Unlike a drilling jig (Section 3.4), which guides the drill as well as locating the work, a milling fixture only locates the work. Figure 3.29(a) shows a component which has been flame cut from steel plate. The faces of the component have been ground flat and parallel, and the two holes have been bored in a previous operation. The lug is now to be milled on the face XX prior to drilling.

A suitable milling fixture is shown in Fig. 3.29(b). Primary location is on the larger bore and the location spigot has been form relieved to assist loading. The secondary location is on the smaller bore, and the secondary location spigot has been form relieved to compensate for small variations in the bore centre distance due to manufacturing

Fig. 3.27 *Milling machine: (a) location of work in the vice; (b) location of work on the table*

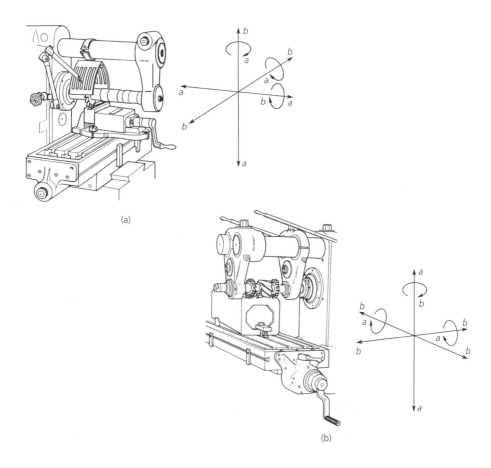

(a)

(b)

Fig. 3.28 *Holding cylindrical work on a milling machine: (a) using a V-block in the vice; (b) using the table*

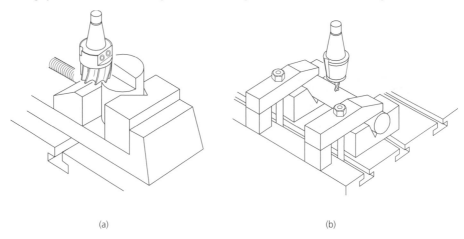

(a)

(b)

Fig. 3.29 *(a) Link. (b) Milling fixture for a link: the setting block in the position cutter is set to the specified feeler gauge between the bottom tooth and the block; this automatically gives the correct machine setting for dimension A; the setting block is removed to prevent wear during machining*

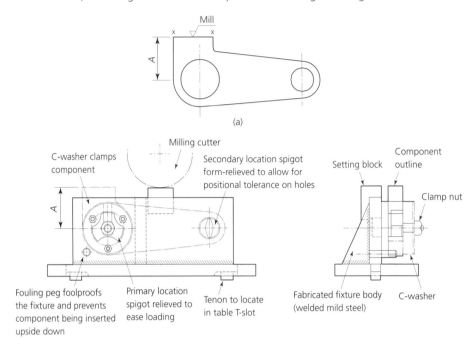

tolerances. Tenons are provided to locate the fixture parallel to the T-slot of the milling machine table, and a fouling peg is provided to prevent incorrect loading of the component; that is, to prevent the component being accidentally loaded upside down. The setting block is used to set the cutter. When the bottom tooth of the cutter is a prescribed distance above the setting block, as checked with a feeler gauge, the machine is set to provide the correct dimension *A* on the component. The component is clamped into position by means of a C-washer and nut.

SELF-ASSESSMENT TASK 3.3

1. Explain why it is preferable to use a slot mill on a horizontal milling machine rather than a long-reach end mill on a vertical milling machine when cutting a deep but narrow slot from a solid. State any factors which limit the depth of the slot.

2. With the aid of sketches, compare the clamping arrangements necessary when machining a large flat surface using a slab mill on a horizontal milling machine and a face mill on a vertical milling machine. Pay particular attention to any positive restraints required.

3.5.4 *Dividing head*

Sometimes it is necessary to make a series of cuts around the periphery of a component, e.g. when gear cutting or cutting splines on a shaft. This operation is called *indexing* and Fig. 3.30 shows an appropriate *dividing head*. This dividing head is called a simple dividing head because its index plate locates the spindle directly; for clarity, only two rows of holes are shown on the index plate. In practice the index plate would have many more rows of holes to provide a wider range of circular division.

Let's now consider the simple component shown in Fig. 3.31(a). Three equispaced slots are to be milled in the flange. The set-up is shown in Fig. 3.31(b). The component is

Fig. 3.30 *Simple dividing head: (a) side elevation; (b) end elevation showing the index plate*

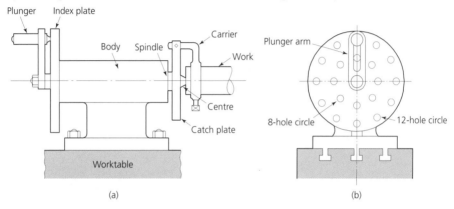

Fig. 3.31 *Example of simple indexing: (a) component requiring indexing; (b) set-up for simple indexing*

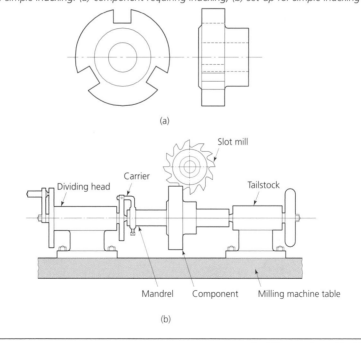

mounted on a mandrel and supported between centres. It is rotated by a carrier and catch plate. Unlike workholding on a lathe, the catch plate on a dividing head is double-sided and has two set screws that can be clamped against the arm of the carrier to remove any backlash. In this example the plunger arm would be adjusted to make the plunger engage with the 12-hole circle. It would then move through four holes between each slot, giving three equispaced slots.

Figure 3.32 shows a *universal dividing head*. This has a worm and worm wheel between the indexing arm and the work spindle. The ratio of this geared drive is standardised internationally at 40:1. The basic layout of the drive is shown in Fig. 3.33. This arrangement together with a range of indexing plates, each having many rows of holes, enables a very wide range of spacings to be achieved.

Fig. 3.32 *Universal dividing head: (a) general arrangement; (b) internal construction*

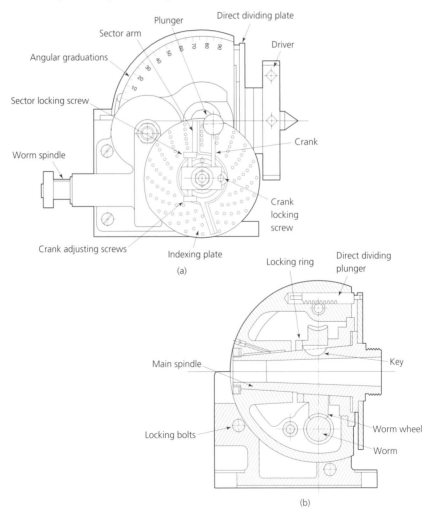

Fig. 3.33 *Spindle drive for a universal dividing head*

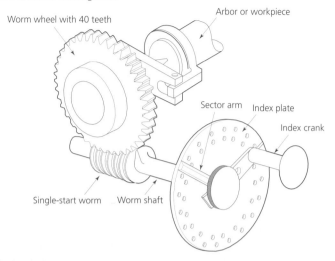

Worm wheel with 40 teeth

Arbor or workpiece

Sector arm   Index plate

Index crank

Single-start worm   Worm shaft

EXAMPLE 3.1

Calculate the index arm setting required to give 13 equally spaced divisions. The index plate has circles with the following numbers of holes: 24, 25, 28, 30, 34, 37, 38, 39, 41, 42, 43.

Index arm setting $= 40/N$ where $N =$ the required number of divisions
$= 40/13$
$= \mathbf{3\,1/13}$

Thus the indexing between each division will be 3 whole turns of the index arm and 1/13 of a turn. Therefore we must look for a row of holes that is divisible by 13. This is the 39-hole circle. So, between each division, the index arm would be moved 3 whole turns and 3 holes in the 39-hole circle ($3 \times 13 = 39$).

EXAMPLE 3.2

Using the same index plate as in the previous example, calculate the index arm setting to give an angular division of the workpiece of $15°\,18'$.

Index arm setting $= \dfrac{\text{required angle}}{9}$

$= \dfrac{15°\,18'}{9}$

$= 918'/(9 \times 60)$

$= 102'/60$

$= \mathbf{1\,21/30}$

The indexing will be 1 whole turn of the indexing arm plus 21 holes in the 30-hole circle of the index plate.

You may wonder where the constant 9 comes from. Remember that, because of the 40 : 1 gear ratio, one whole turn of the index arm results in 1/40 of a turn of the spindle. Since one turn of the spindle is 360°, then one turn of the indexing arm will result in the spindle moving $360°/40 = 9$ (the constant).

3.5.5 *Sector arms*

To save having to count the holes in the index plate every time the dividing head is operated, sector arms are provided (Fig. 3.34). The method of using the sector arms is as follows:

- The sector arms are set so that between arm A and arm B there is the required number of holes plus the starting hole a.
- The plunger and index arm is moved from hole a to hole b against sector arm B.
- The sector arms are rotated so that arm A is now against the plunger in hole b.
- For the next indexing, the plunger is moved to hole c against the newly positioned arm B.
- The process is repeated for each indexing.

Fig. 3.34 *Universal dividing head: (a) how to use sector arms; (b) setting for 152 divisions (10 holes in 38-hole circle)*

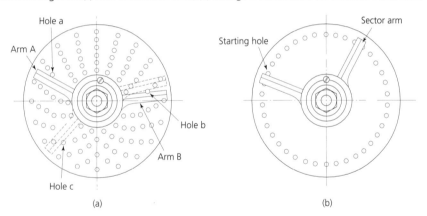

3.5.6 *Differential indexing*

Let's see how we can obtain divisions outside the range of a standard index plate. We can do this by *differential indexing*. Instead of the index plate being clamped to the body of the dividing head, it is coupled to the work spindle by an external geartrain (Fig. 3.35). Thus, as the index arm is rotated, the index plate is advanced or retarded by a small amount automatically. We can obtain the external gear ratio from the following expression:

$$\frac{\text{driver gear}}{\text{driven gear}} = 40 \times \frac{N_1 - N_2}{N_2}$$

where N_1 = required divisions
N_2 = nearest actual divisions available on the index plate

There are two points to note:

- If the answer is *negative*, this has no mathematical significance but indicates that the index plate must rotate in the *same direction* as the index arm.
- If the answer is *positive*, the index plate must rotate in the *opposite direction* to the index arm.

Fig. 3.35 *Differential indexing: the external change gears are set up by the operator to suit the indexing*

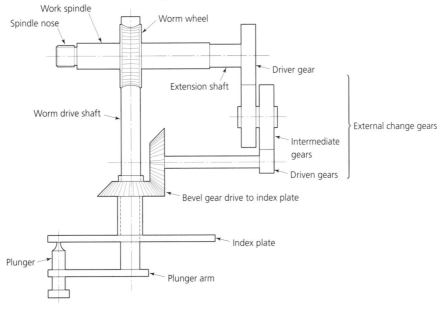

EXAMPLE 3.3

Calculate the geartrain to give an indexing of 113 divisions. The index plate has 24, 25, 28, 30, 34, 37, 38, 41, 42, 43 holes. The gears available have 24(2), 28, 32, 40, 48, 56, 64, 72, 86, 100 teeth.

The procedure in Example 3.1 shows that the required indexing is 40/113, which is not available with the index plate supplied. A near approximation could be 40/120, which indexes as 14/42. This is available by using 14 holes in a 42-hole circle. Applying our new expression gives:

$$\frac{\text{driver}}{\text{driven}} = 40 \times \frac{N_1 - N_2}{N_2}$$
$$= 40 \times (113 - 120)/120$$
$$= 40 \times (-7/120)$$
$$= -7/3$$

The minus sign only indicates the direction of rotation. From the gears available, we would use a 56-tooth driver gear and a 24-tooth driven gear.

3.5.7 *Helical milling*

The universal dividing head can also be used for *helical milling*. For example, helical milling would be used to cut the flutes in a large twist drill. Smaller ones are ground from a solid, prehardened blank but the principle remains the same. A helix is, by definition, the path of a point travelling around an imaginary cylinder so that its axial and circumferential movements maintain a constant ratio. In helical milling the table lead provides the axial movement and the circumferential movement is provided by the dividing head. To maintain the constant ratio, the spindle of the dividing head is geared to the table lead screw by an external geartrain (Fig. 3.36).

Fig. 3.36 *Helical milling*

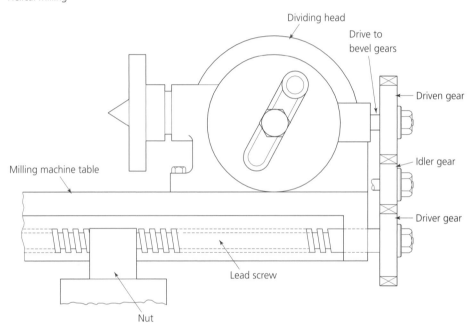

The expression for calculating the geartrain ratio is as follows:

$$\frac{\text{driver gear}}{\text{driven gear}} = \frac{\text{lead of machine}}{\text{lead of helix to be cut}}$$

$$= \frac{40p}{\text{lead of helix to be cut}}$$

where p = the lead of the machine table lead screw

Calculate the geartrain to cut a helix of 480 mm lead on a milling machine with a table lead screw having a lead of 6 mm. The gears available are as for Example 3.3.

$$\frac{\text{driver}}{\text{driven}} = \frac{\text{lead of machine}}{\text{lead of helix to be cut}}$$

$$= (40 \times 6)/480$$
$$= \mathbf{1/2}$$

From the gears available, a 32-tooth gear could be used to drive a 64-tooth gear. Whether or not one or two idler gears would also be needed would depend upon the direction of rotation required to give the helix its specified hand. The flutes of a normal twist drill follow a right-handed helix.

In order to reduce the interference of the cutter with the sides of the slot as much as possible, the table of the milling machine has to be set over by the helix angle (Fig. 3.37). This facility is only available on a *universal* horizontal milling machine; it is not possible on plain drilling machines and vertical-spindle machines.

Fig. 3.37 *Set-over of milling machine table*

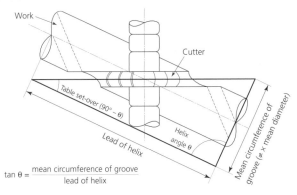

With reference to Fig. 3.37, calculate the angle of set-over when milling a groove with a lead of 480 mm if the mean diameter of the groove is 50 mm.

$$\tan \theta = \frac{\text{lead of work}}{\text{mean circumference}}$$

$$= 480/(50 \times \pi)$$
$$= 3.06$$
$$\theta = \arctan 3.06$$
$$= 71° \, 54'$$

The set over angle of the table $= 90° - 71° 54'$
$$= \mathbf{18° \ 6'}$$

For practical purposes the table would be set over 18°.

3.5.8 *Cam milling*

This is another application of the dividing head. Snail cams of the type shown in Fig. 3.38(a) can be milled using a vertical milling machine as shown in Fig. 3.38(b). As the table feeds the cam blank into the cutter, the dividing head rotates the blank. The dividing head has been set with its spindle axis vertical in Fig. 3.38(b). The gear ratio to provide the required cam *lift* can be calculated from the following expression:

$$\frac{\text{driver}}{\text{driven}} = \frac{\text{lead of machine}}{\text{lift per revolution of cam}}$$

Fig. 3.38 *Simple cam milling: (a) snail cam; (b) set-up for cam milling*

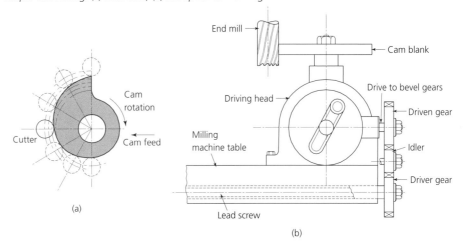

EXAMPLE 3.6

Calculate the gears to cut a cam that has a lift of 15 mm in a 90° rotation of the cam. The table lead screw has a lead of 6 mm.

From Example 3.4 the lead of the machine will be $40 \times 6 \,\text{mm} = 240 \,\text{mm}$. And the lift of the cam per revolution is $15 \,\text{mm} \times (360/90) = 60 \,\text{mm}$. Then:

$$\frac{\text{driver}}{\text{driven}} = \frac{\text{lead of machine}}{\text{lift per revolution}}$$
$$= 240/60$$
$$= \mathbf{4/1}$$

Unfortunately, the ratio of lift to lead rarely works out as conveniently in practice, so some means has to be used to obtain intermediate values. With the dividing head set vertically, as in the previous example, the lift generated on the cam is a maximum for any given gear ratio. However, with the dividing head spindle and the machine spindle set horizontally as shown in Fig. 3.39(a), only a cylindrical surface will be generated and the cam lift will be zero. Thus the required lift will be obtained for some setting between these two extremes.

A vertical-spindle machine with an *inclinable head* is required for cam milling. A convenient gear ratio is then selected that will give a lift slightly *larger* than required, and the machine head and dividing head are inclined to give the actual lift required. This is shown in Fig. 3.39(b). Now we can say:

$$\sin \theta = x/L$$

where x = required lift per revolution of the cam
     L = table movement per revolution of the cam

Fig. 3.39 *Effect of inclined dividing head: (a) horizontal – zero lift is generated; (b) inclined at angle θ, where sin θ = x/L*

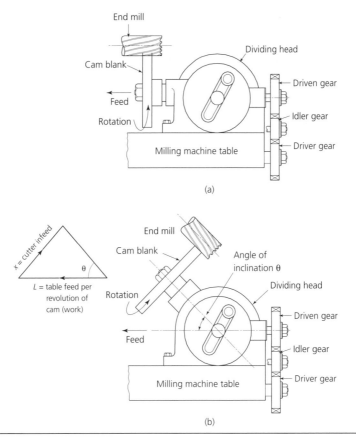

(a)

(b)

Hence $L = x/\sin\theta$. But L is the maximum lift per revolution for any given geartrain, therefore:

$$\frac{\text{driver}}{\text{driven}} = \frac{\text{lead of machine}}{L}$$

$$= \frac{\text{lead of machine}}{x/\sin\theta} \qquad \text{(for inclination } \theta\text{)}$$

$$= \frac{\text{lead of machine}}{x} \times \sin\theta$$

$$\sin\theta = \frac{x}{\text{lead of machine}} \times \frac{\text{driver}}{\text{driven}}$$

EXAMPLE 3.7

Calculate the gear ratio and spindle inclination to cut a cam that has a lift of 23.5 mm in a cam rotation of 83° on a milling machine whose lead is 240 mm.

The lift per revolution of the cam is $x = 23.5\,\text{mm} \times (360°/83°) = 101.9\,\text{mm}$. The nearest convenient gear ratio greater than this lift and available with standard gears would be:

$$\frac{\text{driver}}{\text{driven}} = \frac{\text{lead of machine}}{\text{maximum lift per revolution}}$$
$$= 240/105$$
$$= 48/21$$

To find the angle of inclination θ:

$$\sin\theta = \frac{x}{\text{lead of machine}} \times \frac{\text{driver}}{\text{driven}}$$

$$= \frac{101.9}{240} \times \frac{48}{21} = 0.9705$$

$$\theta = \arcsin 0.9705$$
$$= \mathbf{76°\ 3'}$$

3.5.9 *The rotary table*

Figure 3.40(a) shows a typical rotary table that is used for indexing flat-plate components. The main scale surrounding the working surface is calibrated in degrees of arc (360°). The working surface is driven by a worm and worm wheel, as in a dividing head, but the ratio is not standardised. The illustrated example has a vernier scale for divisions of less than 1°. Some tables are equipped with an indexing plate instead of a vernier scale.

3.5.10 *The dividing chuck*

This is used for holding cylindrical work with the axis of the workpiece in the vertical plane, an example is shown in Fig. 3.40(b). Although a universal dividing head can be inclined through 90°, the dividing chuck is less bulky and is more convenient to use when only simple indexing is required. Some manufacturers supply a chuck and backplate for mounting on a rotary table to avoid the need for two separate pieces of equipment. But this arrangement has a greater overall height and there may not always be sufficient room for it beneath the spindle.

Fig. 3.40 *Miscellaneous indexing equipment: (a) rotary table; (b) dividing chuck*

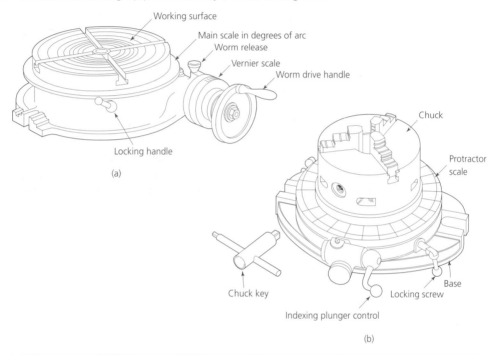

Working surface
Main scale in degrees of arc
Worm release
Vernier scale
Worm drive handle
Locking handle
(a)

Chuck
Protractor scale
Chuck key
Indexing plunger control
Locking screw
Base
(b)

SELF-ASSESSMENT TASK 3.4

1. Calculate the index arm setting required to give 17 equally spaced divisions on the workpiece. Use the index plate from Example 3.1.

2. Explain why it is sometimes necessary to use differential indexing.

3. (a) Calculate the geartrain required to cut a helix of 300 mm lead on a milling machine having a table lead screw with a lead of 6 mm. The gears available are as in Example 3.3.
   (b) Calculate the angle of table set-over if the helical groove has a mean diameter of 60 mm.

4. Explain why it is sometimes necessary to incline the head of the milling machine and the spindle of the dividing head when cam milling.

3.6 Lathe: Toolholding and workholding

The single-point tools used on centre lathes are held in tool posts as shown in Fig. 3.41. The purpose of the tool post is to locate the turning tool in the correct position relative to the machine and the work, to ensure the point of the tool is at the same height as the axis of the workpiece, and to restrain the tool against the forces acting on it during cutting.

The simple clamp tool post in Fig. 3.41(a) is mostly found on small lathes. It relies largely on frictional restraint and it is difficult to adjust the height of the tool point. You need to keep lots of assorted packing handy!

The American pillar tool post in Fig. 3.41(b) is also widely used on light-duty lathes. It is easy to set as the tool height is readily adjustable by the rocking action of the 'boat piece'. Unfortunately adjustment tilts the tool and alters the cutting angles. And this type of tool post tends to lack rigidity due to its inherent tool overhang.

The four-way turret tool post in Fig. 3.41(c) is one of the most widely used on modern industrial centre lathes. It is robust and holds up to four tools which can be quickly swung into position as required. Unfortunately there is no height adjustment, so packing has to be used in order to bring the tool point into line with the workpiece axis.

Figure 3.41(d) shows a quick-change tool holder. The tools are premounted in quick-change holders. These in turn are located on a dovetail slide (machined in the face of the mounting block) and restrained by means of a cam type locking device operated by the handle. To change a tool, the handle is released and the tool and holder are slipped off. The next tool and holder are simply slipped into the vacant position and clamped by means of

Fig. 3.41 *Tool posts for a centre lathe: (a) English (clamp); (b) American (pillar); (c) turret (four-way); (d) quick-release*

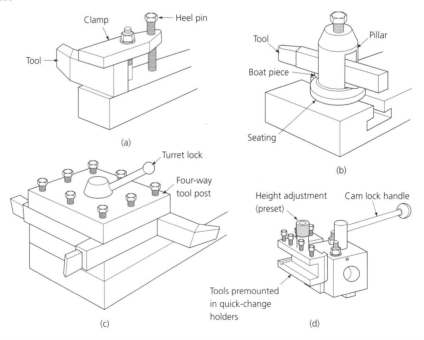

the cam locking lever. The height of the tool can also be preset in a suitable fixture by a micrometer adjustment. This arrangement is costlier than the more common tool posts but the time saved in tool changing during small batch production justifies the cost.

Workholding on the lathe can be achieved in various ways depending on the geometry of the workpiece.

3.6.1 *Workholding between centres*

This is the traditional method of workholding from which the centre lathe gets its name; it is shown in Fig. 3.42(a). Notice that the workpiece is located between centres and is driven by a catch plate and carrier. Since the driving mechanism can float, it has no influence on the locational accuracy of the workpiece. The centres themselves have Morse taper shanks and are located in taper sockets in the spindle and the tailstock barrel. The use of taper locations ensures axial alignment irrespective of variations due to manufacturing tolerance and wear. Figure 3.42(b) shows the restraints acting on the workpiece held between centres.

Fig. 3.42 *Workholding: (a) between centres; (b) restraints, a positive and b frictional*

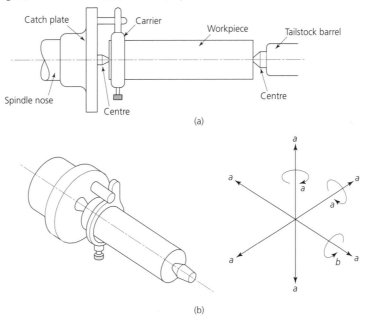

For parallel turning, the axis of the spindle centre and the axis of the tailstock centre must be coincident with each other and parallel to the bed slideways. The tailstock is provided with lateral (sideways) adjustment to achieve this alignment. If the axes of the centres are not coincident, the turned component will be conical (tapered). This will be considered further in Section 4.7.

3.6.2 *Workholding in a self-centring chuck*

Figure 3.43(a) shows the construction of a typical three-jaw self-centring chuck. Notice that the scroll not only restrains the workpiece, but locates it as well. This is fundamentally bad practice, since any wear in the scroll and/or the jaws impairs the accuracy of location. And there is no available means of adjustment to compensate for this wear. However, because of its convenience, the self-centring chuck is widely used; providing it is treated with care, it retains its initial accuracy for a considerable time. To maintain the initial accuracy:

- Never try to hammer the work true.
- Never hold the work on an irregular surface.
- Never hold the work on the tips of the jaws.

The jaws of this type of chuck are not reversible; separate internal and external jaws have to be used as shown in Fig. 3.43(b). It is essential that the sets of jaws carry the same serial number as the chuck body, otherwise they will not close concentrically on the workpiece. The restraints acting on the workpiece held in a three-jaw self-centring chuck are shown in Fig. 3.43(c).

Fig. 3.43 *The three-jaw self-centring chuck: (a) construction; (b) external and internal jaws; (c) restraints, a positive and b frictional*

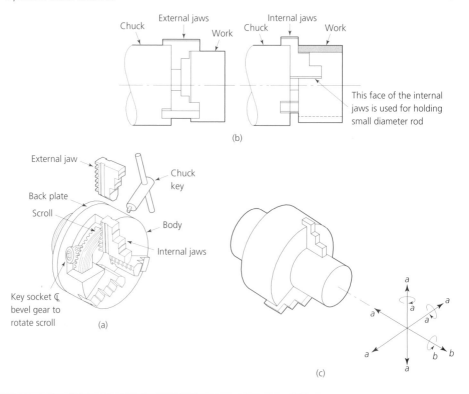

3.6.3 *Workholding in a collet chuck*

This type of chuck is shown in Fig. 3.44. Its advantage over the self-centring chuck is that it provides a higher level of concentricity; but compared with a three-jaw self-centring chuck, it has a limited range of sizes and can only hold on bright-drawn or centreless-ground bar stock or blanks cut from such bar stock. The restraints acting on the workpiece are similar to those for a three-jaw self-centring chuck.

Fig. 3.44 *Collet chucks: (a) chuck for a simple plain nose spindle (typical of small instrument lathes) – tightening the collar forces the collet back into the taper bore of the sleeve which closes the collet down onto the workpiece; (b) split (spring) collet – the four slots allow the collet to be closed onto the work as the collet is drawn into the sleeve taper; (c) draw-bar chuck for taper nose spindles*

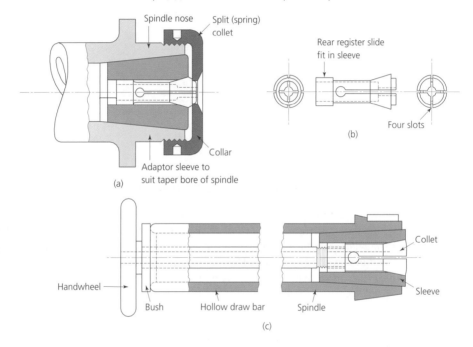

3.6.4 *Workholding in an independent-jaw chuck*

Figure 3.45 shows the construction of a four-jaw independent chuck. It is much more heavily built than the self-centring chuck and has much greater holding ability. It gets its name from the fact that each jaw can be moved independently. Each jaw is moved independently by a square-thread screw, and each jaw is reversible. The restraints acting on a rectangular workpiece are shown in Fig. 3.45(b). This type of chuck is used for holding:

● Work where the number of sides is 4, 8, 12 or any other multiple of 4.
● Irregularly shaped work.
● Work which must be trued up to run concentrically.
● Work which must be deliberately offset to run eccentrically.

Three methods of setting work in a four-jaw independent chuck are shown in Fig. 3.46.

Fig. 3.45 *The four-jaw chuck: (a) construction; (b) restraints, a positive and b frictional*

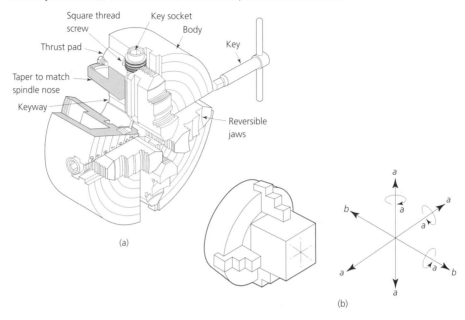

(a)

(b)

3.6.5 *Workholding on the face plate*

The workholding devices described so far enable a diameter to be turned parallel to an existing datum surface. The face plate enables diameters to be turned perpendicular to an existing datum surface, as shown in Fig. 3.47(a). Sometimes the work must be balanced, as shown in Fig. 3.47(b), to avoid vibration which could damage the machine bearings and affect the surface finish. The restraints acting on work held on a face plate are shown in Fig. 3.47(c).

SELF-ASSESSMENT TASK 3.5

1. (a) State which type of lathe tool posts give easy adjustment of tool height and explain with the aid of sketches how this is achieved.
    (b) State which type of lathe tool posts allow the quickest tool changing and explain with sketches how this is achieved.
    (c) Describe which type of tool post combines both the above facilities.

2. Sketch suitable components that are most suitable for holding:

    (a) in a self-centring three-jaw chuck
    (b) between centres
    (c) in an independent four-jaw chuck
    (d) on a face plate

Fig. 3.46 *The four-jaw chuck: (a) truing up a four-jaw chuck with a dial test indicator (DTI) – the DTI will show a constant reading when the component is true; (b) setting work in the chuck using a DTI and centre – the chuck is adjusted until the DTI maintains a constant reading whilst the chuck is revolved; (c) setting work in the chuck using a height gauge (scribing block) – the chuck is adjusted until the scriber point just touches each opposite edge or corner as the chuck is revolved*

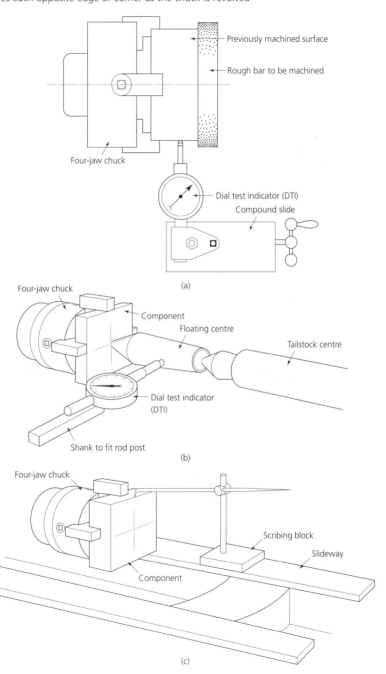

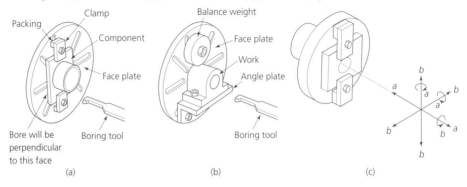

3.7 Grinding machine: Toolholding and workholding

The abrasive wheels for precision grinding machines are not mounted directly onto the machine spindle, as is the case with simple off-hand grinding machines, but are mounted on hubs which have built-in adjustable balance weights (Fig. 3.48). These balance weights enable the wheel-and-hub assembly to be statically balanced before it is mounted on the machine spindle. The hub has a taper bore that locates on the taper of the machine spindle to ensure axial alignment and true running.

Fig. 3.48 *Balancing a grinding wheel: (a) balance weights; (b) balancing fixture*

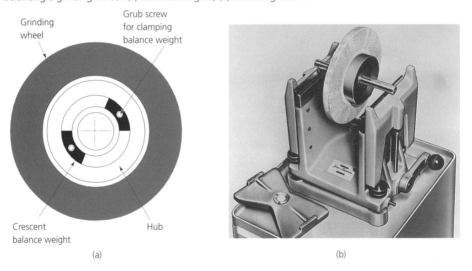

Workholding on surface grinding machines is usually on a magnetic chuck. Modern magnetic materials enable permanent magnetic chucks to have very great holding ability. The magnetic field attracts the component against the face of the chuck so strongly that the component is restrained by friction between itself and the chuck face. Figure 3.49(a) shows a magnetic chuck in the 'on' position with the magnetic flux passing through the component, which must be made from a ferromagnetic material such as steel. The magnets are located in a grid which can be moved sideways by the operating handle. When the handle is moved to the 'off' position, as shown in Fig. 3.49(b), the magnets are moved sideways and their flux is short-circuited through the pole pieces. No longer magnetised by the flux, the component stops being attracted to the chuck and can now be removed.

Fig. 3.49 *The permanent magnet chuck: (a) chuck 'on', lines of flux pass through component; (b) chuck 'off', lines of flux bypassed by pole pieces, which act as keepers; as flux lines no longer pass through the component, the component is no longer attracted to the chuck*

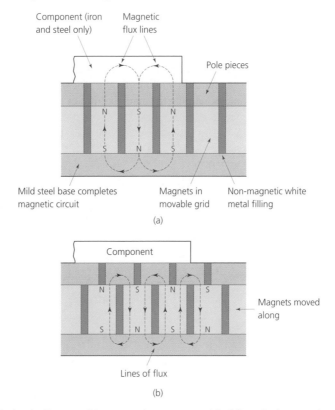

(a)

(b)

Cylindrical grinding machines use the same workholding devices as the centre lathe. The work is held between centres or in chucks and collets. Flat work can be held on a rotary magnetic chuck.

EXERCISES

3.1 (a) With the aid of sketches, describe the difference between restraint and location as applied to workholding on machine tools.
 (b) With the aid of sketches, describe the difference between frictional and positive restraints as applied to workholding on machine tools.

3.2 (a) With the aid of sketches, explain what is meant by the six degrees of freedom as applied to workholding on machine tools.

(b) With the aid of sketches, explain how these freedoms may be restrained when work is held:

(i) between centres on a lathe

(ii) in a three-jaw self-centring chuck on a lathe

(iii) in V-blocks on a drilling machine

(iv) in a machine vice on a vertical-spindle milling machine

3.3 Figure 3.50 shows a component that is to be set up for milling the slot A. The surfaces PQRS have been previously machined and may be used for location purposes.

(a) With the aid of sketches, show how a single component could be located, aligned and clamped on a milling machine table for the production of the slot.

(b) Sketch a suitable milling fixture for machining the slot in a batch of components, naming the main features of the fixture.

Fig. 3.50

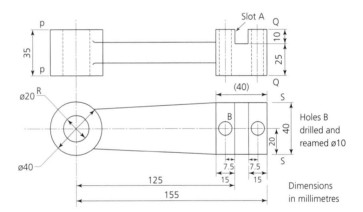

3.4 The holes marked B in the component shown in Fig. 3.50 are to be drilled. The surfaces PQRS have been previously machined and may be used for location purposes. Sketch a suitable drilling jig and name the main features.

3.5 Referring to Fig. 3.50, sketch a suitable workholding fixture for batch production on a CNC milling machine so that slot A can be milled and the holes marked B can be drilled and reamed at the same setting.

3.6 Discuss the advantages and limitations of magnetic chucks as workholding devices on grinding machines.

4 Kinematics of manufacturing equipment

4.1 Generating and forming surfaces

Figure 4.1 shows a component which consists of a number of surfaces that are basic geometrical shapes. These surfaces can be produced by selecting the appropriate manufacturing process. For example:

- The circular plane surface on the end of the stem could be produced on a lathe at the same setting as the conical and cylindrical surfaces.
- The cylindrical surface would be produced on a lathe.
- The conical surface could be produced on a lathe using a taper turning technique.
- The rectangular plane surface adjacent to the cylindrical surface could also be produced on a lathe at the same setting.
- The plane surfaces which are mutually parallel or perpendicular could be produced on a milling machine.
- The inclined plane surfaces could also be produced on a milling machine.
- The hole could be produced on a drilling machine.

Should a high level of surface finish be required, then the turned surfaces could be finished on a cylindrical grinding machine and the plain surfaces could be finished on a surface grinding machine. Most engineering components are designed as combinations of geometrical surfaces with specific manufacturing processes in mind. Where more complex

Fig. 4.1 *Basic geometric shapes*

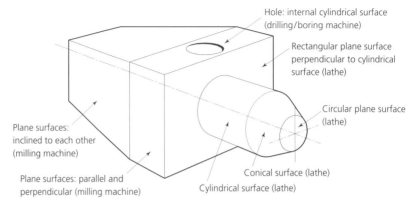

Hole: internal cylindrical surface (drilling/boring machine)

Rectangular plane surface perpendicular to cylindrical surface (lathe)

Circular plane surface (lathe)

Plane surfaces: inclined to each other (milling machine)

Plane surfaces: parallel and perpendicular (milling machine)

Conical surface (lathe)

Cylindrical surface (lathe)

Fig. 4.2 *Generation of a cylindrical surface by turning: (a) straight nose roughing tool; (b) bar turning tool; (c) right-hand facing tool*

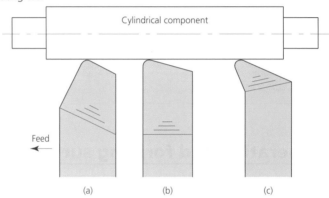

Cylindrical component

Feed

(a)   (b)   (c)

surfaces are required, e.g. in the press tools for stamping out car body panels, copy-machining techniques or computer-controlled (CNC) machining techniques can be employed as these processes have greater flexibility (Chapter 5).

Conventional machine tools were chosen in the above example because they generate the geometrical surfaces and solids required. A surface is *generated* when its geometry is wholly dependent upon the relative movements of the cutting tool and the workpiece and is wholly independent of the shape of the cutting tool.

Figure 4.2 shows a piece of metal being turned using a lathe. Irrespective of the profile of the turning tool, the outcome will be a cylindrical component. That is, a *cylindrical surface* will be *generated*.

Figure 4.3 shows surfaces being produced by a slab milling cutter and a side and face milling cutter such as would be used on a horizontal-spindle milling machine. It also shows surfaces being cut by a face milling cutter and a long-reach shell-end milling cutter such as would be used on a vertical-spindle milling machine. Irrespective of which cutter is used, and despite the fact they are circular and rotating about an axis, plane surfaces are produced by all the cutters. That is, plane surfaces are being *generated*.

Fig. 4.3 *Generation of plain surfaces using milling cutters: (a) slab mill; (b) side and face milling cutter; (c) face mill; (d) long-reach shell-end milling cutter; (a) and (b) are suitable for a vertical milling machine; (c) and (d) are suitable for a horizontal milling machine*

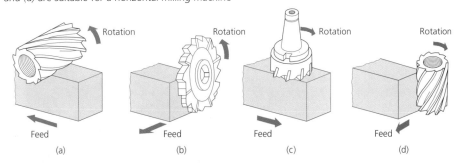

Figure 4.4 shows a selection of surfaces being produced by a variety of cutters. In each case the outcome depends on the shape of the tool. These surfaces are *formed*. Notice that the formed surface is superimposed upon the surface normally generated by the machine. That is, you can generate without forming but, when forming, generation also occurs at the same time.

Forming operations are strictly limited by the rigidity of the workpiece, the cutting tool and the machine tool. A forming operation increases the length of the cutting edge in contact with the workpiece and can cause chatter. Where an extensive forming operation is required, copy machining is preferred to a forming tool. An example is shown in Fig. 4.5. A template or 'master' is produced to the shape of the finished component. In this example the master is mounted between centres on the back of the lathe. As the saddle traverses the bed of the lathe, a stylus mounted on the copying attachment fixed to the cross-slide traces the profile of the master. Through a system of valves, hydraulic cylinders and pistons, the stylus controls the path of the cutting tool so that its cutting point follows the outline of the required component. A succession of cuts are taken with the tool being fed further into the workpiece each time. A single-point turning tool is used; and because no forming tool is required, the cutting conditions and metal removal rates are normal.

Fig. 4.4 *Forming surfaces: (a) forming on a lathe; (b) forming on a milling machine; (c) countersinking on a drilling machine*

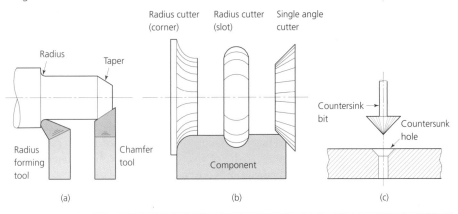

Fig. 4.5 *Copy turning*

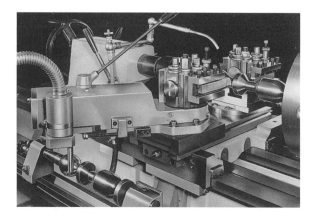

Although suitable for large batch production, the cost of producing a template or master precludes the use of copy machining for one-off and small batch production. The use of a computer-controlled (CNC) turning centre gives the same advantages of cutting with a conventional single-point tool, but has the added advantage that no template or master is required (Chapter 5). Modern CAD/CAM techniques use computer graphics to produce a virtual component that can be downloaded, via suitable software, directly into the machine control system ready for manufacture.

SELF-ASSESSMENT TASK 4.1

1. With the aid of sketches and in your own words, explain the difference between forming and generating surfaces.

2. Explain why forming tools are not required when producing components with complex profiles on copy-turning and CNC lathes.

4.2 Requirements of a machine tool

The basic requirements of a machine tool may be summarised as follows:

- It must provide the geometrically interrelated movements of tool and workpiece necessary to generate the surfaces and shapes for which it has been designed.
- The machine must have a suitable level of accuracy and repeatability for the class of work for which it is intended. That is, once set, all the components in the same batch of work must lie within the dimensional and geometrical tolerances specified.
- The machine must provide a surface finish appropriate for the class of work for which it is intended, providing the tooling, tool maintenance and tool setting are adequate.
- The machine must be capable of an economic rate of material removal.
- The machine must be safe and convenient to use and set up.
- Maintenance must be minimal and easy to carry out.

To achieve these aims, the machine should follow the principles of good kinematic design. That is, each moving element of the machine involved in the generation of the required surface should be designed to carry out a specific function without duplicating functions or interfering with the function of an associated machine element. For example, the conventional use of the lead screw and nut of a milling machine table is an example of bad kinematic design. Not only must the lead screw and nut provide linear motion to the worktable as it is driven against the thrust of the cutter, but it is usual to fit a micrometer dial to the screw so that it can be used as a measuring and positioning device as well. Any wear and strain resulting from its driving function will affect the accuracy of its measuring function.

Good kinematic design is found on many modern machines where these functions are separated. The lead screw and nut merely provide the table movement; a separate scale and transducer provide a digital read-out of the table position. This means the measuring and positioning system is unaffected by any wear in the lead screw and nut. The basic constructional features of a number of machine tools will now be considered; notice they have several features in common (Fig. 4.6).

Fig. 4.6 *Fundamental requirements of a machine tool: since an axis is not a physical entity, it is represented in practice by centres, a chuck or the taper nose of a spindle*

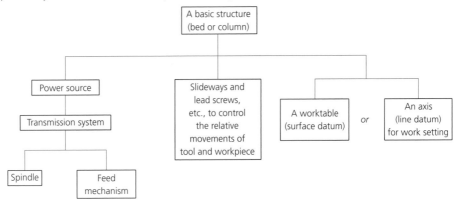

4.3 The basic structure

Figure 4.7 shows examples of the basic structures upon which all machine tools are built. Those which lie in the horizontal plane are called beds, whereas those which stand in the vertical plane are called columns. These beds and columns should have the following properties:

- Strength to resist the cutting forces, and to support the weight of the workpiece and any workholding devices (e.g. jigs and fixtures), and the weight of any subassemblies such as the headstock and gearbox in the case of a lathe.
- Rigidity to resist deflection resulting in misalignment when subject to the weight of the component and to the thrust of the cutting forces.

- Stability so that the basic structure will not distort due to the slow release of internal stresses resulting from its manufacture. Castings should be made from high-stability alloys and stabilised after rough machining by heat treatment or by weathering out of doors.
- The chosen material should have damping properties so that vibrations produced by the cutting process are absorbed before they can affect the accuracy and finish of the workpiece. That is, the material used for machine beds and columns should be non-resonant.

Figure 4.7 shows that most of these requirements are satisfied by the use of high-quality iron castings which are heavily ribbed or have a cellular construction. The high rates of material removal associated with CNC machine tools have necessitated improvements in bed and column design. Steel has also been tried for machine-tool beds and columns, but it has the following disadvantages:

- Stronger and more expensive than cast iron, steel tempts the designer to use thinner sections which are less rigid.
- It is more difficult to cast steel since it has a much higher melting point than cast iron. It is also more viscous (like treacle) than cast iron when molten and will not flow into intricate moulds.
- When fabricated from plate, steel can only be formed into square box shapes; this limits the design possibilities.

Fig. 4.7 *Machine beds and columns: (a) typical cast-iron lathe bed – notice that a rectangle can deflect under load but a triangle cannot; (b) box section casting of a drilling machine column – this type of construction is also used for columns in milling machines and shaping machines. (c) shaping machine column – ribbed box cellular construction; (d) milling machine column – ribbed box cellular construction*

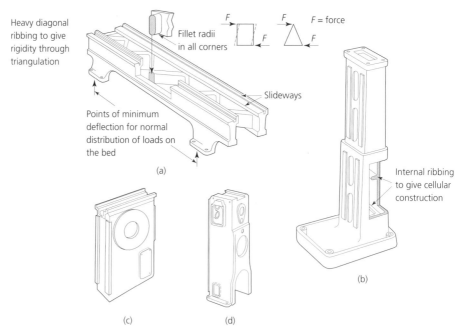

- Steel is not a good bearing material, so antifriction pads have to be fastened to the basic fabrication to form slideways.
- Steel is a resonant material and does not damp vibrations.

However, steel is widely used for welded cabinet bases, supporting the beds and columns of small machines at a convenient working height. One application where steel is superior to cast iron is in the frames of large power presses used for stamping sheet metal. These press frames experience enormous tensile forces which steel is capable of resisting very much better than cast iron.

Fig. 4.8 *Bed and column designs for CNC machines: (a) poor kinematic design, lacking rigidity; (b) improved kinematic design, overhangs reduced and whole head moves up and down, not just the spindle; (c) further improved kinematic design, bifurcated column resists twisting and improves rigidity; (d) most improved kinematic design, slant bed increases rigidity and facilitates swarf removal by conveyor*

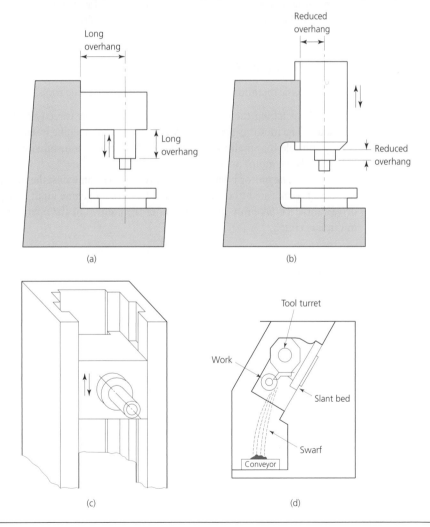

4.4 Power transmission

Power has to be transmitted from the drive motor to the spindle and feed mechanisms of the machine tool. Provision has to be made for connecting and disconnecting the drive from the various machine elements easily and safely. Provision has also to be made for varying the speed of the drive to suit various cutting conditions.

4.4.1 *Starting and stopping*

There are two basic methods of starting and stopping machine tools:

- *Electrical* Switching the drive motor on and off either directly or through a relay or contactor. The motor is permanently coupled to the machine transmission system. This is only suitable for small machines with a low-inertia transmission system which only lightly loads the motor at switch-on.
- *Mechanical* The motor is left permanently switched on whilst the machine is in use and the drive is connected to or disconnected from the machine input by a clutch. Since the drive should be taken up gradually under the control of the operator, this is usually a multiplate clutch.

4.4.2 *Speed control*

There are various methods of speed control. The following examples are the more common ones:

- *Pole-changing motors* These are somewhat limited in that they can only be switched to give a limited number of set speeds (usually four). Unfortunately the torque decreases with the speed.
- *Electronic motor control* This provides an infinite variation in speed within the range of the motor. The torque decreases as the speed decreases. Electronic speed control is widely used on computer-controlled machine tools.
- *Belt drive* A V-belt with stepped pulleys, as shown in Fig. 4.9(a), is used on small machines, especially sensitive bench drilling machines.
- *Gearbox* This is the most widely used method of speed control on conventional machine tools. A schematic diagram for a typical milling machine gearbox is shown in Fig. 4.9(b).

Fig. 4.9 *Speed control: (a) belt drive with stepped pulleys (sensitive drilling machine); (b) typical machine tool gearbox (milling machine)*

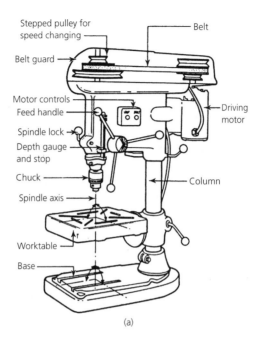

Stepped pulley for speed changing

Belt

Belt guard

Motor controls

Feed handle

Driving motor

Spindle lock

Depth gauge and stop

Chuck

Column

Spindle axis

Worktable

Base

(a)

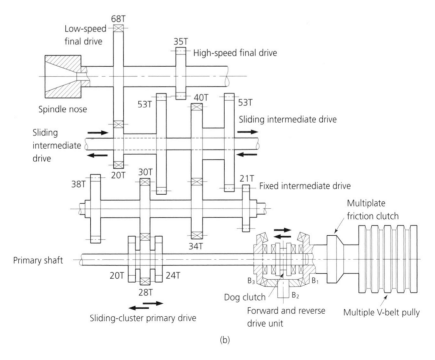

68T

Low-speed final drive

35T

High-speed final drive

Spindle nose

40T

53T

53T

Sliding intermediate drive

Sliding intermediate drive

20T 30T

38T

21T

Fixed intermediate drive

Multiplate friction clutch

34T

Primary shaft

20T

24T

28T

B_3

B_1

Dog clutch

B_2

Forward and reverse drive unit

Multiple V-belt pully

Sliding-cluster primary drive

(b)

4.4.3 *Electrical power transmission*

To avoid complex mechanical drives, separate electrical motors are employed for each separate drive, e.g. the table traverse mechanism of a milling machine. The power to the motor can be supplied through flexible conductors, which is less complex than using telescopic drive shafts fitted with universal joints.

4.4.4 *Hydraulic power transmission*

Rotary hydraulic motors tend to be smaller than electric motors for the same power output, and they have greater starting torque. Energy can also be transmitted by flexible hose, thus giving flexibility of movement to the system. However, hydraulic transmission systems are more costly than electrical systems and are less easily linked to electronic control systems. Hydraulic drives are used in the worktable traverse of large machine tools.

4.4.5 *Belt drives*

Long-centre drives can be achieved with belts or roller chains (similar to bicycle chains). Roller chain drives have the advantage that they are synchronous (non-slip) but require regular maintenance, regular lubrication and a clean environment to operate successfully. They have been largely superseded by toothed belt drives, as shown in Fig. 4.10(a).

Fig. 4.10 *Belt drives: (a) toothed belt and pulley (for non-slip drive); (b) V-belt drive*

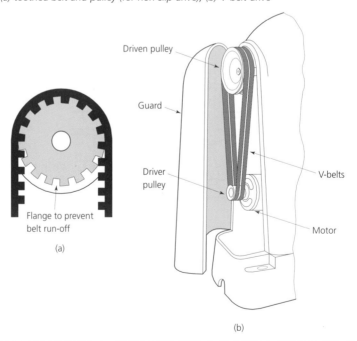

Cheaper and quieter, toothed belts do not require lubrication and can work in a more hostile environment than chain drives. They are now widely used in machine tools where a synchronous drive is required, e.g. transmitting the drive from a stepper motor to a lead screw in a computer-controlled machine tool. V-belts are widely used for transmitting mechanical energy from the drive motor of a machine to the main drive pulley of the gearbox, as shown in Fig. 4.10(b).

4.4.6 *Gear drives*

These are used for short-centre drives. They are compact, reliable and synchronous (non-slip) but relatively expensive and require continuous lubrication. They are used in machine-tool gearboxes and for such applications as the transmission of energy from the spindle gearbox of lathes to the feed gearbox, where slip cannot be tolerated and the velocity ratio sometimes needs to be changed. Block diagrams of some typical machine-tool power transmission systems are shown in Figs 4.11 to 4.13.

4.4.7 *Bearings*

Plane and rolling bearings are used in machine tools. Plain bearings are the simplest and the cheapest to manufacture. However, they depend upon adequate lubrication if they are

Fig. 4.11 *Drilling machine power transmission systems: (a) cone pulley drive (sensitive drilling machine); (b) all-geared drive; (c) variable-speed electrical system*

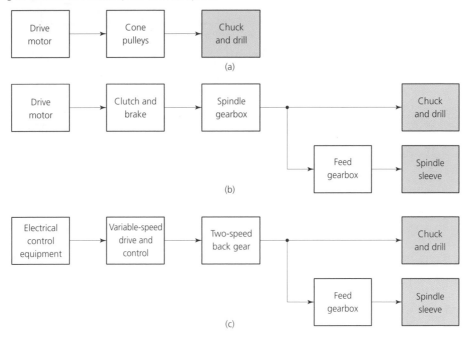

Fig. 4.12 *Power transmission systems: (a) for centre lathe; (b) for milling machine, all-geared drive; (c) for milling machine, variable-speed electrical system*

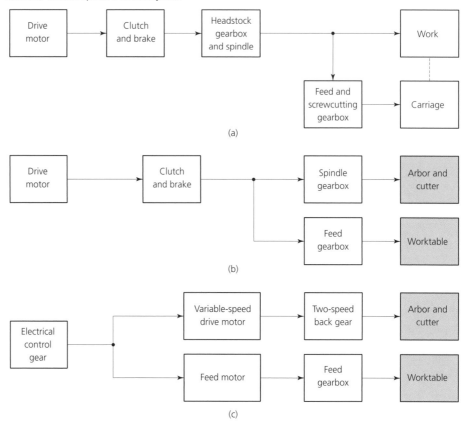

(a)

(b)

(c)

not to fail prematurely. Bearing bushes cannot be adjusted for wear or reloaded to increase radial stiffness. They also exhibit relatively high frictional losses. Figure 4.14(a) shows a typical plane bearing. At rest there is little lubricant between the shaft and the bearing shell, and on start-up the wear is very heavy. Once the shaft is rotating it drags a wedge of oil between itself and the bearing; it depends on this to provide lubrication and prevent wear. The wedge of oil is called a hydrodynamic wedge as its presence depends on the movement (rotation) of the shaft.

Figure 4.14 also shows that the axis of the shaft moves about the centre point of the bearing, causing inaccuracies if it is used for the spindle of a machine tool. Figure 4.14(b) shows a plane bearing with a constant centre line. Instead of a single pressure lobe, the tilting pads 'float' until a system of pressure lobes are built up symmetrically around the shaft and it is centred in the bearing. Any disturbing force upsets the balance of the system, causing a local pressure increase to oppose the disturbance and return the shaft to the centre of the bearing.

Modern practice favours the use of antifriction, rolling bearings. A typical machine spindle of good kinematic design is shown in Fig. 4.15. The minimum number of bearings

Fig. 4.13 *Power transmission systems: (a) for a surface grinding machine; (b) for a cylindrical grinding machine*

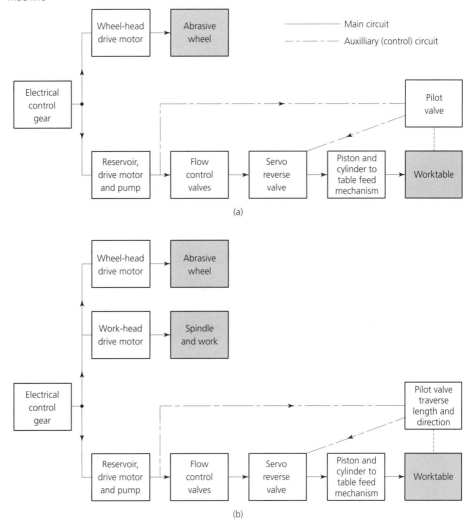

(a)

(b)

are used and they do not act against each other. The spindle nose is at A and is supported in a pair of opposed taper rolling bearings B. These bearings provide axial and radial location of the nose end of the spindle and are self-centring. They are preloaded against any disturbing forces by the screwed rings C, which can also be used to take up wear. The drive to the spindle is through the gear D, and the 'tail' of the spindle is supported in a roller bearing at E. The shaft is free to slide axially in the bearings at E, so that as it warms up and expands there is no tendency for the shaft to bend and any expansion does not disturb the plane of the spindle nose. Oil seals are provided at F to prevent lubricating oil getting out and dirt getting in.

Fig. 4.14 *Plane journal bearings: (a) simple bearing, movement of centre line; (b) bearing with constant centre line*

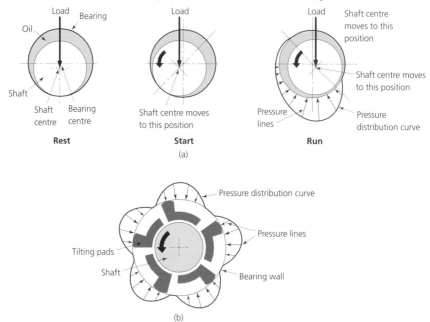

Rest   Start   Run

(a)

(b)

Fig. 4.15 *Roller bearings: (a) geometry of a taper roller bearing; (b) a machine tool spindle*

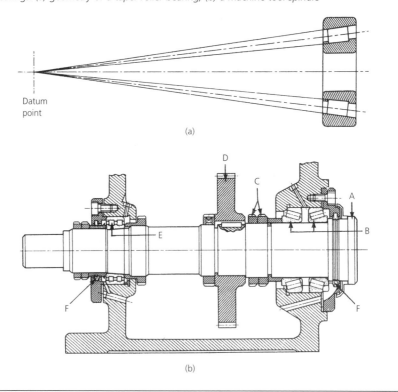

(a)

(b)

4.5 Guidance systems

Slideways are linear bearings which support and guide the sliding members of machine tools. They should provide the easiest possible movement in the required direction and the maximum possible resistance to movement or deflection (stiffness) in all other directions. Many different types of slideways are used on machine tools and some of them will now be considered.

4.5.1 *Flat slides*

The flat slide was the earliest form of slideway to be used extensively. It is simple to produce, robust and accurate when new. However, when subject to uneven wear it is virtually impossible to keep in adjustment. For example, most of the wear on a lathe bed occurs in the vicinity of the chuck, and if the gib strip of the carriage is adjusted to take up the wear in this area, it will be too tight when the saddle is moved to less worn areas of the bed slides (Fig. 4.16). Flat slideways are still used where the wear is uniform or limited by only occasional use.

Fig. 4.16 *Flat slideway: (a) section; (b) adjustment of a gib strip*

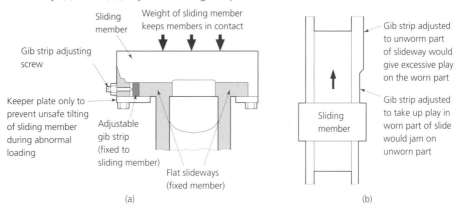

(a)

(b)

4.5.2 *Inverted V-slides*

This type of slide is self-adjusting for wear and is found on most modern lathe beds. As long as the contact between the carriage and the bed is only on the flanks of the V-slide, any wear causes the carriage to settle lower onto the V but there is no lateral displacement. This design is largely kinematic and resists the disturbing forces without redundant restraint. Figure 4.17 shows how a single V-slide provides support for the front of the carriage as well as lateral restraint. Only one V-slide is provided as only one slide is required to provide lateral restraint and also because it is impossible to manufacture two perfectly parallel V-slides. Any deviance from perfect parallelism would cause the slides to pull against each other. Instead the rear of the carriage is supported on a flat slide; this provides support but no lateral control. In practice a second inverted V-slide and flat slide are provided for the support and guidance of the tailstock. Figure 4.17 also shows the various forces acting on the carriage and how the V-slide resists them.

Fig. 4.17 *Inverted-V slideway*

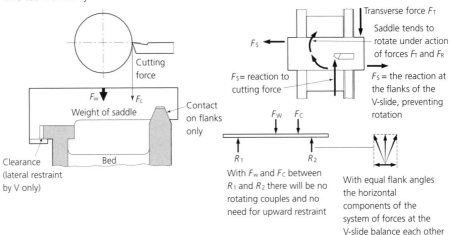

4.5.3 *Dovetail slide*

This type of slideway is used where the applied forces form a couple which tries to rotate the sliding member. Figure 4.18 shows a dovetail slide and the forces acting upon it. Because of its ability to provide a restoring couple, the dovetail slide is used where the applied force is 'outside' the sliding member. Notice that the cutting force F_c and the reaction force R_1 form a disturbing couple, and the reaction forces R_1 and R_2 form a restoring couple. Another advantage of this type of slideway is that adjustment of the gib strip to compensate for wear, restores the vertical and horizontal restraints simultaneously.

4.5.4 *Antifriction slides*

Ball and roller bearings are used for shafts rotating in journal bearings because the rolling friction of a journal bearing is very much less than the sliding friction of a plane bearing. Similarly, the slideways of ball and roller bearing exhibit less friction than the slideways of

Fig. 4.18 *Dovetail slideway: a couple is a system of two equal but oppositely directed parallel forces*

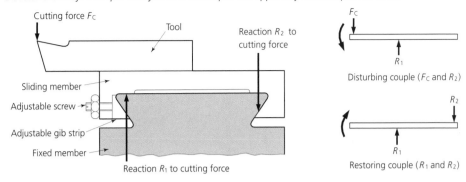

Fig. 4.19 *Antifriction slideways: (a) ball-bearing; (b) roller bearing*

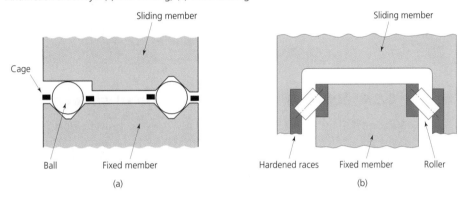

the plane bearings described so far. The layout of typical antifriction slideways is shown in Fig. 4.19. Since the forces acting on the sliding members are supported on much smaller surfaces, these slideways cannot be used on heavy-duty machine tools.

4.5.5 *Hydrostatic slideways*

To understand the function of the hydrostatic slide, consider the principle of the hydrostatic bearing. Figure 4.20(a) shows a typical hydrostatic journal bearing. Notice that the bearing shell is perforated with a symmetrical pattern of fine holes through which air or oil is forced under high pressure. Compared with a conventional bearing, there is considerable clearance between the shaft and the bearing shell. Since the pressure acts equally all round the shaft, it is forced into the centre of the bearing. Any deflection of the shaft causes an imbalance in the fluid forces acting on the shaft in such a direction that the shaft is restored to the centre of the bearing. This type of bearing has a very low friction but its stiffness to deflection is very much greater than a conventional bearing.

Figure 4.20(b) shows the application of this principle to a hydrostatic slide. The sliding member floats over the slideway like a hovercraft. Because the slide is self-centring it exhibits a much higher degree of lateral stiffness than a conventional slideway. At the same time, because there is no metal-to-metal contact, both friction and wear are much reduced.

Fig. 4.20 *Hydrostatic slideways: (a) hydrostatic bearing; (b) hydrostatic bearing principle applied to a linear slideway*

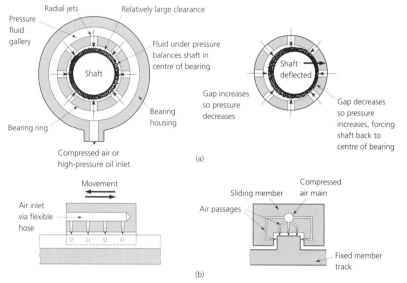

(a)

(b)

4.5.6 *Lubrication and protection of slideways*

Except for hydrostatic slideways it is necessary to provide efficient lubrication of the contact surfaces if wear and friction are to be kept to a minimum. Also dirt and swarf must be prevented from entering between the sliding members and causing excessive and premature wear. Examples of slideway lubrication and protection are shown in Fig. 4.21.

Fig. 4.21 *Lubrication and protection of slideways: (a) to ensure an even distribution of oil to the slideway, there is a pattern of oil grooves on the underside of the sliding member; (b) oil well and roller; (c) slideway wiper; (d) slideway covers*

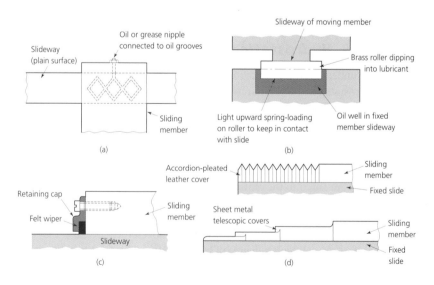

(a)

(b)

(c)

(d)

4.6 Positioning systems

All machine tools require a means of moving the sliding members – the carriage of a lathe or the worktable of a milling machine – along the slideways in a controlled manner, not only for setting purposes but also to oppose the thrust of the cutting forces during machining. The two most common means are a screw and nut, and a rack and pinion. Both mechanisms convert rotary motion into linear motion.

4.6.1 *Screw and nut*

A typical screw-and-nut assembly for traversing a milling machine table is shown in Fig. 4.22(a). Since the screw is being used to transmit power with minimum friction, the locking action of a V-thread is undesirable, so a square thread form or an Acme thread form is used. The lead screw is made of alloy steel, toughened and thread-ground to reduce

Fig. 4.22 *Screw and nut: (a) milling machine table. (b) half-nut and lead screw*

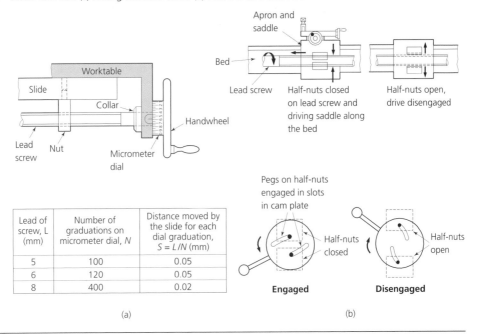

Lead of screw, L (mm)	Number of graduations on micrometer dial, N	Distance moved by the slide for each dial graduation, $S = L/N$ (mm)
5	100	0.05
6	120	0.05
8	400	0.02

(a)

(b)

wear and ensure a high degree of accuracy. The nut is made from a hard-wearing antifriction material such as phosphor bronze. Preferably, the nut should wear out first as it is easier and cheaper to replace than the screw. The lead screw of a lathe is always given an Acme thread form. This thread form has slightly tapered flanks which ease the engagement and disengagement of the half-nut when screw cutting, as shown in Fig. 4.22(b).

Recirculating ball screws and nuts are used on computer-controlled machines to reduce the friction, hence reducing the torque required by the drive motor. These screw-and-nut systems have a higher accuracy and less backlash than conventional screws and nuts. Because there is less wear, they maintain their accuracy longer. Figure 4.23 shows the principle of such a mechanism.

Fig. 4.23 *Recirculating ball screw*

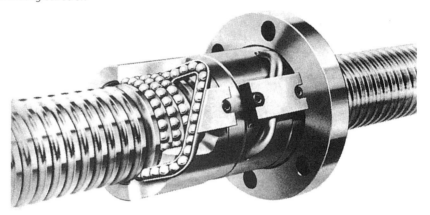

4.6.2 *Rack and pinion*

This mechanism is simple and relatively cheap. It also has the advantage that rapid rates of traverse are easily obtained with little effort. However, it has a lower accuracy of control than a screw and nut. The rack-and-pinion mechanism is conventionally used to move a lathe carriage during normal cylindrical turning, as shown in Fig. 4.24(a). Figure 4.24(b) shows how a rack and pinion is used for the feed mechanism of sensitive drilling machines.

4.6.3 *Hydraulics*

Hydraulic systems are used to provide the table traverse for grinding machines. They are smoother than mechanical drives and are less likely to leave a vibration pattern on the ground surface. A simple example is shown in Fig. 4.25.

4.6.4 *Measurement*

To produce accurate work from a machine tool, it is necessary to be able to position the various sliding elements which control the work and tool movements accurately, and suitable measuring systems have to be built into the machine. The use of a lead screw and nut to impart linear motion to the sliding members of a machine has already been

Fig. 4.24 *Rack and pinion: (a) lathe traverse mechanism; (b) spindle feed mechanism (drilling machine)*

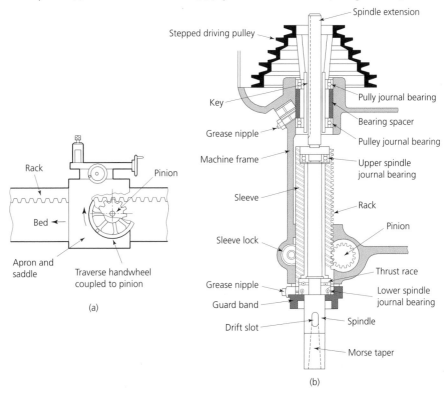

Spindle extension

Stepped driving pulley

Key

Grease nipple

Machine frame

Sleeve

Sleeve lock

Grease nipple

Guard band

Drift slot

Pully journal bearing

Bearing spacer

Pulley journal bearing

Upper spindle journal bearing

Rack

Pinion

Thrust race

Lower spindle journal bearing

Spindle

Morse taper

Rack

Pinion

Bed

Apron and saddle

Traverse handwheel coupled to pinion

(a)

(b)

Fig. 4.25 *Hydraulic circuit*

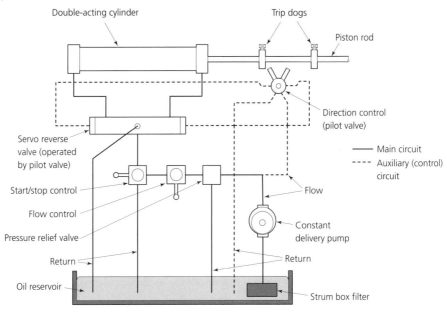

Double-acting cylinder

Trip dogs

Piston rod

Direction control (pilot valve)

Servo reverse valve (operated by pilot valve)

Start/stop control

Flow control

Pressure relief valve

Return

Oil reservoir

—— Main circuit

--- Auxiliary (control) circuit

Flow

Constant delivery pump

Return

Strum box filter

introduced. By the addition of a calibrated dial to the screw, the same screw and nut can also be used to position the sliding member, e.g. the table of a milling machine (Fig. 4.22) or the cross-slide of a lathe. Each revolution of the dial moves the nut along the screw a distance equal to the lead of the nut. The movement is relative; it does not matter whether the nut is fixed and the screw moves, or the screw is fixed and the nut moves.

Consider a nut of lead 5 mm and a dial with 50 divisions:

- Rotation of the dial through 2 complete rotations would move the nut along the screw through $2 \times 5\,mm = 10\,mm$.
- Rotation of the dial through 8 divisions would move the nut along the screw through $5\,mm \times 8/50 = 0.8\,mm$.
- Rotation of the dial through 4 complete rotations and 18 divisions would move the nut along the screw through:

$$(4 \times 5\,mm) + (5\,mm \times 18/50) = 20\,mm + 1.8\,mm = 21.8\,mm$$

It is bad kinematic design to use the same mechanism for measurement and for traversing sliding members of the machine, since wear in the traverse mechanism can lead to inaccuracy when it is used as a measuring device. It is better practice to use a separate measuring system to control positioning of the work and cutting tool. Various optical and mechanical measuring systems have been used, but modern practice favours electronic systems because of the ease of providing digital read-out and of linking the positioning system to a computer numerical control (CNC) unit.

SELF-ASSESSMENT TASK 4.5

1. For any milling machine with which you are familiar:

    (a) measure the lead of the table traverse screw
    (b) count the number of divisions on the micrometer dial on the table traverse handle
    (c) from this information calculate the distance moved by the table for each division on the dial

---

4.7 Machine-tool alignments

4.7.1 *Lathe*

Figure 4.26(a) shows the basic alignment of the headstock spindle and the tailstock barrel on a common axis, itself parallel to the bed slideways in both the vertical and the horizontal plane. This common axis is the datum of the machine to which all other alignments are referred. The bed slideways are also called the 'shears' in some parts of the United Kingdom. The rotation of the workpiece and the movement of the carriage alone are shown in Fig. 4.26(b). Since they move the tool in a path parallel to the datum axis, these movements and alignments will generate a true cylinder. The cross-slide on top of the carriage is aligned at 90° to the spindle axis, as shown in Fig. 4.27(a). This slide has two purposes:

- To control the depth of cut (in-feed of the tool) when cylindrical turning; for this purpose the cross-traverse screw is fitted with a micrometer dial.
- To move the tool in a path at right angles to the datum axis when generating a plain surface; this operation is called facing and is used for machining the ends of components and shoulders.

The compound slide (top slide) is mounted on the cross-slide and can be set at an angle to the datum axis, as shown in Fig. 4.27(b). It has two purposes:

- To control the depth of cut (in-feed of the tool) when a facing operation is being performed; for this purpose the compound slide is set parallel to the datum axis and the traverse screw is fitted with a micrometer dial.
- To move the tool in a path at an angle to the datum axis in order to generate a conical surface (taper) as shown in Fig. 4.27(b); taper turning is considered more fully in Section 4.9.

Fig. 4.26 *Centre lathe: (a) basic alignment; (b) the carriage or saddle provides the basic movement of the cutting tool parallel to the work axis; the bed is the basic structure of the lathe*

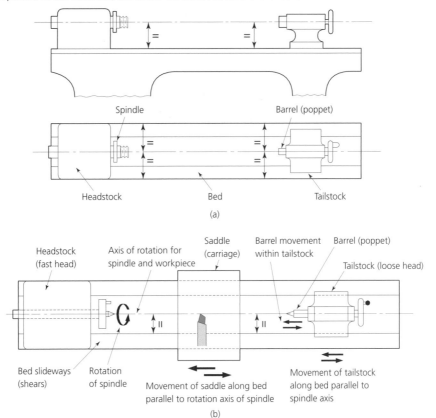

Fig. 4.27 *Centre lathe: (a) cross-slide; (b) compound slide*

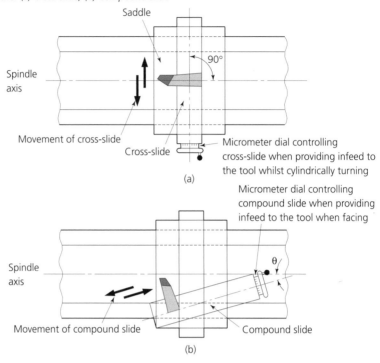

Saddle

Spindle
axis

90°

Movement of cross-slide

Cross-slide

Micrometer dial controlling
cross-slide when providing infeed to
the tool whilst cylindrically turning

(a)

Micrometer dial controlling
compound slide when providing
infeed to the tool when facing

Spindle
axis

θ

Movement of compound slide

Compound slide

(b)

4.7.2 *Drilling machine*

Figure 4.28 shows the basic alignments of the drilling machine. Here the working surface of the worktable is the machine datum. The spindle locates and rotates the drill or other devices such as reamers or countersinking cutters and counterboring cutters. The spindle

Fig. 4.28 *Drilling machine: basic alignments*

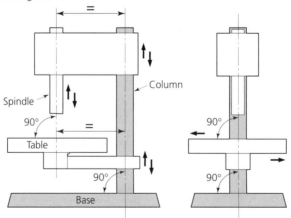

Column

Spindle

90°

Table

90°

90°

90°

Base

itself is located in bearings within a sleeve and can move in the head of the drilling machine. The complete assembly of spindle, bearings and sleeve is known as the quill. A section through such an assembly is shown in Fig. 4.23.

To produce accurate work, the fundamental alignment of the machine requires the spindle axis to be perpendicular to worktable and to remain perpendicular throughout its travel as it is fed into or out of the workpiece. The head itself is mounted on a column, which in turn is perpendicular to the base of the machine. To compensate for drills of differing lengths and work or differing thicknesses, the head and worktable can be raised or lowered on the column without upsetting the fundamental alignment of spindle axis and worktable. Provision is made to tilt the table when holes need to be drilled at an angle to the datum surface.

4.7.3 *Horizontal milling machines*

These machines generate plane surfaces using rotating multi-tooth cutters. The basic movements and alignments required are:

- A spindle with a horizontal axis to rotate and locate the cutter(s) in a given plane.
- A worktable which will locate and move the workpiece in a given plane beneath and relative to the cutter.

The worktable needs to traverse longitudinally to feed the work into the cutter as cutting proceeds. For setting purposes, it needs to move at right angles to the direction of table traverse. It also needs to move up and down in order to control the depth of cut and to compensate for different thicknesses of work. All these movements and alignments are brought together to build up the outline of a horizontal-spindle milling machine (Fig. 4.29).

Fig. 4.29 *Horizontal milling machine: movements and alignments*

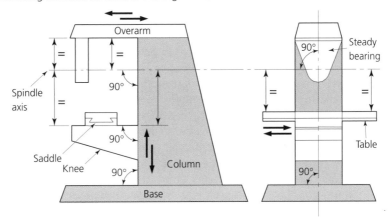

4.7.4 *Vertical milling machine*

This machine only differs from the horizontal machine in that the spindle rotating and locating the cutter has its axis in the vertical plane. The movements and alignments of the worktable and work are the same as for the horizontal-spindle milling machine. All these movements and alignments are brought together to build up the outline of a vertical-spindle milling machine (Fig. 4.30). In the horizontal and vertical machines the working surface of the worktable provides the basic datum surface from which all other alignments are taken. The vertical machine may also have an inclinable head so that the spindle axis may be inclined to the worktable during cam milling or when angular surfaces are required.

Fig. 4.30 *Vertical milling machine: movements and alignments*

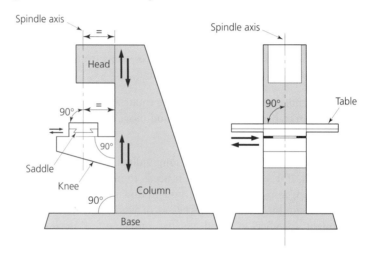

4.7.5 *Precision grinding machines*

Figure 4.31(a) shows the relative alignments and movements of a surface grinding machine. Notice how they are very similar to those of the horizontal-spindle milling machine. This is because both machines are designed to generate plane surfaces using a cylindrical rotating cutter whose axis is horizontal. The precision surface grinding machine produces work of greater accuracy and superior surface finish than the milling machine, but at a lower rate of material removal. It is used to finish-machine precision components previously roughed out on machines with higher rates of material removal.

Figure 4.31(b) shows the relative alignments and movements of the cylindrical grinding machine. Notice how they are similar to those of the centre lathe. This is not surprising since both machines are designed to generate surfaces of revolution. The cylindrical grinding machine produces work of greater accuracy and superior surface finish than the centre lathe but at a lower rate of material removal. It is used to finish precision components previously roughed out on the lathe.

Fig. 4.31 *Movements and alignments: (a) surface grinding machine; (b) precision grinding machine*

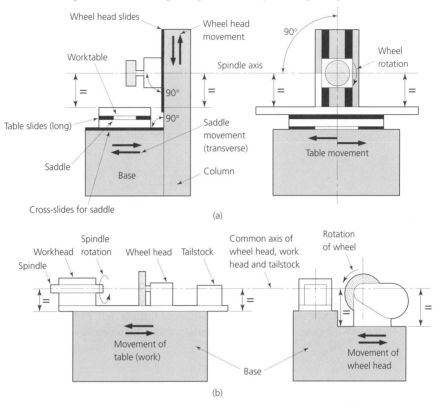

Wheel head slides

Wheel head movement

90°

Wheel rotation

Worktable

Spindle axis

90°

=

=

=

=

Table slides (long)

90°

Saddle movement (transverse)

Table movement

Saddle

Base

Column

Cross-slides for saddle

(a)

Spindle rotation

Common axis of wheel head, work head and tailstock

Rotation of wheel

Workhead

Wheel head

Tailstock

Spindle

=

=

=

Movement of table (work)

Base

Movement of wheel head

(b)

4.8 Machine-tool accuracy

So far this chapter has been concerned with the constructional features common to a number of machine types and to the alignments and movements necessary for the production of accurate work. However, no matter how carefully the principles of good kinetmatic design are observed, no matter how carefully the various subassemblies making up the machine are manufactured, the machine will not function effectively unless all the subassemblies are correctly aligned during the final assembly of the machine. After assembly all machine tools must be tested for alignment to the standards which are laid down and accepted internationally. Machines should be retested from time to time throughout their lives to ensure they are correctly adjusted and maintained. Misalignment can occur through incorrect handling during transportation and through incorrect installation.

Before any new machine can be tested for accuracy and alignment, it must be levelled up to ensure the basic structure is not distorted. Figure 4.32 shows how a lathe should be levelled. Wedges under the base of the machine, or jacking screws built into the base of the machine are adjusted until the machine is level and free from distortion. After levelling, the appropriate standard alignment test should be carried out and reported.

Fig. 4.32 *Levelling a machine tool*

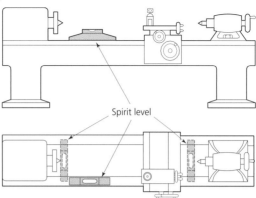

Spirit level

Figure 4.33 shows an abstract from the standard test chart for a centre lathe. The results of the test are entered in the final column and a copy is supplied with the machine. Notice that the tolerance is applied over a specified length or diameter. This compensates for the fact that it is easier to maintain a high degree of accuracy over a short distance than over a long distance. The tolerances are biased so that any possible deflection of the member being tested – due to its own weight, the weight of the workpiece and the cutting forces – improves the alignment accuracy rather than detracts from it.

4.9 Setting accuracy

No matter how accurately a machine tool is manufactured, the ultimate factor in controlling the quality of the work produced is the alignment of the workholding device and workpiece. The alignment of workholding devices and work will now be considered for some typical machines.

4.9.1 *Centre lathe*

For parallel turning, the axis of the headstock spindle and the axis of the tailstock barrel must be coincident and parallel with the bed slideways. The tailstock has lateral adjustment to achieve this alignment. A trial cut should be taken along the workpiece when turning between centres. Its diameter is then measured at each end with a micrometer caliper. If the readings are the same, a true cylinder is being generated and turning may proceed. If the readings are different, the tailstock has to be adjusted as shown in Fig. 4.34(a). Adjustment proceeds largely by trial and error, taking further trial cuts until parallel work is produced. A more convenient method of bringing the tailstock axis into alignment is shown in Fig. 4.34(b). It is essential that the test bar is accurately centred and ground parallel.

Fig. 4.33 *Alignment testing*

Test diagram	Test to be applied	Gauge and methods	Tolerances	Test results
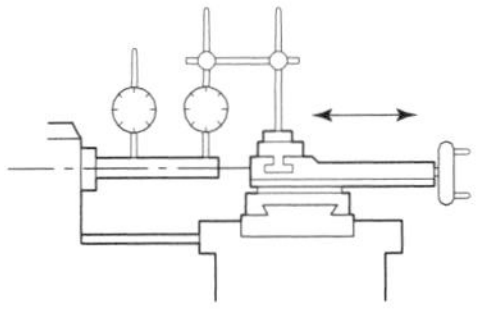	**1.** Saddle: movement of upper slide parallel with spindle (vertical plane)	Stationary mandrel. DTI in tool post. Test over mandrel. Set at zero and traverse top slide by hand	0.02 mm in its movement	
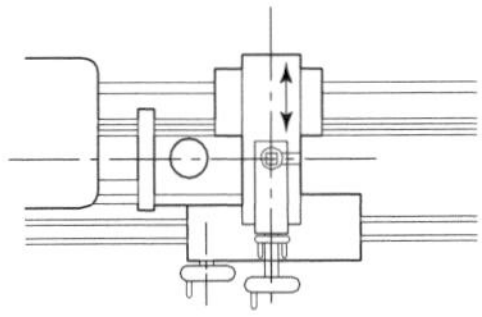	**2.** Movement of lower slide at 90°	DTI in tool post. Test across straight edge on face plate	0 to 0.2 mm per D 250 mm concave only. DTI reading to be minus at centre	
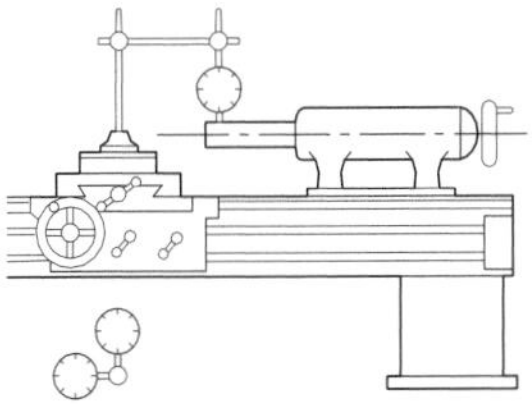	**3.** Tailstock: quill movement parallel with bed (vertical plane)	DTI and stand. Test over quill which is clamped. Centre must rise	0 to 0.02 mm in its movement	
	Tailstock: quill movement parallel with bed (horizontal plane)	As above. Test side of quill. Inclination towards tool pressure	0 to 0.1 mm in its movement	
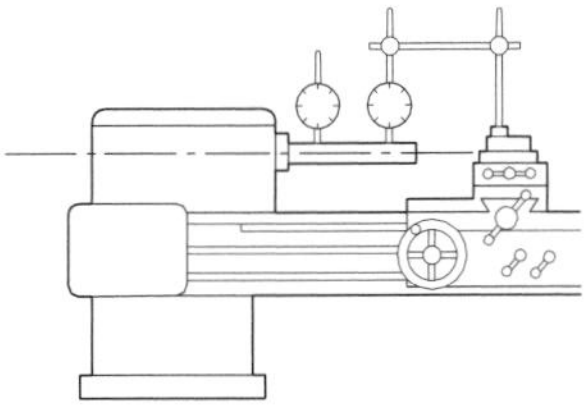	**4.** Saddle: bore true size to gauge (internal taper)	Mandrel 250 mm long, one end is a gauge for spindle DTI and stand	0.02 mm maximum eccentric error	
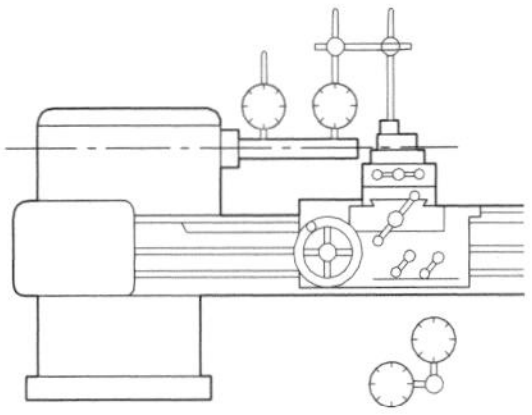	**5.** Axis parallel with bed in a vertical plane	Stationary mandrel (as above). DTI and stand	0 to +0.02 mm per 250 mm from end of mandrel	
	Axis parallel with bed in a horizontal plane	Free end of mandrel inclined towards tool pressure	0 to 0.02 mm per 250 mm from end of mandrel	

When a conical (tapered) surface is required, it is necessary to disturb the basic alignments deliberately to produce the required amount of taper. Figure 4.35 shows three methods of taper turning. The advantages and limitations of each method are summarised in Table 4.1. Notice that methods (a) and (c) require the basic alignments of the machine to be disturbed. After taper turning is finished, the alignments should be restored for normal parallel turning and checked against a parallel test bar using a dial test indicator (DTI) as shown in Fig. 4.34.

Fig. 4.34 *Parallel cylindrical turning: (a) the axis of the headstock spindle must be in alignment with the tailstock barrel, if this is so then diameter A will be the same as diameter B; (b) use of a test bar*

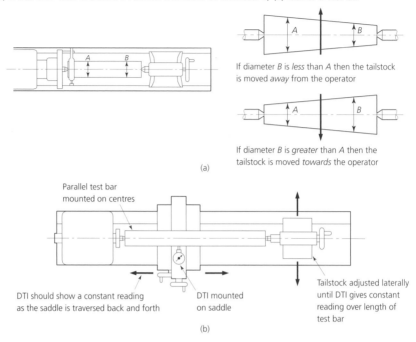

If diameter B is *less* than A then the tailstock is moved *away* from the operator

If diameter B is *greater* than A then the tailstock is moved *towards* the operator

(a)

Parallel test bar mounted on centres

DTI should show a constant reading as the saddle is traversed back and forth

DTI mounted on saddle

Tailstock adjusted laterally until DTI gives constant reading over length of test bar

(b)

Fig. 4.35 *Taper turning: (a) set-over of centres; (b) the taper turning attachment; (c) the compound slide, 2θ is the included angle of the taper*

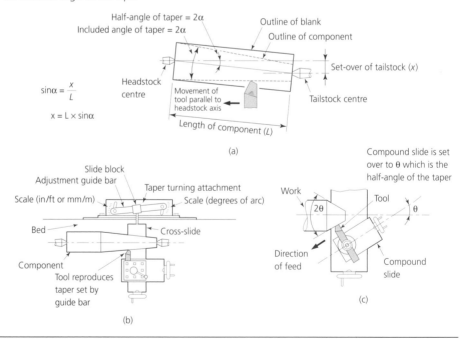

Half-angle of taper = 2α
Included angle of taper = 2α

Outline of blank
Outline of component

$\sin\alpha = \dfrac{x}{L}$

$x = L \times \sin\alpha$

Headstock centre

Movement of tool parallel to headstock axis

Set-over of tailstock (x)

Tailstock centre

Length of component (L)

(a)

Slide block
Adjustment guide bar
Scale (in/ft or mm/m)

Taper turning attachment
Scale (degrees of arc)

Bed

Cross-slide

Component
Tool reproduces taper set by guide bar

(b)

Compound slide is set over to θ which is the half-angle of the taper

Work

Tool

Direction of feed

Compound slide

(c)

Table 4.1 *Comparison of taper turning techniques*

Method	Advantages	Limitations
Set-over of tailstock	1. Power traverse can be used 2. The full length of the bed is used	1. Only small angles can be accommodated 2. Damage to the centre holes can occur 3. Difficulty in setting up 4. Only applies to work held between centres
Taper turning attachments	1. Power traverse can be used 2. Ease of setting 3. Can be applied to chucking and centre work	1. Only small angles can be accommodated 2. Only short lengths can be cut (304–457 mm, 12–18 in) depending on the make
Compound slide	1. Very easy setting over a wide range of angles: normally used for short, steep tapers and chamfers 2. Can be applied to chucking and centre work	1. Only hand traverse available 2. Only very short lengths can be cut. Varies with machine but is usually limited to about 76–101 mm (3–4 in)

To ensure all the turned diameters are concentric, as many of them as possible should be turned at one setting, as shown in Fig. 4.36(a). There are two options when the work has to be turned around or reset for finishing: (1) it should be held in a self-centring chuck using soft jaws which have been bored out true to hold the workpiece, or (2) it should be held in a four-jaw independent chuck. If a four-jaw chuck is used, the component is checked on a previously turned diameter using a dial test indicator (DTI), as shown in Fig. 3.46. Alternatively the work may need to be deliberately offset in the four-jaw chuck to turn an eccentric diameter, as shown in Fig. 4.36(c). When turning eccentric diameters, remember that the throw of an eccentric diameter is twice the offset.

The height of the tool point can also affect the accuracy of the surface being produced when taper turning, screw cutting, copy turning and facing to the centre of the work. The height has little effect on the accuracy of parallel work but it does affect the cutting efficiency of the tool. The tool point should always be aligned with the axis of the workpiece. Failure to do this can cause the effective rake and clearance angles to be very different from their manufactured or reground values. Figure 4.37 shows the effect of tool height on the cutting angles for external work, and Fig. 4.38 shows the effect of tool height when boring.

4.9.2 *Drilling machine*

Holes are normally drilled so their axes are perpendicular to the surface of the workpiece. Thus the spindle axis must be perpendicular to the surface of the worktable of the machine.

Fig. 4.36 *Concentric and eccentric turning: (a) maintaining concentricity – both the bore and the outside diameter are turned at the same setting, i.e. they are turned without removing the component from the chuck; (b) concentric diameters – both diameters have the same centre; (c) eccentric diameters – each diameter has a different centre*

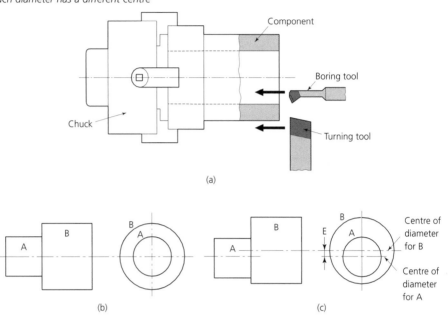

Fig. 4.37 *Effect of tool height on cutting angle for external turning: (a) tool set correctly on centre height; (b) tool set below centre height; (c) tool set above centre height*

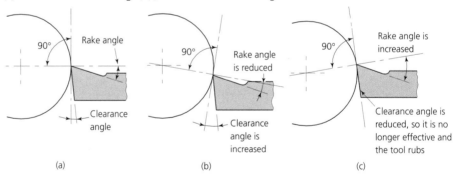

It is possible for this alignment to be disturbed on some machines fitted with an inclinable table. Therefore the table should be checked before commencing work (Fig. 4.39). Where work is held in a machine vice, care should be taken to clean the table and the underside of the vice, as any dirt or swarf under the vice will tilt it and disturb the alignment of the drill and the work. The same applies when the work is clamped directly to the machine table. If through-holes are to be drilled, the work should be supported on parallel strips so that neither the table nor the vice is damaged as the drill breaks through the workpiece, since this would damage and eventually destroy the basic datum surfaces required for setting.

Fig. 4.38 *Effect of tool height on cutting angle for boring: (a) tool set correctly on centre height; (b) tool set below centre height; (c) tool set above centre height*

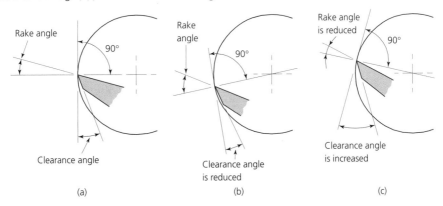

(a)   (b)   (c)

Fig. 4.39 *Setting a drilling machine table: when the spindle axis is perpendicular to a datum surface of the worktable, the DTI reading will be constant in all positions*

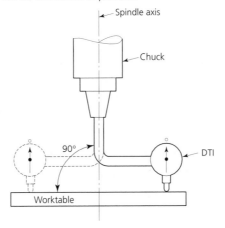

4.9.3 *Milling machine*

A DTI is used to set the fixed jaw of a milling machine vice parallel to the spindle axis of a horizontal milling machine, as shown in Fig. 4.40(a). Alternatively a ground parallel strip can be held in the vice and clocked up with a DTI. This method is more accurate because the length of the parallel strip is greater than the length of the vice jaw. When the vice is correctly set, the DTI reading should be constant as the cross-slide is wound back and forth. Figure 4.40(b) shows a component being set parallel with a milling machine table.

In the case of a vertical-spindle milling machine with an inclinable head, it is essential that the spindle axis is perpendicular to the machine table, otherwise a hollow surface will be machined when a face cutter is used. The method of checking the alignment of the machine head with the table is the same as for checking a drilling machine (Fig. 4.38).

Fig. 4.40 *Setting work on a milling machine table: (a) setting a vice; (b) checking a surface for parallelism – a constant reading of the DTI indicates the upper surface of the component is parallel with the machine table*

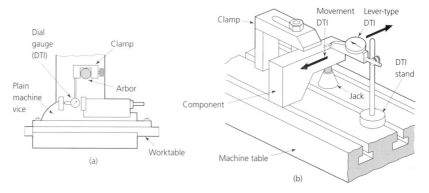

4.9.4 *Surface grinding machine*

The basic datum surface of a surface grinding machine is the worktable. However, most work is held on the magnetic chuck, and this provides a secondary datum which must have a plane surface parallel to the worktable. Therefore, when a new chuck is fitted to a precision surface grinding machine, it is usual to take a cut over the working surface of the chuck with the grinding wheel. The space between the pole pieces is filled with white metal; this clogs the grinding wheel and prevents it cutting properly. So, before grinding, the white metal should be cut away to just below the surface using a narrow chisel. Once ground the chuck should not be disturbed.

4.9.5 *Cylindrical grinding machine*

The alignments for this machine are similar to those of the centre lathe and its standard workholding devices. But there is one essential difference. Instead of setting over the tailstock for parallel or tapered components, the whole worktable (carrying the workhead and the tailstock) is pivoted about its centre (Fig. 4.41) and has a fine screw adjustment. If parallel grinding is required, the table is adjusted until the work measures the same diameter at each end.

Fig. 4.41 *Grinding tapered components*

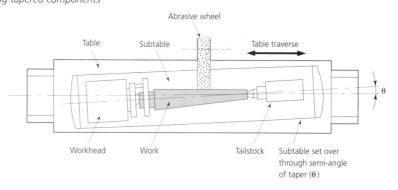

1. (a) Explain why it is important to level a machine tool when installing it in a new position.
   (b) With the aid of sketches, show how you would use a precision level when installing a milling machine.
   (c) Suggest several means of levelling the machine.

2. Use sketches to explain how the tailstock of a lathe should be adjusted to ensure parallel turning of work supported between the centres.

3. With the aid of a sketch, show how you would check that the spindle axis of a vertical milling machine is perpendicular to the worktable.

EXERCISES

4.1 With the aid of sketches, explain the difference between generating, forming and copying surfaces when machining.

4.2 Discuss the essential requirements of a machine tool and explain how they are satisfied in a typical vertical milling machine.

4.3 List the essential requirements of machine-tool beds and columns, and discuss how they are satisfied in a typical centre lathe.

4.4 (a) With the aid of sketches, compare and contrast the power transmission systems of a centre lathe and a milling machine; suggest reasons for any differences.
   (b) Discuss the advantages and limitations of belt drives compared with gear drives as applied to the transmission systems of machine tools.

4.5 Discuss the advantages and limitations of the following types of slideway:
   (a) dovetail
   (b) inverted-V
   (c) hydrostatic

4.6 With the aid of sketches, develop the fundamental geometrical movements and alignments of **two** of the following:
   (a) horizontal-spindle milling machine
   (b) vertical-spindle milling machine
   (c) centre lathe
   (d) column type drilling machine

4.7 (a) Discuss the advantages and limitations of plane bearings compared with antifriction (ball and roller) bearings for machine spindles.
   (b) Sketch a typical machine spindle and its bearings; indicate how your design allows the spindle to expand without distortion or misalignment as it warms up.

4.8 (a) Compare and contrast the advantages and limitations of the following machine-tool positioning systems:
  (i) lead screw and nut
  (ii) rack and pinion
  (iii) hydraulic piston and cylinder

(b) State the advantages of using a recirculating ball screw and nut rather than a conventional lead screw and nut on CNC machine tools.

4.9 (a) Explain why it is bad practice to use the same mechanism (e.g. lead screw and nut) for both positioning and measuring in a machine tool, and suggest how this may be avoided.

(b) List some causes of misalignment resulting from lack of care and attention when setting up a workpiece on a machine tool.

5 Numerical control part programming

The topic areas covered in this chapter are:

- Background.
- Applications.
- Advantages and limitations.
- Axis nomenclature.
- Control systems and data input.
- Programming terminology.
- Program formats and canned cycles.
- Offsets and compensation.
- Tooling for CNC.
- Workholding.
- Simple part programming.

5.1 Background

When a machine tool is manually operated as shown in Fig. 5.1(a), the operator directly controls the relative movements of the tool and workpiece. The amount and accuracy of these movements are controlled by reference to some form of measuring device fitted to the machine slide or its lead screws. The operator will also perform such functions as starting and stopping the machine, turning the cutting fluid on or off, changing the spindle speed and feed rate, and changing the tools if required. Thus, with manual control, the quality of the final workpiece and the time required to manufacture it will depend upon the skill, judgement and concentration of the operator.

Where batches of identical parts are required, it is preferable to use methods which are not so dependent upon the skill of the operator, e.g. jigs, fixtures and templates. Automatic machine tools are also used in order to eliminate the high cost and variable quality of manual operation. Until the advent of numerical control, automatic machine tools had to be controlled mechanically by such devices as cams. Although fast and reliable and still used for very large batches of components, such as nuts and bolts, they are relatively time-consuming and expensive to reset when a new and different workpiece is required. For each new component, a single-spindle automatic lathe not only requires a new set of cams but, if the component has a complex profile, form tools will also be required. Such cams and form

Fig. 5.1 *Control of machine tools: (a) operator control; (b) numerical control*

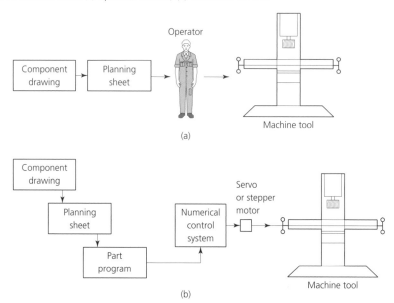

tools are not only expensive, they cannot be produced quickly, and this increases the lead time for bringing a new component into production.

On a numerically controlled machine tool, the decisions which govern the operation of the machine are not made directly by the operator or by a set of cams, but by a series of numbers in binary code which are interpreted by an electronic system. The electronic system converts these numerical commands into the physical movement of the machine elements, as shown in Fig. 5.1(b). Thus each component is an exact replica of this stored data, and high levels of repeatability and consistent quality are achieved.

Early numerically controlled machines had very limited memories, and the information had to be fed into the controller, block by block, from a punched paper tape as required. Computer numerical control (CNC) is now used exclusively. CNC controllers have dedicated built-in computers. Their very powerful memories can store sophisticated system management software, including frequently used standardised programming data (canned cycles), and they can store entire part programs. The system management software interprets the alphanumeric data of the part program and feeds the information for each operation into the electronic system which operates the machine movements.

5.2 Applications

Computer numerical control is used throughout industry. For example, it is used in wood machining, weaving, carpet making, as well as engineering. Some typical engineering applications involving machine tools are as follows:

- Milling machines and machining centres.
- Centre lathes and turning centres.
- Drilling machines.
- Precision grinding machines.
- EDM (spark erosion) machines.

Computer numerical control is also widely used for sheet metalworking and fabrication. For the forming of sheet metal, CNC fills the middle ground between handmade prototypes and small batch production, and volume production involving the use of expensive press tools, where greater rates of production are required. Typical applications of CNC to sheet metal and fabrication equipment are as follows:

- Turret punching machines.
- Flame cutting machines.
- Forming machines.
- Welding machines.
- Tube bending machines.

CNC-controlled automatic inspection machines for checking two- and three-dimensionally contoured components are also available. This improves the frequency and accuracy of part inspection, with a corresponding improvement in quality control.

This chapter is only concerned with a basic introduction to CNC for milling and turning, together with simple part programming. More sophisticated programming techniques are dealt with in *Manufacturing Technology*, Vol. 2.

5.3 Advantages and limitations

Although numerically controlled machines may appear to be costly and complicated, their increasing use by industry suggests their advantages far outweigh their limitations.

5.3.1 *Advantages*

High productivity
Although there is no difference in cutting speeds and feed rates between CNC machines and manually operated machines of similar power and rigidity, production time is saved by rapid traversing and positioning between operations, and greater flexibility. For example, milling, drilling and boring can all be achieved on a CNC machine at the same setting. This avoids expensive jigs and fixtures; it reduces the amount of work in progress stored between operations; and it reduces the time spent in resetting work as it is passed from one machine to the next. CNC machines do not become tired and maintain their high rate of productivity. Their electronic computerised control systems allow them to be easily linked to a robot for loading and unloading, so lights-out production can continue when people have gone to bed.

Tool life
Since the tool approach and cutting conditions are controlled by the program and are constant from one component to the next, tool wear is more even and tool life is extended.

Since profiles and contours are generated by the program-controlled workpiece and tool movements, there is no need for complex and delicate form tools which are susceptible to damage.

Workholding

Since each workpiece merely has to be positioned relative to the same datum point on the machine table, as dictated by the program, then securely clamped, complex jigs and fixtures are no longer required (Section 5.13).

Component modification

When using jigs and fixtures on manually operated machines, even small changes to the design of the workpiece can lead to costly modifications of the jigs and fixtures, perhaps even replacement. However, when using CNC machines, small modifications to the part program take only a few minutes, then production of the modified component can begin.

Design flexibility

Components with complex profiles and components requiring three-dimensional contouring can be more easily and accurately produced on CNC machine tools than on manually controlled machines. It is also possible to produce components on CNC machines which cannot be produced on manually operated machines.

Reduced lead time

CNC machining reduces the lead time required to bring new components into production, compared with the use of conventional automatic machining. There are several reasons for this:

- Writing a part program for a CNC machine is very much quicker than producing the cams for an automatic machine.
- Complex form tools are not required since CNC machines generate the required profile from the program data.
- Workholding is simplified, so complex jigs and fixtures do not have to be made.

Management control

Since machine performance is controlled by the program not the operator, there is greater management control over the cost and quality of production when using CNC machines.

Quality

Since there is less operator involvement, CNC machines produce components of more consistent quality than manually operated machines. If the machine is fitted with *adaptive control*, the machine will always run the tooling at the optimum production rate. It will also sense tool failures or other variations in performance; then it will either stop the machine, or if fitted with automatic tool changing, it will select back-up tooling from the tool magazine. Automatic gauging can also be fitted.

5.3.2 *Limitations*

Capital cost

The initial cost of CNC machine tools is substantially higher than for similar manually operated machines. However, in recent years the cost differential has narrowed somewhat.

Depreciation

As with all computer-based devices, CNC controllers rapidly become out of date. Therefore CNC machine tools should be written off over a relatively short period of time, and they should be replaced more frequently than manually controlled machines.

Tooling Costs

Specialised tooling has been developed to exploit the potential of CNC machine tools. Although the initial cost is high, this largely reflects the tool shanks and toolholding devices, which do not have to be replaced frequently. The cost of replacement tool inserts is no higher than for manually operated machines. The controlled cutting conditions also mean that insert replacement is less frequent, and the tooling can be run continuously at its optimum performance levels.

Maintenance

CNC machine tools and their controllers are extremely complex, and it is unlikely that small and medium-sized companies will have the expertise to maintain and repair them, except for routine lubrication and adjustments. Therefore an approved maintenance contract for each machine or group of machines is required. These contracts are expensive – for one year they usually cost about 10 per cent of the original purchase price.

Training

Comprehensive programmer and operator training is required to convert the workforce from being expert in the use of conventional, manually operated machines to being expert in the programming and operation of CNC machines. Since there is little standardisation between different makes of controller or even between different types and generations of the same make, extensive type-specific familiarisation training is required for each new machine. Although familiarisation training is provided by the equipment supplier within the purchase cost of the equipment, it is time-consuming and therefore costly in terms of lost production.

SELF-ASSESSMENT TASK 5.1

Discuss the reasons behind the widespread adoption of computer numerical control (CNC) in recent years, even though the machines are costly and complex.

5.4 Axis nomenclature

An important feature of the information supplied to the control system is slide displacement. Most machines have two or more slides (usually perpendicular to each other) and they can be moved in one of two directions. It is therefore essential that the control system knows:

- Which slide is to be actuated.
- Which direction the slide is to move in.
- How far the slide is to move.

BS 3635 axis and motion nomenclature is intended to simplify programming and to standardise machine movements. Figure 5.2(a) shows there are three basic axes of movement:

- *Z-axis* This is always the main spindle axis and, as a safety feature, it is positive in a direction towards the tool holder (away from the work). Should the programmer omit the directional sign (in this case negative) from in front of the positional data, the tool will always move away from the work.
- *X-axis* This is always horizontal and parallel to the surface of the work.
- *Y-axis* This is perpendicular to both the X and Z axes.

Once the positive Z-direction has been found, the positive X and Y directions can be found from the right-hand rule, as shown in Fig. 5.2(b). Figure 5.3 shows examples of axis and motion nomenclature for typical machine tools, in accordance with BS 3635.

The programmed movements of a CNC machine tool can be described in three ways:

- *Point-to-point systems* These are designed to position the tool at a series of different points. The path of the tool between the points and the traverse rate between the points are neither under the control of the programmer or the operator. These systems are suitable for simple drilling operations or for a sheet-metal turret punching machine; they are not suitable for profile machining or contouring. A point-to-point system always takes the shortest path, so do not obstruct it.

Fig. 5.2 *Axis rotation: (a) notation; (b) right-hand rule*

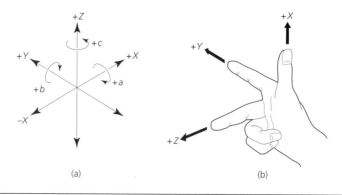

(a)

(b)

Fig. 5.3 *Examples of axis notation: (a) axes for vertical milling machines and drilling machines; (b) axes for lathes, horizontal boring machines and cylindrical spindles*

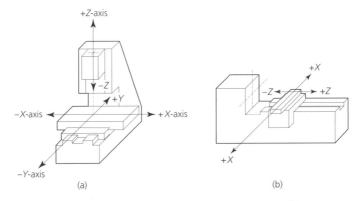

(a)   (b)

- *Linear systems* Also known as parallel-path systems, they are designed to move the tool from one position to the next using a tool path that is a series of straight lines, hence the name. Each line may be parallel to the X-axis or the Y-axis. The traverse rate is under the control of the programmer, and machining operations such as milling can take place between tool positions. A linear system is unsuitable for profile machining or contouring.
- *Continuous-path systems* These are the most widely used systems. The path taken by the tool in moving from one point to the next is fully under the control of the programmer, along with the traverse rate. Angular and curved movements can be made in two or three axes simultaneously, and complex profiles (two-dimensional) and contoured components (three-dimensional) can be generated.

For convenience when programming, it is always assumed that the cutter moves and follows the profile of the workpiece, even though it is the worktable and work that move. The computer in the controller automatically makes the translation from the programmed cutter movement to the actual work movement.

SELF-ASSESSMENT TASK 5.2

Explain why the numerical positional data for the approach of a cutting tool relative to the Z-axis is always given a negative sign.

5.5 Control systems

The importance of antifriction slideways and antifriction lead screws for CNC machine tools is discussed in Chapter 4. Here we consider the method of rotating the lead screws so as to move and position the tool and/or work accurately and under the control of the program.

5.5.1 *Open-loop system*

In an open-loop control system the machine slides are displaced – according to information loaded from the part program into the control system – without their positions being monitored. Hence there is no measurement of slide position and no feedback signal for comparison with the input signal. The correct movement of the slide entirely depends upon the ability of the drive system to move the slide through the exact distance required. Such a system is shown in Fig. 5.4(a).

The most common method of driving the lead screw is by a stepper motor, either directly or via a toothed belt drive. A stepper motor is an electric motor energised by a train of electrical pulses rather than by a continuous electrical signal. Each pulse causes the motor to rotate through a small discrete angle. Thus the motor rotates in a series of steps according to the number of pulses supplied to it. The direction of rotation depends upon the polarity of the pulses, and the feed rate depends upon the number of pulses per second. Unfortunately stepper motors have only limited torque compared with servomotors and are only suitable for small and medium-sized machines. If overloaded, stepper motors may stall and miss one or more pulses. Since there is no feedback in open-loop systems, any slip produces a dimensional inaccuracy.

Fig. 5.4 *Control systems: (a) open-loop control, no feedback loop; (b) closed-loop control, the transducer is part of a feedback loop*

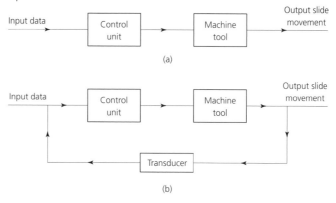

5.5.2 *Closed-loop system*

A closed-loop system, as shown in Fig. 5.4(b), sends back a signal to the control unit from a measuring device (transducer) attached to the slideways. The signal indicates the actual movement and position of the slides. The control unit continues to adjust the position of the slide until it arrives at its destination; this system has feedback. Although more complex and costly than open-loop systems, closed-loop systems give more accurate positioning, especially on medium and large machines, where the forces involved in moving the machine elements can be high.

For this type of system, servomotors are used to drive the lead screws. Unlike stepper motors, they are continuously running DC motors with a high starting torque to overcome the inertia of the machine slides and workpiece. They can be used on all sizes of machine up to the largest.

5.5.3 *Transducers and encoders*

Transducers are devices for converting one physical variable into another physical variable. For example, a microphone converts sound waves into electrical signals. CNC machines use transducers for a variety of purposes:

- Monitoring the position of the worktable, hence the work.
- Measuring the speed and angular displacement of the machine spindle.
- Measuring tool-tip temperature.
- Measuring cutting power being transmitted to the machine spindle.
- Measuring oil pressure in the hydraulic and lubricating systems.
- Measuring the volumetric flow of the coolant.

In a closed-loop system, linear transducers attached to the machine table provide the necessary feedback required for the servomotors accurately to position the worktable and work accurately in accordance with the requirements of the program.

Rotary transducers are used to measure the angular displacement of a machine's rotating elements, e.g. the lead screw and spindle of a lathe. This is essential for synchronising the rotation of the work and the axial movement of the tool when screw cutting on a CNC lathe.

Transducers provide incremental data – the distance moved from one position to the next. However, they do not know where they are relative to the machine datum, and they 'lose their memory' if the power supply to the machine fails or they are turned off. Therefore most systems use *encoders*. Encoders are transducers that use more complex scales to provide additional information, so the control system knows not only how far the worktable and work have moved, but the exact position of the work relative to the machine datum. What is more, at switch-on the machine immediately identifies the position of all the machine elements. Therefore encoders are able to provide both *incremental* and *absolute* information for the control system. Refer ahead to Fig. 5.19 which shows the difference between incremental and absolute information.

SELF-ASSESSMENT TASK 5.3

Discuss the essential differences between servomotors and stepper motors, and compare the advantages and limitations of both devices as a means of positioning and traversing the machine elements.

5.6 Data input

5.6.1 *Manual data input*

This can be used for entering complete programs, for editing the program or for setting the machine by manually pressing the keys on the control console. Loading of complete programs is rarely used nowadays and should be limited to simple programs to keep machine idle time to a minimum. With CNC machines the manually loaded program can then be recorded on a suitable storage medium, e.g. magnetic tape, disk or punched tape, to produce a further batch of components at a later date.

5.6.2 *Conversational data input*

The operator loads the program into the control console by manually depressing the appropriate keys. However, instead of writing it out in machine code before loading, the program is entered in response to questions (prompts) appearing on a visual display unit (VDU) in everyday, conversational English.

The computer is preprogrammed with standard data stored in files within the computer memory. Each item of data is numerically identified and called into the program by the operator's response to a question. To reduce idle time, most modern conversational control units allow a new program to be entered whilst an existing program is still operating the machine.

5.6.3 *Punched tape*

Punched tape was the original method of loading data into numerically controlled machines. In fact, numerically controlled machining was once known as tape-controlled machining. Punched tape is still commonly used for data input. It has several advantages: it can be read visually; any damage is easily identified by visual inspection, and it is not corrupted under normal working conditions. Unfortunately, the paper tape is easily torn, so current practice favours the use of stronger plastic (Mylar) and metallised plastic tapes. The tapes have to be prepared on an electromechanical tape punch and read by an electromechanical or optoelectronic tape reader.

Two tape standards are in current use and examples are shown in Fig. 5.5. Notice that the numbers, letters and symbols are represented by rows of holes punched across the width of the tape. The tape width for both standards is 25 mm. The Electrical Industries Association (EIA) system shown in Fig. 5.5(a) requires that each row has an odd number of holes. If the required character is expressed by an even number of holes, an extra hole has to be punched in track 8, which is known as the parity track. If the required character is expressed by an even number of holes, there is no hole in the parity track. Thus the EIA standard is called an odd-parity system.

The International Standards Organisation (ISO) system shown in Fig. 5.5(b) requires that each row has an even number of holes, and track 8 is used to make up an even number of holes in the row if a character is expressed by an odd number of holes. This is the opposite of the EIA system, so the ISO standard is called an even-parity system. Only those characters of immediate use when part programming have been shown.

The tape format is shown in Fig. 5.6. Each row of holes across the tape is called a character, each set of characters is called a word, and each set of words making a complete statement is called a block.

Fig. 5.5 *Punched tape codes: (a) EIA (odd parity) code; (b) ISO (even parity) code*

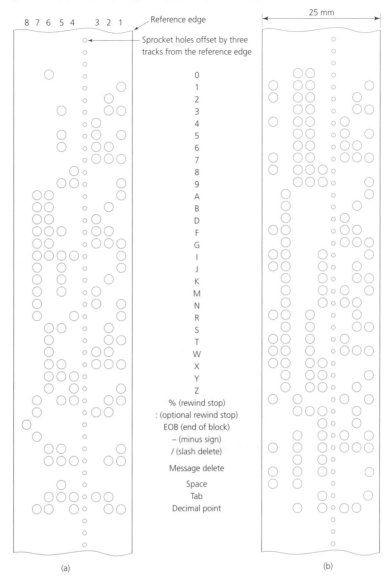

5.6.4 *Magnetic tape*

The data is recorded on magnetic tape in much the same way as recording music onto an audio cassette. Although the magnetic tapes are compact and easily stored, and recording and reading is quicker than for punched tape, they have two major disadvantages:

- The recorded data cannot be read visually.
- The recorded data is easily corrupted by electromagnetic interference under workshop conditions.

Fig. 5.6 *Punched tape format (ISO)*

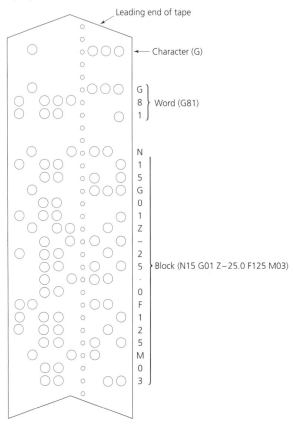

5.6.5 *Magnetic disk*

Since computer-aided programming software is now widely available and widely used, it is most convenient to record the program on computer disks. This software has the advantage that the computer does most of the calculations, removing a major source of programming errors. The shape of the component is displayed on the screen, so the programmer can check the program before loading it into the machine. The programmer only has to be familiar with one entry language; the software will convert the program into the language of the machine controller. This is of great importance when a firm has many machines and controllers of different makes and languages.

5.6.6 *Direct numerical control*

The program is prepared on a remote computer and stored in its memory. It can be checked on the computer using simulation software as previously described. But this time the program is downloaded directly into the CNC machine; a copy is retained on magnetic disk. The computer software has to be compatible with the machine controller's management software.

5.7 Program terminology

5.7.1 *Characters and words*

A *character* is a number, a letter or a symbol that is recognised by the controller. A *word* is an associated group of characters that define one complete element of information, e.g. N150. There are two types: dimensional words and management words.

5.7.2 *Dimensional words*

These are any words related to a linear dimension; that is, any word commencing with the letters X, Y, Z, I, J, K, or any word in which the above characters are implied. The letters X, Y, Z, refer to the machine axes defined in Section 5.4. The letters I, J, K, refer to circles and arcs of circles. The start and finish positions of arcs are defined by X, Y, Z coordinates, whereas the centre and radius of the arc or circle are defined by I, J, K coordinates with:

- I dimensions related to X dimensions.
- J dimensions related to Y dimensions.
- K dimensions related to Z dimensions.

Older systems, many of which are still in use, did not use a decimal point but used leading and trailing zeros to indicate the position of the decimal point. For example, 35.6 mm would be written as 0035600. Current practice favours the use of a decimal point in specifying dimensional words, and a machine manual may stipulate that an X-axis dimension word has the form X4,3. That is, the X dimension may have up to 4 digits in front of the decimal point and up to 3 digits behind the decimal point. This is standard practice when the dimensions are in millimetres. For inch dimensions the form is X3,4; that is, up to 3 digits in front of the decimal point and up to 4 digits behind it.

Besides the movement axis and the movement distance, the specification needs to include a sign, positive or negative. If there is no sign in front of the digits, they will be taken as positive. If there is a minus sign in front of the digits, they will be taken as negative. This convention provides a safety feature for the Z-axis; if a minus sign is accidentally omitted, the tool will move away from the work (Fig. 5.7).

5.7.3 *Management words*

These are any words which are not related to a dimension; that is, any word commencing with the character N, G, F, S, T, M, or any word in which these characters are implied. Here are some examples:

- *N4 is a block or sequence number* The character N followed by up to 4 digits (i.e. N1 to N9999). Block words are usually the first words which appears in any block and identifies that block. Blocks are usually numbered in steps of 5 or 10, so that additional blocks can be inserted if needs be.
- *G2 is a preparatory code or function* The character G followed by up to 2 digits (i.e. G0 to G99). These are used to *prepare* or inform the machine controller of the functions required for the next operation. The preparatory codes as specified in BS 3635 are listed

Fig. 5.7 *Z-axis movement*

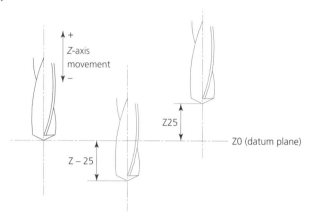

in Table 5.1. Unfortunately they vary slightly between different makes and types of controller, so the maker's programming manual should always be consulted. Many preparatory codes are *modal*; once they have been selected, they remain in operation until changed or cancelled.

- *F4 is a feed rate command* The character F followed by up to 4 digits (i.e. F1 to F9999). They indicate to the controller the desired feed rate for machining and may be defined in millimetres per minute, inches per minute, millimetres per revolution, inches per revolution or a feed rate number. Feed rate numbers are an older system in which typical feed rates, in say millimetres per minute or millimetres per revolution, are predetermined by the manufacturer and selected by an appropriate F-code, together with a G-code to tell the controller which system is being used.

- *S4 is a spindle speed command* The character S followed by up to four digits (i.e. 1 to 9999). There are various ways of defining the spindle speed: the rotational speed in revolutions per minute, the cutting speed in metres per minute and the constant cutting speed in metres per minute. The constant cutting speed is used when facing across the end of a component; as the effective diameter gets smaller, the spindle speed is increased to a predetermined safe maximum so that the surface cutting speed is maintained at a constant value.

- *T2 is a tool number* The character T followed by up to 2 digits (usually 1 to 20) identifies which tool is to be used. Each tool will have its own tool number and, as well as memorising the tool number, the computer also memorises additional data such as the tool length offset and/or the tool diameter (radius) compensation for each tool. In a machine with automatic tool changing, the position of the tool in the tool magazine is also memorised by the computer under the tool number file.

- *M2 is a miscellaneous command* The character M followed by up to two digits (i.e. 0 to 99). Apart from the preparatory functions (G-codes), there are several other commands which are required throughout the program; for example, starting and stopping the spindle, turning the coolant on or off, changing speed and changing tools. The miscellaneous codes are listed in Table 5.2.

Table 5.1 *Preparatory functions*

Code number	Function		
G00	Rapid positioning, point to point		(M)
G01	Positioning at controlled feed rate	} Normal	(M)
G02	Circular interpolation	} dimensions	(M)
G03	Circular interpolation CCW*		(M)
G04	Dwell for programmed duration		
G05	Hold: cancelled by operator		
G06 }			
G07 }	Reserved for future standardisation		
G08	Programmed slide acceleration		
G09	Programmed slide deceleration		
G10	Linear interpolation (long dimensions)		(M)
G11	Linear interpolation (short dimensions)		(M)
G12	3D interpolation		(M)
G13–G16	Axis selection		(M)
G17	XY plane selection		(M)
G18	ZX plane selection		(M)
G19	YZ plane selection		(M)
G20	Circular interpolation CW* (long dimensions)		(M)
G21	Circular interpolation CW (short dimensions)		(M)
G22	Coupled motion positive		
G23	Coupled motion negative		
G24	Reserved for future standardisation		
G25–G29	Available for individual use		
G30	Circular interpolation CCW (long dimensions)		(M)
G31	Circular interpolation CCW (short dimensions)		(M)
G32	Reserved for future standardisation		
G33	Thread cutting, constant lead		(M)
G34	Thread cutting, increasing lead		(M)
G35	Thread cutting, decreasing lead		(M)
G36–G39	Available for individual use		
G40	Cutter compensation, cancel		(M)
G41	Cutter compensation, left		(M)
G42	Cutter compensation, right		(M)
G43	Cutter compensation, positive		
G44	Cutter compensation, negative		
G45	Cutter compensation +/+		
G46	Cutter compensation +/−		
G47	Cutter compensation −/−		
G48	Cutter compensation −/+		
G49	Cutter compensation 0/+		
G50	Cutter compensation 0/−		
G51	Cutter compensation +/0		
G52	Cutter compensation −/0		
G53	Linear shift cancel		(M)
G54	Linear shift X		(M)
G55	Linear shift Y		(M)
G56	Linear shift Z		(M)
G57	Linear shift XY		(M)
G58	Linear shift XZ		(M)
G59	Linear shift YZ		(M)
G60	Positioning exact 1		(M)
G61	Positioning exact 2		(M)
G62	Positioning fast		(M)
G63	Tapping		
G64	Change of rate		
G65–G79	Reserved for future standardisation		
G80	Fixed cycle cancel		(M)
G81–G89	Fixed cycles		(M)
G90–G99	Reserved for future standardisation		

*CCW = counter-clockwise; CW = clockwise

Notes: Functions marked (M) are modal. These codes are based on BS 3635; but codes vary between makes and types of controller, and the manufacturer's programming manual should always be consulted. Most controllers use G90 to establish the program in *absolute* dimensional units, G91 to establish the program in *incremental* dimensional units. FANUK controllers use G20 in place of G90, and G21 in place of G91.

Table 5.2 Miscellaneous functions

Code number	Function
M00	Program stop
M01	Optional stop
M02	End of program
M03	Spindle on CW*
M04	Spindle on CCW*
M05	Spindle off
M06	Tool change
M07	Coolant 2 on
M08	Coolant 1 on
M09	Coolant off
M10	Clamp slide
M11	Unclamp slide
M12	Reserved for future standardisation
M13	Spindle on CW, coolant on
M14	Spindle on CCW, coolant on
M15	Motion in the positive direction
M16	Motion in the negative direction
M17 M18	Reserved for future standardisation
M19	Oriented spindle stop
M20–M29	Available for individual use
M30	End of tape
M31	Interlock bypass
M32–M35	Constant cutting speed
M36	Feed range 1
M37	Feed range 2
M38	Spindle speed range 1
M39	Spindle speed range 2
M40–45	Gear changes
M46–49	Reserved for future standardisation
M50	Coolant 3 on
M51	Coolant 4 on
M52–54	Reserved for future standardisation
M55	Linear tool shift, position 1
M56	Linear tool shift, position 2
M57–M59	Reserved for future standardisation
M60	Workpiece change
M61	Linear workpiece shift, position 1
M62	Linear workpiece shift, position 2
M63–M67	Reserved for future standardisation
M68	Clamp workpiece
M69	Unclamp workpiece
M70	Reserved for future standardisation
M71	Angular workpiece shift, position 1
M72	Angular workpiece shift, position 2
M73–M77	Reserved for future standardisation
M78	Clamp slide
M79	Unclamp slide
M80–M99	Reserved for future standardisation

*CW = clockwise; CCW = counter-clockwise
Notes: N, G, T, M commands *may* require a leading zero to be programmed on some older but still widely used systems. For example, G0 becomes G00, G1 becomes G01, M2 becomes M02.

The N, G, T, M, commands may require a leading zero to be programmed on some older but still widely used systems; for example, G0 becomes G00, G1 becomes G01, and M2 becomes M02.

A CNC program using these codes and commands could look like this:

N5 G90 G71 G00 X35.4 Y25.5 T01 M06
N10 X15.0 Y15.0 S1250
N15 G01 Z-25.0 F120 M03

5.8 Program formats

Different control systems use different formats for the assembly of each block of data. Always consult the programming manual for the machine being programmed. A block of data consists of a complete line on a program containing a complete set of instructions for the controller.

5.8.1 *Word (or letter) address system*

Currently this is the most widely used system. Each word commences with a letter character called an address. Hence a word is identified by its letter, not by its position in the block (unlike the fixed block format). The word (or letter) address system has the advantage that instructions which remain unchanged from a previous block may be omitted from the subsequent blocks until a change becomes necessary.

A typical letter address structure (format), as given by a maker's handbook, could be:

- *Metric* N4 G2 X4,3 Y4,3 Z4,3 I4,3 J4,3 K4,3 F3 S4 T2 M2.
- *Inch* N4 G2 X3,4 Y3,4 Z3,4 I3,4 J3,4 K3,4 F3 S4 T2 M2.

It breaks down as follows:

- The letter signifies the function of the word.
- The number determines the maximum number of digits which may follow the letter.
- The comma determines the position of the decimal point.

5.8.2 *Fixed block (sequential) format system*

This is an outdated format which is still widely used on older machines. The instructions in the blocks are always written and recorded in the same fixed sequence. No letter commences each word, but the letter is implicit from the position of the word in the block or sequence. For example, a block could read:

20 1 25.0 37.550 80 1500 2 6

Since the first word in the sequence is 20, the controller reads this as N20 because of its position. Since the third word in the sequence is 25.0, the controller reads this as X25.0 because of its position, and so on. Thus all instructions have to be given in every block, including those instructions which have not changed from the preceding block, to ensure each word is in its identifying position within the sequence.

5.9 Canned cycles

Standardised fixed cycles, or *canned cycles* as they are usually known, save the repetitive programming of common operations. The sequence of events for a standard cycle of operations is embedded in the memory of the controller's computer at the time of manufacture and is called up, when required, by an appropriate G-code. One of the most frequently used canned cycles is the drilling cycle in Fig. 5.8. Called up using a G81, the cycle of events is as follows:

1. Rapid traverse to centre of first hole.
2. Rapid traverse to clearance plane height.
3. Feed to depth of hole.
4. Rapid up to clearance plane height.
5. Rapid traverse to centre of next hole.
6. Repeat for as many holes as required.

The only data the programmer has to provide is:

- The positions of the hole centres.
- The spindle speed.
- The feed rate.
- The tool number.

The machine setter will provide the tool length offset for the drill and any other information that needs to be recorded under the tool number file in the computer memory.

Another important canned cycle used on milling machines is the rectangular pocket milling cycle in Fig. 5.9. This cycle is called up by G78 and its movements are as follows:

- Rapid traverse to position 1 inside the pocket boundary.
- Rapid down to the clearance plane height.
- Feed down in the Z-axis to the roughing depth. (Since the cutter is plunging into solid metal, a slot drill must be used, not an end mill.)

Fig. 5.8 *Canned drilling cycle (G81)*

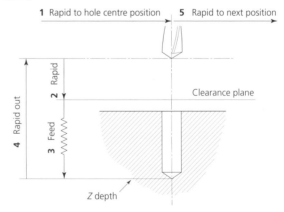

Fig. 5.9 *Canned pocket milling cycle (G78)*

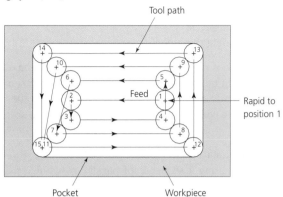

- Machine out the pocket as the cutter path moves sequentially through positions 2 to 15 (minus a finishing allowance on the profile).
- Rapid up to the clearance plane height in the Z-axis, and return to position 1.
- Feed down to finish-pass depth.
- Repeat movements 2 to 15 to finish-machine bottom and sides of pocket. Note the number of movements when roughing and finishing will depend upon the diameter of the cutter and the size of the pocket. In practice the paths may overlap by as much as 60 per cent of the cutter diameter.
- At position 15, rapid up to the clearance plane height and return to position 1.

The programmer has to provide the positional data, the dimensions for the pocket, and the cutter and cutting data.

Circular pockets can be generated using G77. The ability to produce circular pockets and bores immediately shows a major advantage in using CNC milling machines. Using conventional, manually operated machines, any circular pocket in a milled surface would have to be produced on a lathe or a boring machine, so the work would have to be removed from the milling machine and reset in another machine. Using a CNC milling machine, all the operations can be performed at one setting: milling, drilling and tapping; production of circular and rectangular pockets; production of circular and rectangular islands; pockets and islands with complex profiles. Here are some other examples of canned cycles used on CNC milling machines:

G80	cancels all canned cycles
G83	deep-hole drilling cycle
G85	boring cycle
G82	drilling cycle with dwell
G84	tapping cycle
G87/88	deep-hole boring cycle

Canned cycles can also be used with CNC lathes:

G66/67	contouring cycles
G81	turning cycle
G83	deep-hole drilling cycle

G88   autogrooving cycle
G68/69   rough turning cycles
G82   facing cycle
G84/85   straight threading cycles

Note how most of these canned cycles for use on a lathe differ from the canned cycles for use on a milling machine, even though they have the same code. Figure 5.10 shows the sequence of events activated by G68 on a lathe. After the roughing cuts have been made, the profile of the component would be finish-turned to size. Here is the sequence of events for G68:

- Rapid to the start point.
- Rough out using a sequence of parallel roughing passes. The number of passes will depend upon the depth of cut and the profile.
- A profiling pass, leaving a finishing allowance on the component.
- Rapid return to start point.

Fig. 5.10   *Rough turning cycle (G68)*

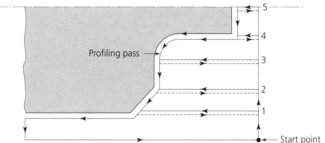

5.10 Tool length offset

Tool length offset (TLO) allows tools of various lengths to be used with a common datum (Fig. 5.11), without having to alter the program. Setting commences with tool T01, which is 'touched' onto the work surface or a setting block, depending upon the Z-axis datum height, and the Z-axis read-out is set to zero.

T01 is now the master tool. Each subsequent tool is then touched onto the Z-axis datum and its Z-axis read-out is noted. Using this information, the Z-axis offset is then applied to each tool in turn, to compensate for differences in length compared with T01. The tool length offsets are recorded in the memory of the machine's computer under the tool number file. Each time a tool is called up by its T-code, the correct length offset will automatically be applied. The tool number file is altered if the offset changes for a particular tool (perhaps it is reground), but no change has to be made to the program.

The application of tool length offsets to turning tools is shown in Fig. 5.12. Notice that offsets are required in both the X and the Z axes relative to a common datum. The lathe turret usually contains several different tools; each tool protrudes by a different distance, so each tool will require its own offset. The offset for a tool becomes operative as soon as that tool is called into the program by its T-code.

Fig. 5.11 *Tool length offset: milling*

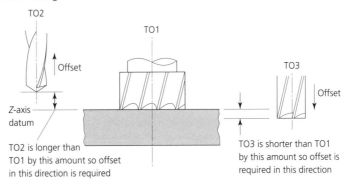

Fig. 5.12 *Tool length offset: turning*

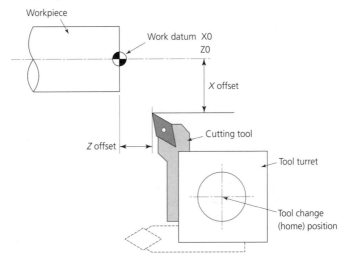

5.11 Cutter diameter compensation

Like tool length offsets, cutter diameter compensation (milling) and tool-nose radius compensation (turning) are also facilities to aid programming. Not only do they allow tools of different sizes to be interchanged without altering the program, they simplify the writing of the program. The tool centreline (axis) can be assumed to travel round the profile being machined, and the controller automatically makes allowance for the actual diameter of the cutter. For turning on a lathe, the programmer can assume the tools have a sharp nose; the controller automatically compensates for the nose radius of the tool.

Cutter diameter compensation for milling machines is controlled by the following preparatory codes:

G41   compensates cutter to the left of the workpiece (Fig. 5.13(a))
G42   compensates cutter to the right of the workpiece (Fig. 5.13(b))
G40   compensation cancelled

At first sight, the 'handing' of the compensation is a little difficult to interpret. Consider Fig. 5.13(a). Start at any point and face in the direction of the cutter travel by following the arrows. Notice that the path of the cutter is always to the left of the surface being machined. Similarly, in Fig. 5.13(b) the path of the cutter is always to the right of the surface being machined. The path of the cutter traverse is determined by whether upcut or downcut (climb) milling techniques are being used.

Whenever you activate the G41 or G42 code, the cutter diameter compensation is always applied at the next move in the X and Y axes, as shown in Fig. 5.14(a), and always cancelled at the next X and Y move after you activate the G40 code, as shown in Fig. 5.14(b). Diameter compensation can never be applied or cancelled whilst cutting is taking place.

Fig. 5.13 *Milling cutter diameter compensation: (a) to the left (G41); (b) to the right (G42)*

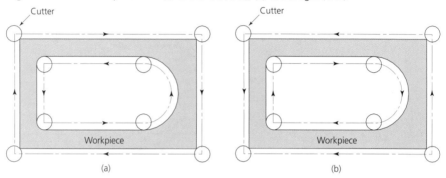

Fig. 5.14 *Cutter compensation: (a) ramping on, G41 initiates cutter compensation; (b) ramping off, G40 cancels cutter compensation*

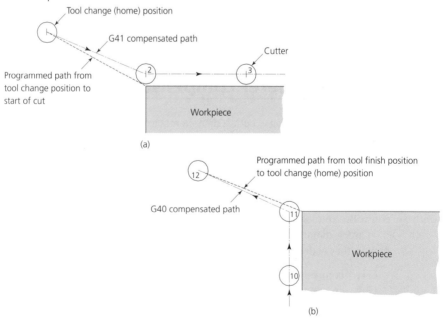

Notice that preliminary movement of the cutter takes place clear of the work so that compensation is fully effective and the cutter has achieved its required feed rate by the time the cutter is in contact with the workpiece. Known as ramping on, it allows the feed servomotor to accelerate up to speed. Similarly, at the end of the cut, the cutter feeds clear of the work as the feed servomotor decelerates when the compensation is cancelled; this is called ramping off.

Lathe tool-nose radius compensation is similar to diameter compensation for milling; it uses the same preparatory codes:

G41   tool-nose radius compensation to the left (Fig. 5.15(a))
G42   tool-nose radius compensation to the right (Fig. 5.15(b))
G40   compensation cancelled

Tool-nose radius compensation simplifies programming; it allows the programmed tool-nose path to follow the profile of the component and it allows the programmer to assume that a sharp-nosed tool is being used. Like tool length offsets, diameter compensation and tool-nose radius compensation are set on the machine itself when the actual tool parameters are known. After the first trial component has been made, the compensation settings may need to be 'tweaked' to allow for cutter deflection and other variables, in order to bring the component within its design tolerances.

Fig. 5.15 *Lathe tool-nose radius compensation: (a) G41, facing tool always moves to the left of the surface being turned; (b) G42, facing tool always moves to the right of the surface being turned*

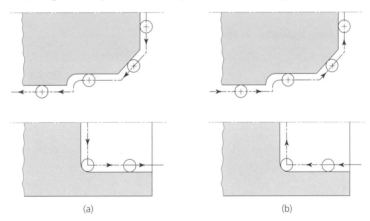

(a)   (b)

SELF-ASSESSMENT TASK 5.5

Explain how the operator of a CNC machine can make allowance for variation in cutter sizes resulting from regrinding.

5.12 Tool retrieval

CNC machines may have their cutting tools changed manually or automatically. In either case, rapid tool changing is required to match the performance of the machine.

5.12.1 *Manual tool changing*

The tools are kept in a tool crib adjacent to the machine as an aid to rapid identification and retrieval. Each tool is kept in a numbered position within the crib (T01, T02, etc.).

5.12.2 *Automatic tool changing*

There are several systems of automatic tool changing, but the most popular are indexible tool turrets and chain magazines. The tools are kept in specific positions in the turret or magazine, and the appropriate tool is selected and changed as required by the program. As an aid to setting and replacement, tooling is usually kept ready for use in a tool library from which it may be withdrawn when required by a specific program.

5.12.3 *Qualified tooling* (ISO 1832)

This is manufactured with the dimensions from its datum faces to the tool tip guaranteed to within ±0.08 mm, as shown in Fig. 5.16(a).

Fig. 5.16 *Tooling: (a) qualified; (b) preset*

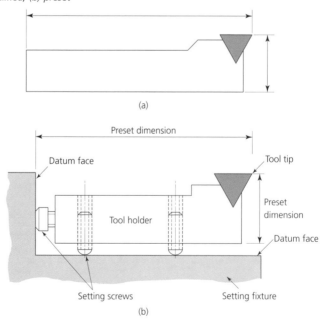

5.12.4 *Preset tooling*

This is used where qualified tooling is unavailable, inappropriate or insufficiently accurate. Presetting is performed away from the machine in a suitable fixture, as shown in Fig. 5.16(b).

5.13 Workholding

The normal principles of location and restraint apply to CNC machine tools; workholding devices must:

- Locate the work in the correct position relative to the machine datum.
- Restrain the work against the cutting forces.

Small work may be held in a machine vice in which the fixed jaw is machined to provide the necessary locations (Fig. 5.17). The fixed jaw has to be aligned with the machine datum.

Fig. 5.17 *Machine vice with formed jaws*

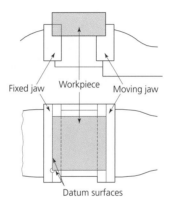

Fixed jaw   Workpiece   Moving jaw

Datum surfaces

Alternatively a grid plate may be used. Figure 5.18(a) shows a typical grid plate; it is usually made from steel or cast iron and has a matrix of holes. Sometimes the matrix consists entirely of tapped holes which may be counterbored for location buttons; sometimes the matrix consists of alternate tapped and reamed location holes. The tapped holes are used for clamping; they are identified by a system of numbers and letters. The grid plate is often permanently fitted to the machine table, allowing a part programmer to identify the exact position of the location holes when establishing datums for use in a program. Figure 5.18(b) shows how a grid plate is used.

Large components may be mounted directly onto the machine table, or traditional milling fixtures may be used if a large batch is being machined. Workholding devices for turning centres (lathes) usually consist of pneumatically or hydraulically actuated plain or stepped collet chucks. Pneumatically or hydraulically actuated self-centring chucks are used for larger work. They are fitted with soft jaws, bored out to run true and to locate the work axially.

Fig. 5.18 *The grid plate: (a) principle; (b) use*

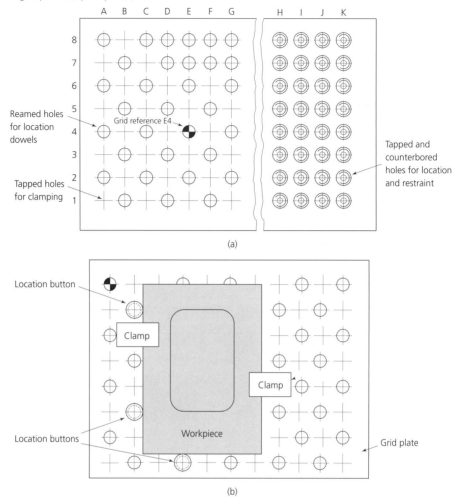

Reamed holes for location dowels

Grid reference E4

Tapped holes for clamping

Tapped and counterbored holes for location and restraint

(a)

Location button

Clamp

Clamp

Location buttons

Workpiece

Grid plate

(b)

5.14 Part programming

Before a part program can be written, one further decision has to be made; that is, to select either absolute or incremental dimensioning. Absolute dimensions are more commonly used as the dimensioning is related to a common datum, as shown in Fig. 5.19(a). Incremental dimensioning is shown in Fig. 5.19(b).

A simple component to be made on a CNC milling machine is shown in Fig. 5.20. A typical part program for milling and drilling this component is shown in Fig. 5.21. A simple turned component is shown in Fig. 5.22, and a typical part program for this component is shown in Fig. 5.23. The computer numerical control of machine tools will be developed further in *Manufacturing Technology*, Volume 2, where macros, subroutines, mirror imaging, scaling and computer-aided programming will be considered.

Fig. 5.19 *Dimensioning techniques: (a) absolute; (b) incremental*

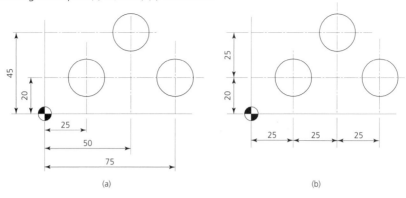

(a)   (b)

Fig. 5.20 *Component to be milled and drilled on a CNC machine*

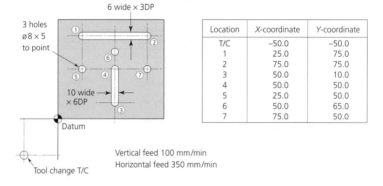

Location	X-coordinate	Y-coordinate
T/C	−50.0	−50.0
1	25.0	75.0
2	75.0	75.0
3	50.0	10.0
4	50.0	50.0
5	25.0	50.0
6	50.0	65.0
7	75.0	50.0

Vertical feed 100 mm/min
Horizontal feed 350 mm/min

Fig. 5.21 *Typical part program for machining the component in Fig. 5.20*

Sequence number	Code program	Explanation
%	G00 G71 G75 G90	Default line
N10	X −50.0 Y −50.0 S1000 T1	
N20	M06	TC posn spindle/tool offset (ø6mm)
N30	X25.0 Y75.0 Z1.0	Rapid posn (1)
N40	G01 Z −3.0 F100	Feed to depth
N50	G01 X75.0 F350	Feed to posn (2)
N60	G00 Z1.0	Rapid tool up
N70	X −50.0 Y −50.0 S800 T2	
	M06	TC posn spindle/tool offset (ø10 mm)
N80	X −50.0 Y10.0 Z1.0	Rapid posn (3)
N90	G01 Z −6.0 F100	Feed to depth
N100	G01 Y50.0 F350	Feed to posn (4)
N110	G00 Z10	Rapid tool up
N120	X −50.0 Y −50.0 S1100 T3	
	M06	TC posn spindle/tool offset (ø8 mm drill)
N130	X25.0 Y50.0 Z1.0	Rapid posn (5)
N140	G81 X25.0 Y50.0 Z −7.0 F100	Drill on restate depth (inc) feed
N150	X50.0 Y65.0	Posn (6)
N160	X75.0 Y50.0	Posn (7) drill
N170	G80	Switch off drill cycle
N180	G00 X −50.0 Y −50.0 M02	TC posn end of prog
E		End of tape

Fig. 5.22 *Component to be turned on a CNC lathe. (Dimensions in millimetres)*

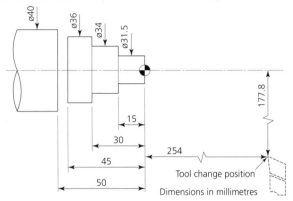

Fig. 5.23 *CNC part program for turning the component in Fig. 5.22 – machine: Hardinge lathe; controller: GE1050*

Sequence number	Code program	Explanation
N10	G71	Metric
N20	G95	Feed in mmr ev
N30	G97 S1000 M03	Direct spindle. 1000 r.p.m. Spindle on CW
N40	G00 M08	Rapid mode. Coolant on
N50	G53 X177.8 Z254 TO	To tool change position
N60	M01	Optional stop
N70	T100	Rotate turret Pos 1
N80	G54 X0 Z2 T101	Move to start with tool 1
N90	G01 Z –0.5 F0.2	Move to depth prior to face end
N100	X31.5	Face end
N110	Z –15	Turn 031.5
N120	X34	Face edge
N130	Z –30	Turn 034
N140	X36	Face edge
N150	Z –50	Turn 036
N160	G53 X177.8 Z254 TO	To tool change
N170	T400	Rotate turret Pos 4
N180	G54 X40 Z –45 T404	To part off position. Tool 4 offset 04
N190	G01 X –1.0 F0.1	Part off
N200	G00 X40	Retract
N210	G53 X177.8 Z254 TO	To tool change
N220	M02	End program

5.1 Despite the fact that CNC machine tools are more complex and costly than conventional machine tools, they are being used increasingly in manufacturing industry. Discuss the reasons why this is so.

5.2 Figure 5.24 shows outline diagrams of a number of machine tools. Redraw these diagrams and add and label the *X*, *Y*, and *Z* axes.

Fig. 5.24 *Exercise 5.2*

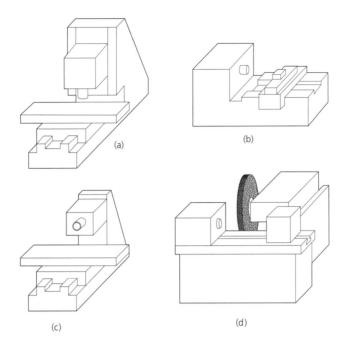

(a)

(b)

(c)

(d)

5.3 Explain the essential differences between point-to-point control systems and continuous-path control systems.

5.4 With the aid of sketches, explain the difference between open-loop and closed-loop control systems, and list the advantages and limitations of each system.

5.5 Discuss the relative merits and limitations of the following methods of data input:
(a) coversational manual data input and punched tape data input
(b) magnetic tape data input
(c) direct numerical control

5.6 (a) Explain what is meant by the following terms used in programming:
   (i) character
   (ii) management word
   (iii) dimensional word
   (iv) block
   (v) preparatory code
   (vi) miscellaneous command
   (vii) modal command
(b) A typical letter address format could be N4 G2 X4,3 Y4,3 Z4,3 I4,3 J4,3 K4,3 F3 S4 T2 M2:
   (i) state whether this format is in inch or metric units and state the reason for your choice
   (ii) explain what this format means

5.7 (a) State what is meant by the term **canned cycle.**

(b) State the advantage of using canned cycles.

(c) With the aid of sketches, show the sequence of a typical canned cycle.

5.8 Explain why CNC machine controllers are provided with tool length offset and cutter diameter or tool-nose radius compensation. Why is it important that the machine operator is fully trained in the use of these facilities?

5.9 (a) Prepare an operation planning sheet listing the tooling, spindle speeds and feed rates for machining the component shown in Fig. 5.25.

(b) Use this data to prepare a part program for machining the component from the solid on a vertical-spindle CNC milling machine. Specify the controller and the machine for which the program has been written.

Fig. 5.25 *All dimensions in millimetres; surfaces and edges of blank previously machined to size (90 × 55 × 20); material is 0.5% plain carbon steel*

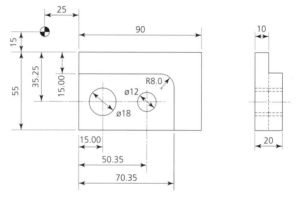

5.10 (a) Prepare an operation planning sheet that lists the tooling, spindle speeds and feed rates for turning the component shown in Fig. 5.26.

(b) Use this data to prepare a part program for machining the component from a solid blank on a CNC lathe. Specify the controller and the machine for which the program has been written.

Fig. 5.26 *All dimensions in millimetres; blank Ø55 × 72 one end already faced; material is Duralumin*

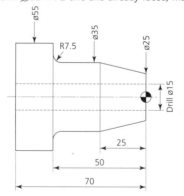

6 Assembly

The topic areas covered in this chapter are:

- Design of components.
- Batch size.
- Accuracy of fit.
- Interchangeability.
- Joining methods.
- Cost of assembly.

6.1 Design of components

Traditionally, ease of assembly has not ranked highly among the design disciplines. However, as labour costs have risen, the need to make assembly quicker (and therefore cheaper) and better suited to automated processes has meant that the assembly of components into a complete unit must now be considered from the earliest stages of design.

Let's consider the electronic circuit shown in Fig. 6.1(a) for a simple small-signal amplifier. There is a transistor, four fixed resistors, one variable resistor and three capacitors – a total of nine components. Compare this with Fig. 6.1(b) which uses an integrated circuit (chip). There is one integrated circuit, three fixed resistors, one variable resistor and one capacitor – a total of six components. The chip circuit not only has fewer components and is easier to assemble, it also has a higher performance.

And it is easier for a robot to pick and place a chip, with its short stiff legs, than the fine wire connections of a transistor.

The chip is built into a protective plastic housing with its legs arranged symmetrically. It can therefore be placed into the holes in the printed circuit board in two positions, as shown in Fig. 6.2(a). However, to reduce the probability of this happening, the designer of the housing has included a distinctive mark at one end and a dot by pin number 1. If the designer of the printed circuit board arranges for this outline to be reproduced on the board, as shown in Fig. 6.2(b), the assembler can immediately see which way round the chip is to be installed.

Fig. 6.1 *Small-signal amplifiers: (a) using discrete components; (b) using an integrated circuit*

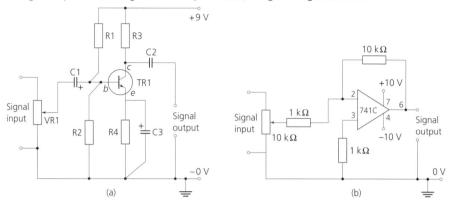

Fig. 6.2 *Designing to avoid assembly errors: (a) alternative positions for integrated circuit (IC) with symmetrical legs; (b) positioning an IC using identification marks*

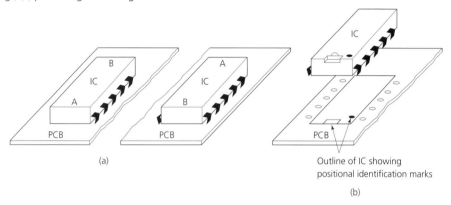

(a)

Outline of IC showing
positional identification marks

(b)

Now let's consider the plastic box and lid shown in Fig. 6.3(a). From this exploded drawing, you can see that the assembler has to fit four screw, nuts, washers and spring-washers to secure the lid to the body. Since they are small, this will be a slow and fiddly job. However, there are two redeeming features:

- The lid can be fitted in either of two directions and either way up; the assembler has few decisions to make.
- Standard screws, nuts, washers and spring-washers are used. Any nut or washer will fit on any screw, which can be used in any hole. The use of standardised components speeds up the assembly process. It enables non-selective assembly to take place.

The box-and-lid assembly can be simplified by modifying the moulding so that self-tapping screws can be used, as shown in Fig. 6.3(b). This also reduces the number of components and their cost..

Finally, the box and lid can be redesigned so as to take advantage of the elastic properties of thermoplastics. Figure 6.3(c) shows how a snap fastening is used. The snap fastener may provide a permanent joint or it may be designed so it can be dismantled. These alternatives are shown in Fig. 6.3(d).

Fig. 6.3 *Exploiting material properties: (a) original design; (b) alternative design with fewer parts and easier assembly – the properties of the moulded plastic allow the screws to be driven into plain holes; the plastic grips the screws so they will not work loose; (c) alternative design with moulded-on snap fasteners and no loose parts; (d) snap fasteners*

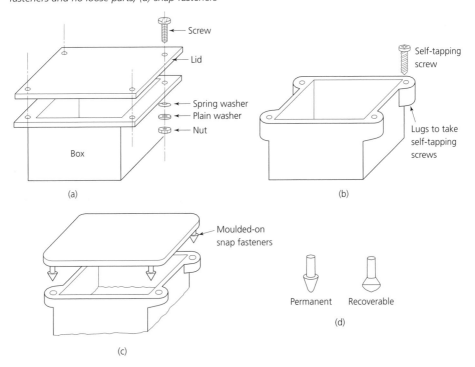

6.2 Batch size

The batch size has a considerable influence on the component design and the method of assembly. Although the snap-fit lid in Section 6.1 was the cheapest and easiest to assemble, the cost of the complex moulding tools required to make it could only be justified by producing large quantities.

In the early days of car manufacturing, each car was individually hand built. Therefore the standardisation of components and the ease of assembly was of little consequence. However, in today's volume car assembly lines, every part must be designed and manufactured to fit easily and quickly into place. There is no room for selective assembly and 'fitting': the part must bolt on and work straight away.

Preferably it must be capable of being picked, placed and fixed automatically by an industrial robot. Only when large batches of a product are being assembled can the design work and manufacturing accuracy be justified to ensure such a high level of interchangeability and ease of assembly. However, even in smaller batch production, a high level of interchangeability can be achieved relatively cheaply by using standard components that are available off the shelf.

1. Give **three** reasons why it is advisable to consider assembly at the earliest stages of any new design.
2. Explain in your own words the advantages of using standard components wherever possible.

6.3 Accuracy of fit

Non-selective assembly reduces assembly costs by speeding up production. Non-selective assembly is only possible if there is complete *interchangeability* between mating components in the assembly. Figure 6.4 shows a shaft which has been dimensioned so that its diameter lies between 74.981 mm and 75.000 mm. The designer has decided, in his or her wisdom, that any shaft lying between these sizes will function satisfactorily in a given assembly. These dimensions are called *limits of size* and the algebraic difference between these limits of size is called the *tolerance*. Toleranced dimensions are provided because it is not possible to work to an exact size, and even if this were, there is no means of measuring an exact size.

The terms *maximum metal condition* and *minimum metal condition* are often met in connection with limit systems and the gauging of workpieces. Gauging will be introduced in Chapter 10. Figure 6.5 explains what is meant by maximum and minimum metal conditions.

The class of fit between two mating components, such as a shaft and a bearing, may be obtained in two ways:

- By having a constant size of hole and varying the diameter of the shaft to suit; this is called a *hole basis system*.
- By having a constant size of shaft and varying the diameter of the hole to suit; this is called a *shaft basis system*.

The hole basis system is the more usually employed since it is easier to maintain a standard hole size using standard drills and reamers. The shaft diameter is more easily machined to the size required by the fit during a turning or grinding operation. Both these systems are shown in Fig. 6.6. Let's now consider the classes of fit we can obtain between any two components, depending upon their relative dimensions.

Fig. 6.4 *Application of limits. (Dimensions in millimetres)*

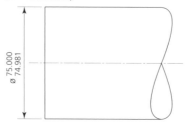

Fig. 6.5 *Maximum and minimum metal conditions*

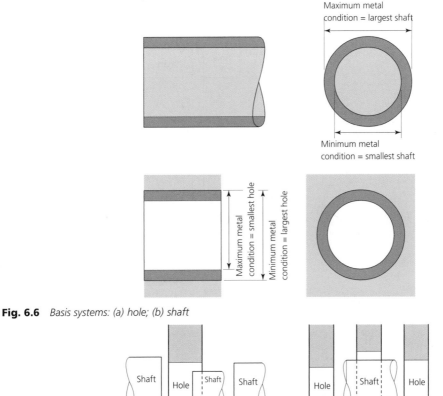

Maximum metal
condition = largest shaft

Minimum metal
condition = smallest shaft

Maximum metal
condition = smallest hole

Minimum metal
condition = largest hole

Fig. 6.6 *Basis systems: (a) hole; (b) shaft*

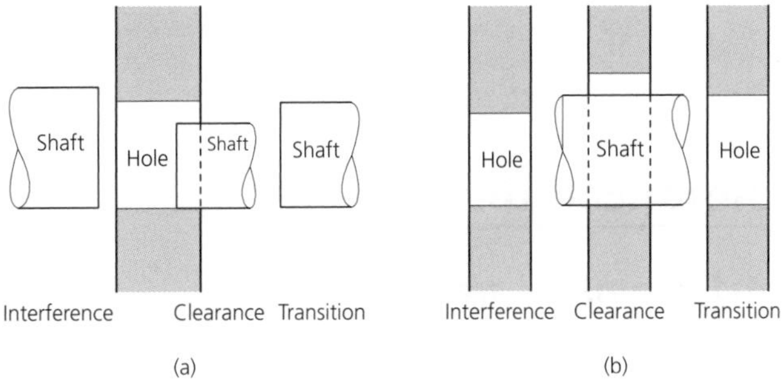

Shaft | Hole | Shaft | Shaft

Interference | Clearance | Transition

(a)

Hole | Shaft | Hole

Interference | Clearance | Transition

(b)

6.3.1 *Clearance fit*

This is achieved when the shaft is always smaller than the corresponding hole it is mating with, even under maximum metal conditions.

6.3.2 *Interference fit*

This is achieved when the shaft is always slightly larger than the corresponding hole it is mating with, even under minimum metal conditions.

6.3.3 *Transition fits*

These fits are achieved when mating shafts and holes which are within limits will give a range of fits from clearance under minimum metal conditions to interference under maximum metal conditions. All these types of fit are shown in Fig. 6.7.

Fig. 6.7 *Classes of fit: (a) clearance; (b) transition; (c) interference*

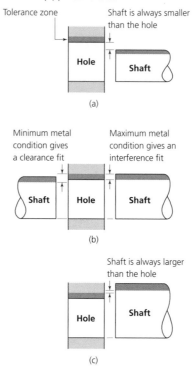

6.4 Toleranced dimensions

6.4.1 *Nominal size*

This is the size by which a component feature or dimension is known (Fig. 6.8). For example, a 75.15 mm diameter hole would usually be known as a 75 mm hole.

Fig. 6.8 *Limit systems: definitions*

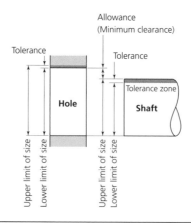

6.4.2 *Basic size*

This is the exact functional size from which the limits are derived by the application of the necessary tolerances and clearance (Fig. 6.8). The nominal size and the basic size may sometimes be the same.

6.4.3 *Limits of size*

These are the maximum and minimum sizes between which it is permissible to manufacture a given workpiece if it is to function correctly (Fig. 6.8).

6.4.4 *Tolerance*

This is the algebraic difference between the upper and lower limits of size (Fig. 6.8). Its size and shape is determined by the functional needs of the component and the economics of production. It can generally be assumed that the smaller the tolerance, the more costly the process to achieve that tolerance.

6.4.5 *Minimum clearance*

Minimum clearance used to be known as 'allowance'; it is the arithmetical difference between the maximum metal condition of the shaft and the maximum metal condition of the hole (Fig. 6.8). That is, the largest shaft and the smallest hole giving the 'tightest fit' which will function correctly.

6.4.6 *Some other terms*

The *actual size* of a component feature is the measured size corrected to 20 °C.

Figure 6.9 shows how tolerances may be applied to the basic size, together with alternative methods of dimensioning. *Unilateral tolerances* are those where the tolerance zones lie to one side only of the basic size. *Bilateral tolerances* are those where the tolerance zones cross the basic size. The basic size is 20.00 mm in all the examples.

Fig. 6.9 *Tolerances: (a) unilateral – tolerance zone does not cross the basic size; (b) bilateral – tolerance zone always crosses the basic size*

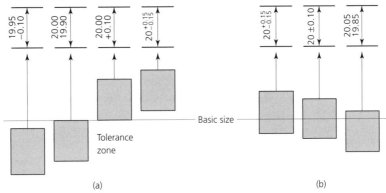

As dimensional tolerances get smaller and the precision of components becomes greater to satisfy present-day quality requirements and facilitate assembly, geometrical tolerancing also becomes of importance. A slightly oval shaft can be within the dimensional limits if they are relatively coarse. However, the same oval shaft could lie within the dimensional limits when measured across its minor diameter but outside them when measured across its major diameter. It is no longer adequate to specify a process and rely on the machine to give an acceptable level of geometrical tolerance. Geometrical tolerancing will be considered in *Manufacturing Technology*, Volume 2.

SELF-ASSESSMENT TASK 6.2

1. Explain why it is necessary to give dimensions a tolerance and why this tolerance should be as large as possible for any given application.

2. With the aid of a sketch, explain the difference between limits and tolerance.

3. Give an example where you would use:

   (a) a clearance fit
   (b) an interference fit

4. Explain why the use of a hole basis system of fits is usually preferable to a shaft basis sytem of fits.

6.5 Standard systems of limits and fits

This section is based upon BS 4500: *Limits and fits (metric units)*, which is suitable for all classes of work from the finest instruments to heavy engineering. It allows for the size of the work and the type of work, and it provides for hole basis and shaft basis systems as required.

The tables provide for 28 types of shaft designated by lower-case letters (a, b, c, etc.) and 28 types of hole designated by upper-case (capital) letters (A, B, C, etc.). For each type of shaft or hole, the grade of tolerance is designated by a number (01, 0, 1, 2, ..., 16), giving 18 grades of tolerance in all.

The letter indicates the position of the tolerance relative to the basic size and is called the fundamental deviation. The number indicates the magnitude of the tolerance and is called the fundamental tolerance. A shaft is completely defined by its basic size, letter and number, e.g. 75 mm h6. Similarly a hole is completely defined by its basic size, letter and number, e.g. 75 mm H7. Figure 6.10(a) shows how a precision clearance fit is specified by using a combination of a 75 mm H7 hole and a 75 mm h6 shaft. Table 6.1 shows the primary selection of limits and fits for a wide range of hole and shaft

Fig. 6.10 *Application of limits and fits: (a) tolerance specification – precision clearance fit; (b) dimensional limits to give precision clearance fit as derived from BS 4500: 1969*

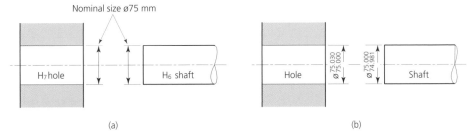

(a)   (b)

combinations for a variety of applications. Let's now see how we can give these tolerances actual sizes.

6.5.1 *Hole*

Enter the table along diameter band 50 mm to 80 mm, and where this band crosses the column H7 the limits are given as +30 and +0. These dimensions are in units of 0.001 mm, where 0.001 mm = 1 micrometer (μm). When applied to a basic size of 75 mm, they give working limits of 75.030 mm and 75.000 mm, as shown in Fig. 6.10(b).

6.5.2 *Shaft*

Enter the table along diameter band 50 mm to 80 mm, and where this band crosses the column h6 the limits are given as −0 and −19. Again these dimensions are in units of 0.001 mm (1 μm). When applied to a basic size of 75 mm, they give working limits of 75.000 mm and 74.981 mm, as shown in Fig. 6.10(b).

SELF-ASSESSMENT TASK 6.3

With reference to Table 6.1, derive the dimensions to give an average clearance fit for a hole and shaft with a basic size of 40 mm diameter. Use a sketch to show how the derived dimensions could be applied in practice.

6.6 Selection of tolerance grades

The closer the limits (the smaller the tolerance) the more difficult and expensive it is to manufacture a component. It is no use specifying very small (close) tolerances if the manufacturing process specified by the designer cannot achieve such a high degree of precision. Alternatively there is no point in choosing a process on grounds of low cost if it cannot achieve the accuracy necessary for the component to function correctly. Table 6.2, based upon BS 4500, shows the standard tolerances from which the tables of limits and fits

Table 6.1 Primary selection of fits

| Normal sizes | | Loose clearance | | Average clearance | | Close clearance | | Precision clearance | | Transition | | Interference | |
Over (mm)	Up to (mm)	H9	e9	H8	f7	H7	g6	H7	h6	H7	k6	H7	p6
—	3	+25 +0	−14 −39	+14 +0	−6 −16	+10 +0	−2 −8	+10 +0	−0 −6	+10 +0	+6 +0	+10 +0	+12 +6
3	6	+30 +0	−20 −50	+18 +0	−10 −22	+12 +0	−4 −12	+12 +0	−0 −8	+12 +0	+9 +1	+12 +0	+20 +12
6	10	+36 +0	−25 −61	+22 +0	−13 −28	+15 +0	−5 −14	+15 +0	−0 −9	+15 +0	+10 +1	+15 +0	+24 +15
10	18	+43 +0	−32 −75	+27 +0	−16 −34	+18 +0	−6 −17	+18 +0	−0 −11	+18 +0	+12 +1	+18 +0	+29 +18
18	30	+52 +0	−40 −92	+33 +0	−20 −41	+21 +0	−7 −20	+21 +0	−0 −13	+21 +0	+15 +2	+21 +0	+35 +22
30	50	+62 +0	−50 −112	+39 +0	−25 −50	+25 +0	−9 −25	+25 +0	−0 −16	+25 +0	+18 +2	+25 +0	+42 +26
50	80	+74 +0	−60 −134	+46 +0	−30 −60	+30 +0	−10 −29	+30 +0	−0 −19	+30 +0	+21 +2	+30 +0	+51 +32
80	120	+87 +0	−72 −159	+54 +0	−36 −71	+35 +0	−12 −34	+35 +0	−0 −22	+35 +0	+25 +3	+35 +0	+59 +37
120	180	+100 +0	−85 −185	+63 +0	−43 −83	+40 +0	−14 −39	+40 +0	−0 −25	+40 +0	+28 +3	+40 +0	+68 +43
180	250	+115 +0	−100 −215	+72 +0	−50 −96	+46 +0	−15 −44	+46 +0	−0 −29	+46 +0	+33 +4	+46 +0	+79 +50
250	315	+130 +0	−110 −240	+81 +0	−56 −108	+52 +0	−17 −49	+52 +0	−0 −32	+52 +0	+36 +4	+52 +0	+88 +56
315	400	+140 +0	−125 −265	+89 +0	−62 −119	+57 +0	−18 −54	+57 +0	−0 −36	+57 +0	+40 +4	+57 +0	+98 +62
400	500	+155 +0	−135 −290	+97 +0	−68 −131	+63 +0	−20 −60	+63 +0	−0 −40	+63 +0	+45 +5	+63 +0	+108 +68

Source: Abstract from BS 4500

Table 6.2 Standard tolerances

Nominal size over (mm)		IT01	IT0	IT1	IT2	IT3	IT4	IT5	IT6[†]	IT7	IT8	IT9	IT10	IT11	IT12	IT13	IT14[‡]	IT15[‡]	IT16[‡]
–	3	0.3	0.5	0.8	1.2	2	3	4	6	10	14	25	40	60	100	140	250	400	600
3	6	0.4	0.6	1	1.5	2.5	4	5	8	12	18	30	48	75	120	180	300	480	750
6	10	0.4	0.6	1	1.5	2.5	4	6	9	15	22	36	58	90	150	220	360	580	900
10	18	0.5	0.8	1.2	2	3	5	8	11	18	27	43	70	110	180	270	430	700	1 100
18	30	0.6	1	1.5	2.5	4	6	9	13	21	33	52	84	130	210	330	520	840	1 300
30	50	0.6	1	1.5	2.5	4	7	11	16	25	39	62	100	160	250	390	620	1 000	1 600
50	80	0.8	1.2	2	3	5	8	13	19	30	46	74	120	190	300	460	740	1 200	1 900
80	120	1	1.5	2.5	4	6	10	15	22	35	54	87	140	220	350	540	870	1 400	2 200
120	180	1.2	2	3.5	5	8	12	18	25	40	63	100	160	250	400	630	1 000	1 600	2 500
180	250	2	3	4.5	7	10	14	20	29	46	72	115	185	290	460	720	1 150	1 850	2 900
250	315	2.5	4	6	8	12	16	23	32	52	81	130	210	320	520	810	1 300	2 100	3 200
315	400	3	5	7	9	13	18	25	36	57	89	140	230	360	570	890	1 400	2 300	3 600
400	500	4	6	8	10	15	20	27	40	63	97	155	250	400	630	970	1 500	2 500	4 000
500	630	–	–	–	–	–	–	–	44	70	110	175	280	440	700	1 100	1 750	2 800	4 400
630	800	–	–	–	–	–	–	–	50	80	125	200	320	500	800	1 250	2 000	3 200	5 000
800	1 000	–	–	–	–	–	–	–	56	90	140	230	360	560	900	1 400	2 300	3 600	5 600
1 000	1 250	–	–	–	–	–	–	–	66	105	165	260	420	660	1 050	1 650	2 600	4 200	6 600
1 250	1 600	–	–	–	–	–	–	–	78	125	195	310	500	780	1 250	1 950	3 100	5 000	7 800
1 600	2 000	–	–	–	–	–	–	–	92	150	230	370	600	920	1 500	2 300	3 700	6 000	9 200
2 000	2 500	–	–	–	–	–	–	–	110	175	280	440	700	1 100	1 750	2 800	4 400	7 000	11 000
2 500	3 150	–	–	–	–	–	–	–	135	210	330	540	860	1 350	2 100	3 300	5 400	8 600	13 500

[†] Not recommended for fits in sizes above 500 mm.
[‡] Not applicable to sizes below 1 mm.

are derived. Notice that as the *international tolerance* (IT) number gets larger, the tolerance increases. The recommended relationship between process and standard tolerance is as follows:

IT16   sand casting, flame cutting
IT15   stamping
IT14   die casting, plastic moulding
IT13   presswork, extrusion
IT12   light presswork, tube drawing
IT11   drilling, rough turning, boring
IT10   milling, slotting, planing, rolling
IT9   low-grade capstan and automatic lathe work
IT8   centre lathe, capstan and automatic lathe work
IT7   high-quality turning, broaching, honing
IT6   grinding, fine honing
IT5   machine lapping, fine grinding
IT4   gauge making, precision lapping
IT3   high-quality gap gauges
IT2   high-quality plug gauges
IT1   slip gauges, reference gauges

EXAMPLE 6.1

Derive the dimensions for a hole and shaft of nominal diameter 40 mm so that an average clearance fit is obtained.

Since the nominal size of the shaft/hole is 40 mm, Table 6.1 is entered at the band 30 mm to 50 mm. The following conditions then apply:

Nominal size	H8	f7
	+39	−25
30–50 mm	+0	−50

The tolerance unit is 0.001 mm, so we obtain the following limits:

- Hole diameter
  upper limit = 40 + 0.039 = 40.039 mm
  lower limit = 40 + 0.000 = 40.000 mm

- Shaft diameter
  upper limit = 40 − 0.025 = 39.975 mm
  lower limit = 40 − 0.050 = 39.950 mm

Since only an *average clearance fit* is required, there is no benefit in working to an accuracy of three decimal places. The designer would use his or her experience and reduce the cost of manufacture by rounding off the limits to:

- Hole diameter
    upper limit = 40.04 mm
    lower limit = 40.00 mm

- Shaft diameter
    upper limit = 39.98 mm
    lower limit = 39.95 mm

Had a close or precision clearance been required, rounding off would not have been permissible and the increase in accuracy would have increased manufacturing costs and therefore the unit component cost. There is no advantage in using greater precision than is required. Figure 6.11 shows how these limits are applied.

Other information that can be obtained from the dimensions:

Hole tolerance = 40.04 − 40.00 = 0.04 mm

Shaft tolerance = 39.98 − 39.95 = 0.03 mm

Maximum clearance = largest hole − smallest shaft
= 40.04 − 39.95 = 0.09 mm

Minimum clearance = smallest hole − largest shaft
= 40.00 − 39.98 = 0.02 mm

EXAMPLE 6.2

Determine a process that is suitable for manufacturing the shaft dimensioned in Example 6.1.

Your answer to the previous self-assessment task should have resulted in a sketch similar to Fig. 6.11 and showing the same dimensions. The diameter of the shaft is 40 mm and the tolerance is 0.03 mm. From Table 6.2 you can see that the IT number which corresponds to these conditions lies between IT7 and IT8. Therefore a good turned finish would be sufficiently accurate. However, to avoid wear, the shaft would probably receive some form of heat treatment, in which case the tolerance could easily be achieved by commercial-quality cylindrical grinding.

Fig. 6.11 *Average clearance fit: toleranced dimensions*

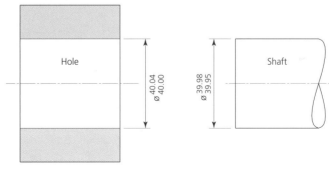

6.7 Interchangeability

We tend to take interchangeability for granted. We expect every ISO M6 nut to fit every ISO M6 bolt. This was not always so and, in the early days of engineering, nuts and bolts were made as matched pairs that were not interchangeable with any other nuts and bolts of the same size, a fitter's nightmare. It was not until the middle of the nineteenth century that Sir Joseph Whitworth developed the standard screw thread system that still bears his name and introduced the concept of standardised, interchangeable components to engineering.

Interchangeability not only leads to reduced assembly cost, it also facilitates maintenance and repairs as standard parts are readily available and can be changed for the original equipment without difficulty or loss of performance. The principles of selective and non-selective assembly are shown in Fig. 6.12. Non-selective assembly can only be achieved by the use of standard specifications and standard dimensioning.

The impression may have been given that there is no room for selective assembly in modern engineering production. However, with modern automatic computer-controlled gauging systems, semi-selective assembly is still justified where workpieces cannot be made to the required accuracy but can be readily measured and graded. This technique is used, for example, when machining pistons and cylinder bores, where the required quality of fit exceeds the accuracy of the available manufacturing process.

Using an automatic system, the bores and pistons are individually measured and graded with a colour code; suppose there are five grades in this example. On assembly a grade 3 piston is used in a grade 3 bore and will give the required class of fit, even though the pistons and cylinders cannot be machined economically to give such a high-quality fit. Figure 6.13 shows how this is achieved.

Fig. 6.12 *Interchangeability: (a) selective assembly; (b) non-selective assembly*

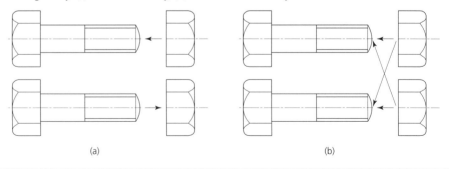

(a)   (b)

Fig. 6.13 *Semi-selective assembly*

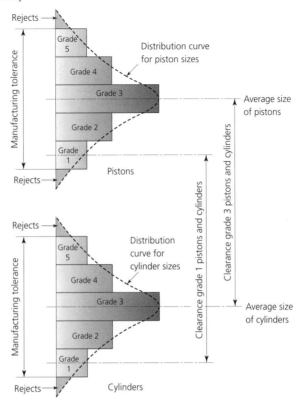

6.8 Materials

The choice of materials can also affect the assembly process. For example, copper conductors can easily be soldered to lugs and terminals. Aluminium conductors, although lighter and cheaper, cannot easily be soldered; they have to be fitted to terminal lugs by mechanical crimping.

Sections of sheet-metal ventilation ducting can be joined by fitting flanges at each end of the section and bolting the flanges of adjacent sections together, or the sections can be oxyacetylene welded together on site. On the other hand, plastic ducting sections can be welded together using a hot-air torch. This is less energy intensive, the fire risk from fusion welding in a confined space is eliminated, and ventilation of the working zone is easier. The plastic ducting is not only lighter and easier to handle, it is also corrosion resistant.

When an assembly is made using compression joints (Fig. 6.14), the inner component is subjected to compressive stress and the outer component to tensile stress. The materials chosen for these components must be capable of withstanding these stresses. Cast iron, with its high compressive strength and low-tensile strength, would be suitable for the inner component but unsuitable for the outer component of a compression joint.

6.9 Joining methods

There are many ways of fastening components together to make an assembly, but all joining processes fall into three main groups: permanent, semi-permanent and temporary joints.

6.9.1 *Permanent joints*

Permanent joints cannot be dismantled without the destruction of the joint, and possibly the joined components. They include:

- Riveting.
- Soldering.
- Silver soldering.
- Brazing.
- Welding.
- Adhesive bonding.

These joining techniques are considered in detail in Chapter 7.

6.9.2 *Semi-permanent joints*

Compression joints can be used for connecting precision components; they rely upon the elasticity of materials to secure one component to another without the use of any additional fastening devices such as bolts, rivets or adhesives. In a lightly compressed joint, such as a drill bush pressed into a bush plate, friction alone maintains the assembly. The principles of this type of pressed or 'staked' joint are shown in Fig. 6.14. However, in cases of extreme interference, such as when the starter-ring gear is shrunk onto the flywheel of a car engine, permanent deformation may take place with one component biting into the other so that positive as well as frictional restraint takes place.

 Figure 6.15(a) and (b) show the principles of making a thermal compression joint in which the outer element is expanded by heating before assembly; Fig. 6.15(c) and (d) show the principle of making a thermal compression joint where the inner element is shrunk by cooling before assembly. In either case the outer element is in tension and the inner element

Fig. 6.14 *Mechanical compression joints: (a) the bush is compressed and the hole in the bush plate is expanded as the bush is pressed home; (b) the elasticity (spring-back) of the assembled components creates high frictional forces which maintain the joint*

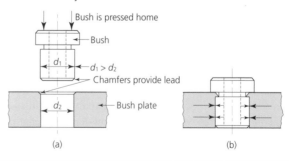

Fig. 6.15 *Thermal compression joints: (a) hot shrink; (b) cold expansion*

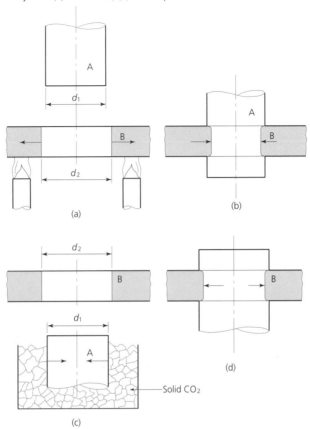

(a)

(b)

(c)

(d)

Solid CO_2

is in compression when the assembly is complete. Therefore the materials have to be carefully chosen to resist these forces. Although heating of the outer element is the simpler process, care must be taken so this does not affect the structure of the material.

6.9.3 *Temporary joints*

Temporary joints may be assembled and dismantled as frequently as required, not only for initial assembly, but also for maintenance and renewal. The more important of these joining techniques will now be considered.

6.10 Screwed fastenings

Figure 6.16 shows some typical applications of screwed joints. The bolted joint in Fig. 6.16(a) shows that the plain shank extends beyond the joint face so that all shear loads are taken on the plain shank, not by the thread.

Fig. 6.16 *Screwed fastenings: (a) section through a bolted joint; (b) stud-and-nut fixing for an inspection cover; (c) cap head socket screw; (d) cheese head brass screws*

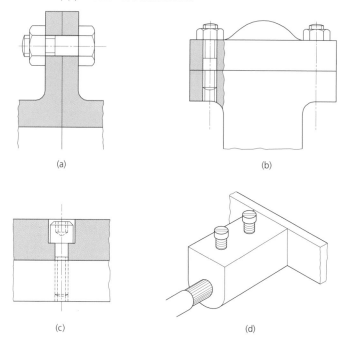

(a)

(b)

(c)

(d)

The stud-and-nut fixing in Fig. 6.16(b) is used where a joint is regularly dismantled. The bulk of the wear comes on the stud, which can eventually be replaced cheaply and easily. This prevents wear falling on the threads of the more expensive casting or forging. If studs cannot be used, the component (female) thread can be lined with a wire thread insert. Studs can have dissimilar threads to suit the material into which they are screwed. For example, one end could have a coarse thread for screwing into a soft aluminium alloy cylinder block, whereas the other end of the stud could have a fine thread to increase the clamping force and resist any loosening of the nut by vibration.

The cap head socket screw in Fig. 6.16(c) is much more expensive than the ordinary hexagon head bolt. However, the cap head socket screw is made from heat-treated high-tensile steel, upset-forged and thread rolled. This makes it very much stronger, tougher and wear resistant than the ordinary low-carbon steel hexagon head bolt. Cap head socket screws are widely used in the manufacture of machine tools. Because the heads can be sunk flush with the surface of the component they are securing, the result is a better appearance, easy cleaning and greater safety.

Cheese head screws can also be used where flush fitting is required, but they are not as strong as cap screws and are only used for lightly stressed joints where low cost is important. Figure 6.16(d) shows small brass cheese head screws being used to clamp electric cable into a terminal.

SELF-ASSESSMENT TASK 6.5

Briefly describe the differences between the following types of joint. Give an example of where each type would be used. Do not repeat the examples in the text.

(a) permanent
(b) semi-permanent
(c) temporary joints

6.11 Locking devices

In order to prevent screwed fastenings working loose due to vibration, various locking devices are employed; some of them are shown in Fig. 6.17. Notice how they can be divided into two main categories: those where the locking action is positive, and those where the locking action is frictional. Positive locking devices are more time-consuming to fit, but they are essential for critical joints where failure could cause a serious accident, as in the controls of a machine tool, motor vehicle or aircraft.

Fig. 6.17 *Locking devices: (a) positive; (b) frictional*

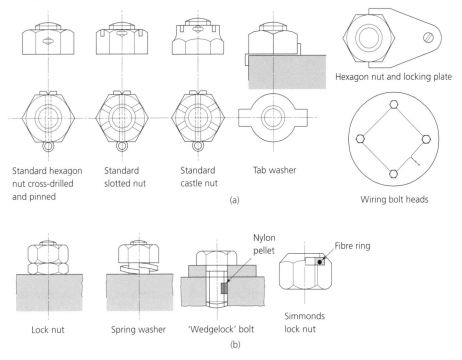

6.12 Spanners

Spanners are used to tighten screwed fastenings. They are carefully proportioned so their length enables a person of average strength to fully tighten the fastenings correctly. A torque spanner should be used to ensure that critical screwed fastenings are tightened correctly. Some torque spanners use a preset slipping clutch so that when the specified torque has been reached, the clutch slips and the fastening cannot be overtightened. Figure 6.18 shows a cheaper and simpler type of torque spanner which has a pointer to indicate when the correct torque has been applied. The torque arm of this type of spanner is made from spring steel.

Fig. 6.18 *Torque spanner: (a) the square tang is designed to fit a standard socket set; (b) the torque scale has a centre zero for left- and right-hand threads; (c) the fastening is tightened until the pointer indicates the prescribed torque has been reached; as the torque increases, the torque arm bends and the scale moves across the pointer*

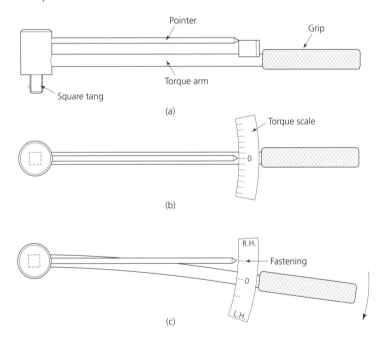

6.13 Keys

Keys are used to connect components such as wheels onto shafts in order to transmit rotary motion without slip. Some typical keys are shown in Fig. 6.19. Taper gib-head keys are only used on shaft ends where they can be withdrawn by means of a drift driven between the key head and the wheel. Woodruff keys are used where axial alignment is required.

Fig. 6.19 *Keys: (a) gib head; (b) Feather; (c) Woodruff*

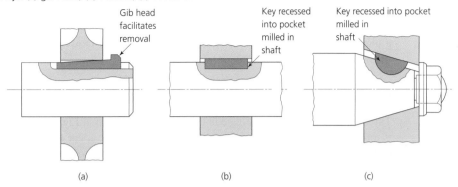

Gib head facilitates removal

Key recessed into pocket milled in shaft

Key recessed into pocket milled in shaft

(a)   (b)   (c)

6.14 Miscellaneous fastenings

Let's now look at a selection of temporary fastenings such as dowels, taper pins, cotter pins and circlips (Fig. 6.20).

6.14.1 *Plain Dowels*

Screwed fastenings are usually fitted through drilled clearance holes. Therefore, lateral location depends largely upon friction and the fastenings do not come into shear until sufficient movement has occurred to take up all the clearance. More precise location may

Fig. 6.20 *Miscellaneous fastenings: (a) plain dowel; (b) taper dowel; (c) cotter pin; (d) circlips*

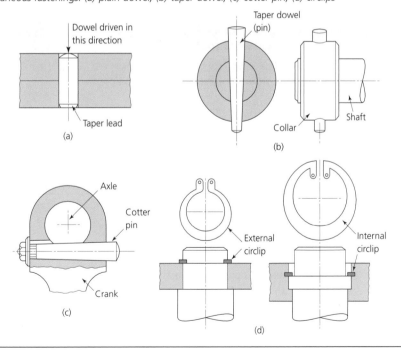

Dowel driven in this direction

Taper dowel (pin)

Taper lead

Collar

Shaft

(a)   (b)

Axle

Cotter pin

Crank

External circlip

Internal circlip

(c)   (d)

be achieved by using bolts with ground shanks fitted through reamed holes or by plain dowels, as shown in Fig. 6.20(a). Dowels should never be fitted in blind holes, since air trapped under a dowel will make it difficult to fit and it is impossible to drive a dowel out of a blind hole when the assembly needs to be dismantled. Dowels should be a light-drive (interference) fit in a reamed hole.

6.14.2 *Taper Dowels (taper pins)*

Taper dowels are used for fitting collars to shafts, as shown in Fig. 6.20(b). When the collar is correctly positioned, a parallel hole is drilled through the collar and the shaft. This is then opened up with a taper pin reamer and the taper dowel (pin) is driven home. Any wear from repeated assembly and dismantling is compensated by merely driving the pin further into the hole. The taper is a Morse self-locking taper, so there is little chance of the pin dropping out.

6.14.3 *Cotter Pins*

Cotter pins are parallel pins with a taper flat that engages with a corresponding flat on the shaft, as shown in Fig. 6.20(c). The nut secures the pin and prevents it working loose, as well as drawing the tapers tightly together.

6.14.4 *Circlips*

Circlips are used to provide a locating shoulder on a shaft or in a hole, as shown in Fig. 6.20(d). They are usually made from spring steel and are fitted or removed with specially shaped pliers which locate in the holes of the circlip. They provide a neat fixing that is easy to assemble or dismantle.

SELF-ASSESSMENT TASK 6.6

With the aid of sketches, show examples where the use of the following devices would be appropriate. Do not repeat the examples in the text.

(a) tab washer
(b) cotter pin
(c) plain dowel
(d) taper dowel
(e) circlip
(f) Woodruff key

6.15 Cost of assembly

The cost of assembly depends upon the time taken, the wage rates and the cost of any special tools and workholding fixtures required. Consider the plastic box and lid introduced in Section 6.1. The cost of assembling the lid on the box manually using self-tapping screws could be determined as follows.

6.15.1 *Manual assembly*

We assess the total assembly time by analysing the time in seconds to assemble the individual elements.

	Time (s)
Pick up box and place in convenient position	2
Pick up lid and position over box	2
Pick up screw, fit washer and position in hole	3
Run screw home using screwdriver	5
Repeat the last two operations for the remaining holes	24
Total time	36

A typical wage rate at the time of publication could be £5.40 per hour. Therefore the cost of assembly would be $(540\text{p} \times 36)/3600 = 5.4\text{p}$ per box.

6.15.2 *Power assisted assembly*

To speed up assembly and reduce costs, the manufacturer decided to provide the assembler with a power-assisted screwdriver. This reduced the overall time of assembly per box to 20 s and the cost to $(540\text{p} \times 20)/3600 = 3\text{p}$ per box.

However, the power-assisted screwdriver cost £280.00 and this is taken into account by using the *break-even diagram* of Fig. 6.21. This shows that below 18,667 boxes it is cheaper to use a manual screwdriver, but above 18,667 boxes it is cheaper to use the power-assisted screwdriver. This simple example neglects operator fatigue, the possibility of long-term wrist injury, the cost of electricity, maintenance and the eventual replacement of the power-assisted screwdriver, all of which would have to be considered in practice.

Fig. 6.21 *Cost of assembly*

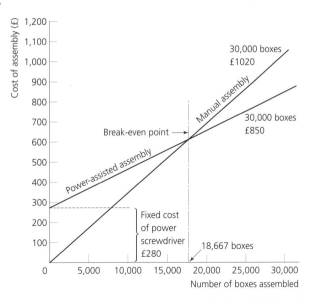

6.1 Discuss the effects of design, batch size and component accuracy on the assembly process.

6.2 Explain the meaning of the following terms using sketches if necessary:
    (a) selective and non-selective assembly
    (b) maximum and minimum metal condition
    (c) hole basis and shaft basis systems
    (d) clearance, interference and transition fits

6.3 Explain the meaning of the following terms using sketches if necessary:
    (a) limits of size
    (b) tolerance
    (c) minimum clearance (allowance)
    (d) unilateral tolerance and bilateral tolerance
    (e) fundamental deviation and fundamental tolerance

6.4 (a) With reference to the primary selection of fits (BS 4500), calculate the dimensions for the combination of a 25 mm H8 hole and a 25 mm f7 shaft.
    (b) With reference to the standard tolerances (BS 4500) and the recommended relationship between process and standard tolerance, select suitable processes for manufacturing the shaft and hole specified in part (a).

6.5 (a) With the aid of sketches, explain the principles of:
      (i) a mechanical compression joint
      (ii) a thermal compression joint
    (b) Name a suitable application for joints (i) and (ii) in part (a), giving reasons for your choice.
    (c) State the essential properties of the materials selected for the inner and outer elements of a compression joint.

6.6 Figure 6.22 shows a simple engineering assembly. Specify suitable fastenings to be used at positions A, B, C and D, giving reasons for your choice.

Fig. 6.22 *Retaining device A is easily removed and causes minimum overhang; device B prevents the spigot from rotating in the crank; screwed fastening C has a locking device; device D fixes the crank rigidly to the shaft yet allows it to be removed (consider a bicycle crank)*

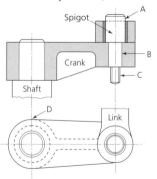

7 Fabrication

The topic areas covered in this chapter are:

- Preparation for fabrication: cutting and forming.
- Location for joining operations.
- Screwed fastenings.
- Riveted joints.
- Soldered joints: soft-soldered, hard-soldered, brazed.
- Fusion welding.
- Resistance welding.
- Adhesive bonding.
- Manipulation and fabrication of sheet plastic.

7.1 Preparation for fabrication

Chapter 6 described assembly as the joining together of individual components and subassemblies to make a single usable device. Fabrication describes the forming and joining of assemblies made largely from sheet metal, thin and thick plate, and sections such as standard angles, British Standard beams (BSBs), and rolled-steel joists (RSJs).

7.1.1 *Cutting processes*

The cutting out of sheet-metal and thin-plate blanks, using a guillotine or a blanking tool in a press, is discussed in Section 1.9. Thick plate is usually flame-cut using an oxyacetylene gas mixture and a special cutting torch. Details of the nozzle for this torch and the method of cutting are given in Fig. 7.1. This process is only successful with ferrous metals; non-ferrous metals tend to conduct the heat away from the cutting zone too rapidly. The metal at the cutting zone is raised to a temperature of 890 °C (bright cherry red) using the oxyacetylene preheating flame. A jet of pure oxygen is then directed onto this spot. The oxygen instantly reacts with the hot steel. The steel actually burns to iron oxide, and as this reaction is exothermic (gives out heat), it becomes largely self-sustaining, providing the oxygen supply is not interrupted. The cutting torch may be controlled manually, or the nozzle may be controlled by a pantograph copying machine using a template, or by computer numerical control.

Fig. 7.1 *Flame cutting: (a) points to be observed when cutting with a hand torch; (b) cutting action with a two-piece nozzle*

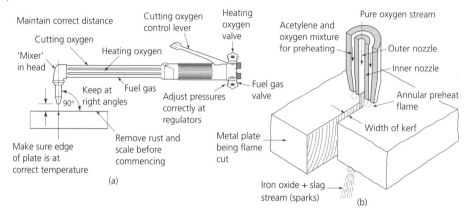

(a)

(b)

For cutting angle sections, T-sections and plate, a universal cropping and notching machine may be used (Fig. 7.2). Typical operations performed by this machine are shown in Fig. 7.3.

Fig. 7.2 *Universal cropping and notching machine: (A) frame: (B) notcher; (C) adjustable hold-down; (D) shearing blades for cropping flat bar and small plates; (E) section cropper for cropping angles and T-shapes; (E) punch and stripper*

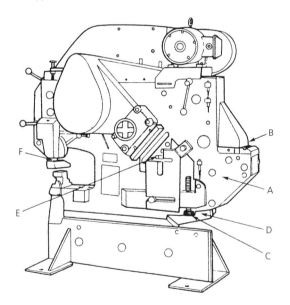

Fig. 7.3 *Cutting operations: (a) shearing plate; (b) provision for holding plate; (c) notching T-section; (d) mitring an angle flange with the notching tool; (e) typical mitre cutting operations performed on a section cropper*

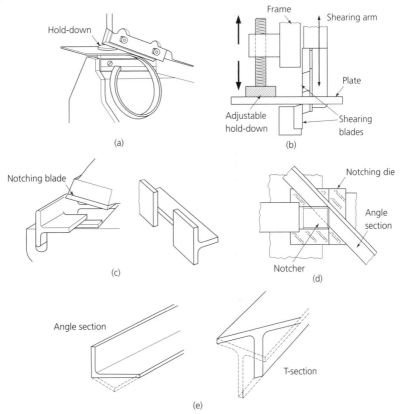

7.1.2 *Forming processes*

The basic principles of V-bending and U-bending for small components are discussed in Section 1.7. Sheet-metal fabrications often require relatively long bends to be made, and they are usually carried out on a press brake. Figure 7.4 shows a mechanical downstroke press brake; it has a capacity of 76 tonnes (the load which can be exerted by the press on the tools) and it can bend a 2.44 m length of 4 mm thick steel. Figure 7.5 shows a selection of forming operations which may be performed on this machine.

Cylindrical components may be formed by the use of sheet and plate rolling machines. Figure 7.6 shows a heavy-duty, power-driven roll bending machine, and Fig. 7.7 shows how it is used to roll a steel pipeline section. The top roller can be 'slipped' for removal of the finished component. The minimum diameter of a cylinder produced on this machine is usually taken as 1.5 to 2 times the diameter of the roller on which it is formed.

Fig. 7.4 *Press brake (mechanical downstroke type)*

Fig. 7.5 *Press brake operations: (a) using gooseneck punches; (b) flattening; (c) channel forming; (d) radius bending; (e) V-bending; (f) acute angle tools*

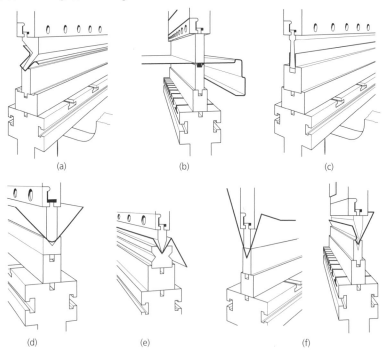

Fig. 7.6 *Plate rolling machine*

Fig. 7.7 *Principle of plate rolling: (a) align plate; (b) set ingoing edge; (c) initial rolling; (d) rolling; (e) set outgoing edge*

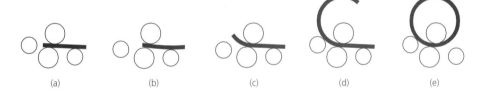

(a)   (b)   (c)   (d)   (e)

7.2 Location for joining operations

Work to be joined by permanent connections, such as welding and riveting, must be carefully located and restrained, since errors cannot be corrected once the connection is made. A wide variety of clamping and locating devices are used and some of the specialised types for fabrication processes will now be considered.

7.2.1 *Skin pins*

Skin pins are frequently used for locating box-section sheet-metal panels before riveting (Fig. 7.8); they are widely used in the aircraft industry. The knurled nut is unscrewed and

Fig. 7.8 *Work holding for riveting: (a) skin pin; (b) work clamped for drilling – work of this nature must be supported to prevent sagging under the drill pressure, which would cause misalignment of the rivet holes; (c) work clamped for riveting – after drilling, the holes are deburred and the assembly is clamped with the aid of location pins (skin pins) ready for riveting*

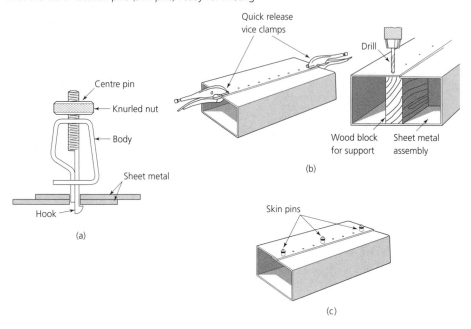

the centre pin is depressed and inserted into the hole in the sheets. Tightening the nut then draws the hooked centre pin upwards, clamping the sheets to be joined between the clamp body and the hook. They are removed by simply releasing the knurled nut and withdrawing the hook. Skin pins are also called location pins; they give quick accurate location and positive grip without distorting the hole, and they allow single-handed operation.

7.2.2 *Clamping devices for welding*

Clamps of many types are needed to assemble parts before welding. G-clamps are most commonly used, but care must be taken to prevent welding spatter from damaging the threads, otherwise the life of an ordinary G-clamp will be very short. The threads can be protected by coating them with antispatter compound. The rack G-clamp (Fig. 7.9), is preferable as its shield protects the screw against damage and spatter.

Many of the clamps used for holding work during welding are quick acting and can be easily and rapidly adjusted for various thicknesses of workpiece. Quick-acting clamps are essential for the batch production of work using welding jigs, where screw clamps would be too slow and uneconomical. Quick-action clamps generally have a toggle action or a cam action; typical examples are shown in Fig. 7.10.

Fig. 7.9 *Rack type G-clamps*

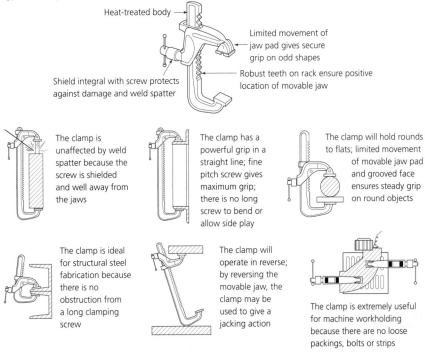

Heat-treated body

Limited movement of jaw pad gives secure grip on odd shapes

Robust teeth on rack ensure positive location of movable jaw

Shield integral with screw protects against damage and weld spatter

The clamp is unaffected by weld spatter because the screw is shielded and well away from the jaws

The clamp has a powerful grip in a straight line; fine pitch screw gives maximum grip; there is no long screw to bend or allow side play

The clamp will hold rounds to flats; limited movement of movable jaw pad and grooved face ensures steady grip on round objects

The clamp is ideal for structural steel fabrication because there is no obstruction from a long clamping screw

The clamp will operate in reverse; by reversing the movable jaw, the clamp may be used to give a jacking action

The clamp is extremely useful for machine workholding because there are no loose packings, bolts or strips

Fig. 7.10 *Quick-acting clamp devices: (a) typical hand-cam operated welding clamp; (b) typical hand-toggle operated welding clamp – quick-acting clamps are essential for holding work in welding jigs; (c) the toggle action – as the links come into a straight line, a considerable clamping force is exerted for a small effort at the handle; no reaction in direction A will cause the toggle to open*

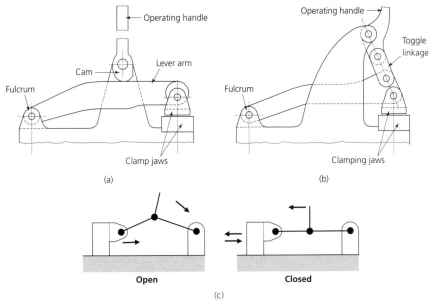

Operating handle

Lever arm

Cam

Fulcrum

Clamp jaws

(a)

Operating handle

Toggle linkage

Fulcrum

Clamping jaws

(b)

Open

Closed

(c)

7.2.3 *Magnetic clamps*

Figure 7.11(a) shows a magnetic workholding device suitable for holding sheet metal and thin plate during a welding operation, and Fig. 7.11(b) shows a magnetic clamp suitable for holding tube or rod. Care must be taken in positioning magnetic clamps when welding. If they become hot, from radiated or conducted heat, they will lose their magnetism.

Fig. 7.11 *Magnetic clamps for welding: (a) adjustable magnetic links; (b) magnetic holder*

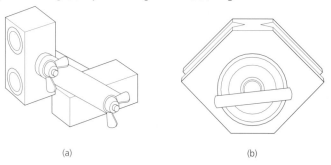

(a)   (b)

SELF-ASSESSMENT TASK 7.1

1. Using appropriate sketches, describe suitable applications for the following cutting equipment or processes:

    (a) guillotine shear
    (b) portable nibbling machine
    (c) oxyacetylene flame cutting
    (d) universal cropping and notching machine

2. Name the forming equipment you would use to make the following items. Give reasons for your choices.

    (a) a sheet-metal channel section 50 mm wide by 150 mm deep by 1000 mm long
    (b) a cylinder from 6 mm thick plate 1500 mm long by 1000 mm internal diameter
    (c) a one-off sheet-metal box with a square base of side length 300 mm and a depth of 100 mm. The thickness of the metal is 1.0 mm.

7.3 Screwed fastenings

7.3.1 *Bolts and nuts*

Screwed fastenings were introduced in Chapter 6. However, the screwed fastenings used in fabrication work tend to differ in detail. For instance, the bolts should be hot forged so as to give them greater strength than bolts machined from hexagon bar. Fitted bolts are sometimes used to provide lateral location; they are hot-forged bolts with machined shanks

---

to fit reamed holes. Friction grip bolts are high-tensile forged bolts used with hardened washers; they are tightened using a torque spanner and provide lateral location by friction that arises from the very high clamping forces they achieve.

7.3.2 *Washers*

Washers should be used between the face of a nut and the part being fastened; they increase the frictional grip of the nut by providing a smooth flat surface for it to bite into, and they prevent damage to the structural member by the scouring action of the corners of the nut as it is tightened. As well as plain washers, tapered washers are also required on structural steelwork to ensure the shank of the bolt is not bent as the fastening is tightened. The use of tapered and flat washers is shown in Fig. 7.12. The tapered washers must be chosen to suit the sections being connected. Two standard angles of taper are available, 5° and 8°.

Fig. 7.12 *Flat, taper and spring washers*

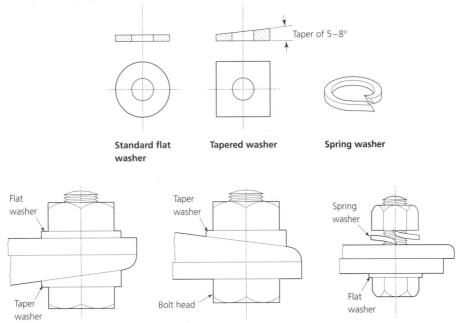

7.3.3 *Fabricated steelwork*

Some typical examples of fabricated steelwork using screwed and riveted connections are shown in Fig. 7.13. Wherever possible, bolted joints should be designed so that major forces acting on the joint are taken across the shank of the bolt, keeping it in shear. For this reason the line of the joint should lie across the shank of the bolt, never across the thread. However, unlike riveted joints, correctly proportioned screwed fasteners should be as strong in tension as they are in shear.

Fig. 7.13 *Bolted and riveted joints in fabricated steelwork: (a) simple riveted and bolted detail; (b) bolted connection; (c) riveted stanchion base*

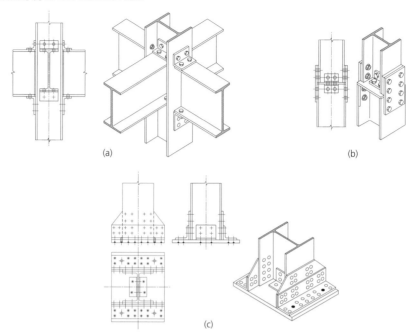

(a) (b) (c)

7.4 Riveted joints

Riveting is a method of making permanent joints. The members to be joined are drilled or punched; the rivets are inserted through the holes and then closed by forming a head on the rivet. Small-diameter rivets are usually closed at room temperature, but large rivets are closed whilst red-hot. Not only does this reduce the force necessary to close the rivet, but as the rivet cools and shrinks it pulls the members being joined tightly together.

A variety of riveted joints are used in constructional and fabrication work, and some of them are shown in Fig. 7.14. In order to close the rivet correctly, the shank of the rivet must swell to take up the clearance required to insert it. The length of the rivet must also be correctly chosen so that it will form the required head. The correct proportions for rivet length and hole clearance are shown in Fig. 7.15. The shank diameter of the rivet depends on the thicknesses of the members being joined and the forces acting upon them. The joint should be designed so that the major forces acting on the rivet place the shank in shear. The heads of the rivets are not intended to be load bearing and should only be lightly loaded. Consider them as merely keeping the rivet in place. Some typical riveted joints used in fabricated steelwork are shown in Fig. 7.13.

Blind riveting ('POP'® riveting) systems are used for fabricating closed, hollow, box sections in sheet metal where a hold-up (dolly) cannot be used to support the rivet head. Figure 7.16(a) shows some types of blind 'POP' rivets and their applications, and Fig. 7.16(b) shows the principle of closing blind rivets. ('POP' is a registered trademark of Tucker Fasteners Ltd.).

Fig. 7.14 *Types of riveted joints: (a) single-riveted lap joint; (b) double-riveted lap joint (chain); (c) double-riveted lap joint (zigzag); (d) single-strap butt joint (chain); (e) double-strap butt joint (zigzag)*

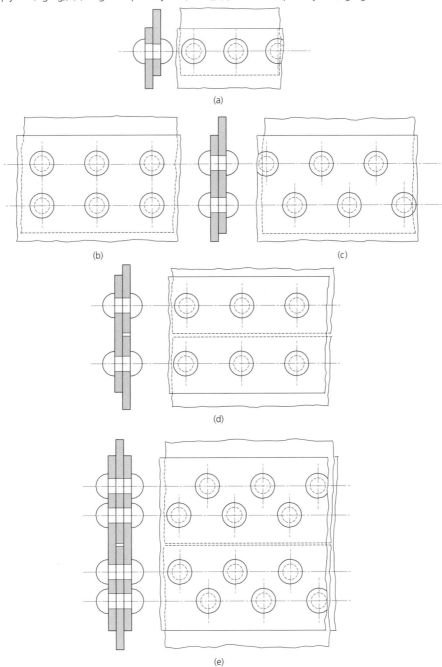

(a)

(b)   (c)

(d)

(e)

Fig. 7.15 *Proportions for riveted joints: (a) snaphead or roundhead rivet shape, L = 2T + 1.5D,
hole diameter = D + D/16; (b) countersunk rivet shape, L = 2T + D, hole diameter = D + D/16*

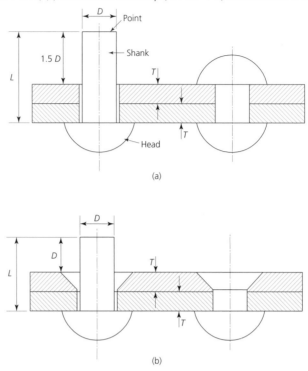

(a)

(b)

Fig. 7.16 *Blind 'POP' riveting: (a) some types of rivet and their application; (b) principle of closing a hollow rivet.
(Reproduced courtesy of Tucker Fasterners Ltd.)*

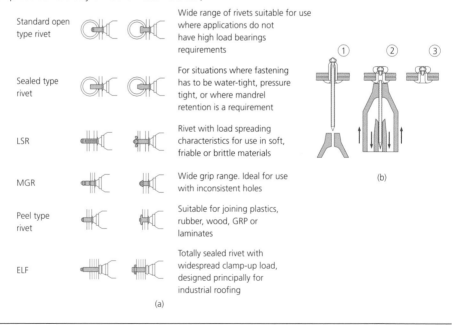

Standard open type rivet		Wide range of rivets suitable for use where applications do not have high load bearings requirements
Sealed type rivet		For situations where fastening has to be water-tight, pressure tight, or where mandrel retention is a requirement
LSR		Rivet with load spreading characteristics for use in soft, friable or brittle materials
MGR		Wide grip range. Ideal for use with inconsistent holes
Peel type rivet		Suitable for joining plastics, rubber, wood, GRP or laminates
ELF		Totally sealed rivet with widespread clamp-up load, designed principally for industrial roofing

(a)

(b)

7.5 Soft-soldered joints

Soft soldering is a low-temperature thermal jointing process in which the metal of the members being joined (parent metal) is not melted. It is essential that each of the joint faces is tinned by a film of solder; these solder films are made to fuse (melt) with the solder filling the space between the joint faces.

Examination under a microscope shows that correct tinning of metals such as brass, copper and steel produces a definite chemical reaction. The surface of the parent metal reacts with the tin content of the solder alloy to form an intermetallic compound, which acts as a key for the bulk of the solder in the joint.

The layer of intermetallic compound will continue to increase in thickness as long as the joint is kept at the soldering temperature. Once tinned, this film of solder cannot be mechanically wiped or prised off the parent metal. This is because, in reacting with the surface of the parent metal, molecules of solder diffuse into the parent metal to form an amalgam.

Soft solders are basically tin–lead alloys. Some solders for special applications have additional alloying elements. The full range of these solders, together with their recommended applications are found in BS EN 29453: 1994; some examples are listed in Tables 7.1, 7.2 and 7.3.

For successful soldering, the joint surfaces must be physically and mechanically clean; that is, they must be free from dirt, grease and oxide films. Once they have been cleaned, the surfaces must be protected from the action of atmospheric oxygen, otherwise oxide film will reform.

When a metal is exposed to air at room temperature it acquires a thin film of oxide within a few minutes of being cleaned. Since oxygen combines with metals even more rapidly at elevated temperatures, it is essential that both the joint surfaces and the molten solder are protected from the atmosphere. This protection is provided by a *flux*. The requirements of a soft-soldering flux are:

- It must remain liquid at the soldering temperature.
- In its liquid state it must cover and protect the joint surfaces from atmospheric oxygen.
- It must act as a wetting agent so that the molten solder flows freely over the joint faces and does not form into droplets.
- It must dissolve the initial oxide film and any that reforms.
- It must be readily displaced from the joint surfaces by the molten solder as soldering proceeds.

Table 7.1 Chemical compositions of tin–lead and tin–lead–antimony solder alloys

Group	Alloy no.	Alloy designation	Melting or solidus/liquidus temperature (°C)	Chemical composition (wt%)† Sn	Pb	Sb	Cd	Zn	Al	Bi	As	Fe	Cu	Sum of all impurities except Bi and Cu	BS 219 equivalent (see Table 7.2)
Tin–lead alloys	1	S-Sn63Pb37	183	62.5 to 63.5	Rem	0.12	0.002	0.001	0.001	0.10	0.03	0.02	0.05	0.08	AP
	1a	S-Sn63Pb37E	183	62.5 to 63.5	Rem	0.05	0.002	0.001	0.001	0.05	0.03	0.02	0.05	0.08	AP
	2	S-Sn60Pb40	183–190	59.5 to 60.5	Rem	0.12	0.002	0.001	0.001	0.10	0.03	0.02	0.05	0.08	KP
	2a	S-Sn60Pb40E	183–190	59.5 to 60.5	Rem	0.05	0.002	0.001	0.001	0.05	0.03	0.02	0.05	0.08	KP
	3	S-Pb50Sn50	183–215	49.5 to 50.5	Rem	0.12	0.002	0.001	0.001	0.10	0.03	0.02	0.05	0.08	–*
	3a	S-Pb50Sn50E	183–215	49.5 to 50.5	Rem	0.05	0.002	0.001	0.001	0.05	0.03	0.02	0.05	0.08	–*
	4	S-Pb55Sn45	183–226	44.5 to 45.5	Rem	0.50	0.005	0.001	0.001	0.25	0.03	0.02	0.05	0.08	R
	5	S-Pb60Sn40	183–235	39.5 to 40.5	Rem	0.50	0.005	0.001	0.001	0.25	0.03	0.02	0.08	0.08	G
	6	S-Pb65Sn35	183–245	34.5 to 35.5	Rem	0.50	0.005	0.001	0.001	0.25	0.03	0.02	0.08	0.08	H
	7	S-Pb70Sn30	183–255	29.5 to 30.5	Rem	0.50	0.005	0.001	0.001	0.25	0.03	0.02	0.08	0.08	J
	8	S-Pb90Sn10	268–302	9.5 to 10.5	Rem	0.50	0.005	0.001	0.001	0.25	0.03	0.02	0.08	0.08	–*
	9	S-Pb92Sn8	280–305	7.5 to 8.5	Rem	0.50	0.005	0.001	0.001	0.25	0.03	0.02	0.08	0.08	–*
	10	S-Pb98Sn2	320–325	1.5 to 2.5	Rem	0.12	0.002	0.001	0.001	0.10	0.03	0.02	0.05	0.08	–*
Tin–lead alloys with antimony	11	S-Sn63Pb37Sb	183	62.5 to 63.5	Rem	0.12 to 0.50	0.002	0.001	0.001	0.10	0.03	0.02	0.05	0.08	A
	12	S-Sn60Pb40Sb	183–190	59.5 to 60.5	Rem	0.12 to 0.50	0.002	0.001	0.001	0.10	0.03	0.02	0.05	0.08	K
	13	S-Pb50Sn50Sb	183–216	49.5 to 50.5	Rem	0.12 to 0.50	0.002	0.001	0.001	0.10	0.03	0.02	0.05	0.08	F
	14	S-Pb58Sn40Sb2	185–231	39.5 to 40.5	Rem	2.0 to 2.4	0.005	0.001	0.001	0.25	0.03	0.02	0.08	0.08	C
	15	S-Pb69Sn30Sb1	185–250	29.5 to 30.5	Rem	0.5 to 1.8	0.005	0.001	0.001	0.25	0.03	0.02	0.08	0.08	L
	16	S-Pb74Sn25Sb1	185–263	24.5 to 25.5	Rem	0.5 to 2.0	0.005	0.001	0.001	0.25	0.03	0.02	0.08	0.08	–*
	17	S-Pb78Sn20Sb2	185–270	19.5 to 20.5	Rem	0.5 to 3.0	0.005	0.001	0.001	0.25	0.03	0.02	0.08	0.08	N

*No related alloy.
† Key to symbols: Sn = tin, Pb = lead, Sb = antimony, Cd = cadmium, Zn = zinc, Al = aluminium, Bi = bismuth, As = arsenic, Fe = iron, Cu = copper.
Notes: All single-figure limits are maxima. Elements shown as Rem (remainder) are calculated as differences from 100%. The temperatures given in column 4 are for information only; they are not part of the alloy specification.
Source: Based on BS EN 29543

Table 7.2 Remarks about BS 219 solders

BS 219 solder	Remarks
A, AP	Free-running solder ideal for soldering electronic and instrument assemblies. Commonly known as *electrician's solder*
K, KP	Used for high-class tinsmith's work. Known as *tinman's solder*
F	Used for general soldering work in coppersmithing and sheet metalwork
G	This is supplied in strip form with a D cross-section 0.3 mm wide. Known as *blowpipe solder*
J	Its wide melting range allows it to become 'pasty', so it can be moulded and wiped. Known as *plumber's solder*

Figure 7.17 shows the essential functions of a soldering flux. Fluxes may be classified as active or passive.

Fig. 7.17 *The essential functions of a soldering flux. (Reproduced courtesy Tin Research Institute)*

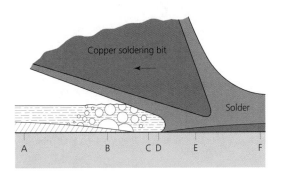

7.5.1 *Active fluxes*

Active fluxes quickly remove any residual oxide film on a metal as well as protecting the deoxidised surface from the atmosphere. Unfortunately, all active fluxes leave a corrosive residue after soldering. This residue absorbs moisture from the atmosphere and corrosion sets in. Therefore, if an active flux is used, the residue must be carefully washed off and the work dried after soldering is complete. Typical active fluxes are Baker's fluid and Fluxite. Baker's fluid is made from zinc dissolved in an excess of hydrochloric acid together with some ammonium chloride, and Fluxite is an acidified resin paste.

Table 7.3 *Alloy cross-references between BS 219 and BS EN 29453*

Solder type	BS 219 alloy code	Nearest related alloy in BS En 29453	
		Alloy designation	Alloy number
Tin	9985	–*	–
	9975	–*	–
Tin–lead and	A	S-Sn63Pb37Sb	11
tin–lead–antimony	AP	S-Sn63Pb37/S-Sn63Pb37E	1/1a
	K	S-Sn60Pb40Sb	12
	KP	S-Sn60Pb40/S-Sn60Pb40E	2/2a
	F	S-Pb50Sn50Sb	13
	R	S-Pb55Sn45	4
	G	S-Pb60Sn40	5
	H	S-Pb65Sn35	6
	J	S-Pb70Sn30	7
	V	No related alloy	–
	W	No related alloy	–
	B	No related alloy	–
	M	No related alloy	–
	C	S-Pb58Sn40Sb2	14
	L	S-Pb69Sn30Sb1	15
	D	S-Pb69Sn30Sb1	15
	N	S-Pb78Sn30Sb2	17
	No related alloy	S-Pb50Sn50	3
	No related alloy	S-Pb50Sn50E	3a
	No related alloy	S-Pb90Sn10	8
	No related alloy	S-Pb92Sn8	9
	No related alloy	S-Pb98Sn2	10
	No related alloy	S-Pb74Sn25Sb1	16
Tin–antimony	95A	S-Sn95Sb5	18
Tin–silver	96S	S-Sn96Ag4	28
	97S	S-Sn97Ag3 (equivalent)	29 (equivalent)
	98S	No related alloy	–
Tin–copper	99C	S-Sn99Cu1	23
	No related alloy	S-Sn97Cu3	24
Tin–indium	No related alloy	S-Sn50In50	27
Tin–lead–silver	5S	S-Pb93Sn5Ag2	34
	62S	S-Sn62Pb36Ag2	30
	No related alloy	S-Sn60Pb36Ag4	31
Tin–lead–copper	No related alloy	S-Sn60Pb38Cu2	25
	No related alloy	S-Sn50Pb49Cu1	26
Tin–lead–bismuth	No related alloy	S-Sn60Pb38Bi2	19
	No related alloy	S-Pb49Sn48Bi3	20
Tin–lead–cadmium	T	S-Sn50Pb32Cd18	22
Bismuth–tin	No related alloy	S-Bi57Sn43	21
Lead–silver	No related alloy	S-Pb98Ag2	32
	No related alloy	S-Pb95Ag5	33

*Tin grades 9985 and 9975 are specified in BS 3252: 1986 *Specification for ingot tin*. BS 3252 is to be superseded by BS EN 610, in which the related grade to 9985 will be Sn 99.85. There will then be no related grade to 9975.

7.5.2 *Passive fluxes*

Passive fluxes are used for soldering joints and connections in electrical and electronic equipment where the acid residue of an active flux could not be tolerated and where washing to remove the residue is not possible. The most widely used passive flux is resin, and solders for electrical purposes are often resin cored for convenience. Although the resin protects the joint surface from oxidation, it does not clean the surfaces like an active flux. Therefore greater care in cleaning the joint surfaces, immediately prior to soldering, is necessary when using passive fluxes.

7.5.3 *Soft soldering*

Figure 7.18 shows the steps in making a simple soft-soldered joint. Before making the joint, the soldering iron has to be heated up and tinned. The hot copper bit is cleaned with a file, flux is applied to prevent the oxide film reforming, and the cleaned bit is then loaded with solder. The metal being joined is then fluxed and tinned. The tinned surfaces are brought into contact and held in place whilst the metal itself is raised to the soldering temperature by the soldering iron. If the parent metal is not at the melting temperature of the solder, a sound joint will not be made. For this reason the metal being joined must be supported on a thermally non-conductive surface such as wood.

Fig. 7.18 *Soft-soldered joint: (a) tinning the metal surface; (b) adding solder to fill the joint; the metal to be soldered is supported on a wooden block (heat insulator) to prevent unnecessary heat loss by conduction*

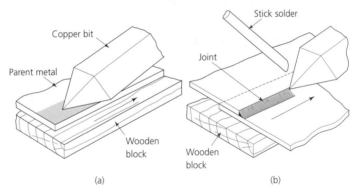

7.6 Hard-soldered joints

Hard soldering is a general term used to cover brazing and silver soldering. In these processes, as in soft soldering, melting or fusion of the metals being joined does not take place. The filler material has a lower melting point than the metals being joined.

Hard solders do not posses the low melting points of the soft solders but they produce very much stronger joints. For example, traditional bicycle frames are made from alloy steel tubes hard soldered into malleable iron brackets. The hard solder used in this instance is a brass alloy, so the process is called *brazing*. The melting temperature of the brazing alloy is 850 °C, much higher than any soft solder. Brazing can be defined as a process of joining metals in which the molten filler is drawn by capillary action into the space between closely adjacent surfaces of the parts to be joined. The filler materials used for hard soldering can be classified into three main types, all of which may be found in BS 1845.

7.6.1 *Silver solders*

Silver solders are alloys containing a high percentage of the precious metal silver. Therefore they are expensive and are only used for fine work. They produce strong, ductile joints at a temperature somewhere between the soft solders and the brazing alloys. Common borax brazing fluxes are not suitable for silver soldering; use a proprietary flux matched to the solder alloy.

7.6.2 *Brazing alloys containing phosphorus*

Usually called self-fluxing alloys, typical filler materials contain silver, copper and phosphorus (lower-melting alloy) or just copper and phosphorus (higher-melting alloy). Their oustanding feature is an ability to form high-quality joints in air without the use of a flux. The products of oxidation react with the phosphorus to form a fluid compound which acts as an efficient flux. These brazing alloys are used for production brazing in furnace and resistance brazing operations. Care must be taken when using them as they have relatively high melting temperatures and, in many applications, there is little difference between the alloy melting temperature and the melting temperatures of the parent metals.

7.6.3 *Brazing brasses*

The oldest and most familiar brazing alloys are the copper–zinc (brass) alloys known as spelters; a typical alloy is 50 per cent copper and 50 per cent zinc. Although hard and brittle, spelters have a shorter and lower melting range than other non-ferrous alloys, alloys which might form the parent metals. Some of the zinc content vaporises during brazing; using a 50/50 alloy the composition of the finished joint is nearer 60 per cent copper and 40 per cent zinc, which is stronger and less brittle. Borax mixed into a paste with water is a suitable flux when brazing with brass alloys.

The metals which can be silver soldered and brazed include:

- Copper and copper-based alloys.
- Stainless steels and irons.
- Nickel-based alloys.
- Plain carbon and alloy steels.
- Malleable cast irons.
- Aluminium alloys (Section 7.7).

7.6.4 *Flame brazing*

Flame brazing can be used to fabricate almost any assembly using a gas–air blowpipe with the air being supplied under pressure. For fine work and also the higher-melting alloys, an oxypropane flame is used as this gives a higher temperature and closer control. No matter what method of heating is used, to successfully braze or silver solder a joint, the parent metal must be raised to a sufficiently high temperature that the solder or spelter will melt on contact with the parent metal whilst the heat source is temporarily removed.

7.6.5 *Furnace brazing*

Furnace brazing is used extensively for quantity production when:

- The parts to be brazed can be preassembled or jigged to hold them in position.
- The brazing alloy (spelter) can be preformed and preplaced as shown in Fig. 7.19(a).
- A controlled atmosphere is required.

The components to be brazed are assembled together with the brazing alloy in the required position and fluxed if necessary. Preplaced brazing alloy inserts are available in a variety of forms such as wire rings, bent wire shapes, washers and foils.

The method of heating varies according to the application; muffle furnaces are generally used because they prevent the flame from directly impinging on the parts being brazed. Furnaces may be heated by gas, oil or electricity, and they are classified into two types.

- Batch furnaces with either air or a controlled atmosphere.
- Conveyor furnaces with a controlled atmosphere.

Fig. 7.19 *Furnace brazing: (a) preplaced brazing alloy (spelter); (b) schematic layout of batch brazing in a sealed container; (c) schematic layout of continuous brazing furnace*

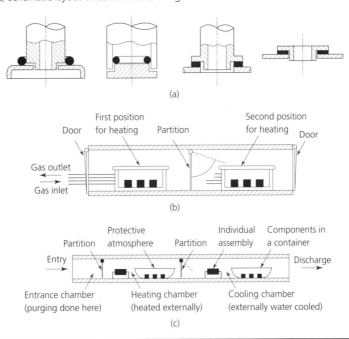

Schematic layouts of brazing furnaces are shown in Fig. 7.19(b). Individual assemblies or components for batch brazing are loaded into the furnace in sealed containers to prevent contamination by the flue gases. After brazing, the sealed containers are removed from the furnace and the components are cooled whilst they are still inside. For continuous brazing the components are mounted on trays and passed through the furnace on a conveyor. The furnace is externally heated so that it can have a controlled atmosphere.

7.6.6 *Dip brazing*

Molten spelter bath

Instead of placing the assembled components into a muffle furnace together with preformed brazing alloy, they are fluxed then dipped into a crucible (bath) of molten spelter. Larger components need to be preheated to avoid cooling the bath.

Molten flux bath

The components are assembled together with the preformed and preplaced spelter. The components and the spelter must be self-locating so they do not come apart in the bath. The molten flux raises the components to the required brazing temperature. This process is suitable for brazing aluminium alloy components. Dip brazing techniques are shown in Fig. 7.20; the work must be dried by preheating to prevent any moisture from causing an explosion.

Fig. 7.20   Dip brazing: (a) the bath is heated externally; (b) baths are usually fitted with an insulated lid or cover to prevent heat loss; (c) the brazing alloy is a preplaced insert and the assemblies are generally self-locating; components for dip brazing can be assembled and retained in position without the use of complicated jigs; parts to be brazed must be dry; immersion of wet parts can produce a violent explosion

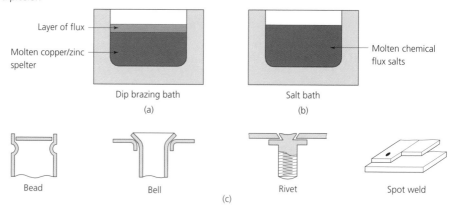

7.6.7 *Induction heating*

Induction heating is used extensively on larger assemblies that are self-locating; the principle is shown in Fig. 7.21. A high-frequency alternating current is passed through a copper induction coil and eddy currents are induced in the work; these eddy currents cause the heating. Since the heat is generated within the work, any heating is rapid and can be closely controlled. A paste flux is used and the brazing alloy is preplaced.

Fig. 7.21 *Electric induction brazing: (a) external coil, more usual but internal coils can be used for certain applications; (b) internal coil, silver solders are used extensively*

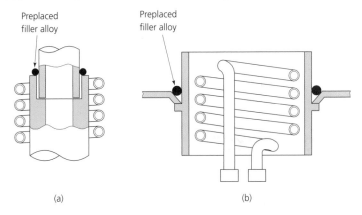

7.6.8 *Resistance heating*

The principle of resistance heating is shown in Fig. 7.22. An electric current is passed through the joint; the joint offers a resistance to the current flow; heat is generated and this raises the temperature of the joint to the appropriate brazing temperature. Heating can be precisely localised and there is no loss of mechanical properties in the parent metal.

Fig. 7.22 *Resistance heating: (a) direct heating; (b) indirect heating; both techniques apply pressure at the brazing temperature; special machines are used which are very similar in operation to spot welders, except the electrodes are usually made of carbon, molybdenum, tungsten or steel*

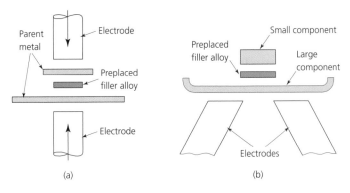

7.7 Aluminium brazing

There is a distinction between the brazing of aluminium and the brazing of other metals. For aluminium and its alloys, the filler material is an aluminium alloy whose melting temperature is lower than that of the parent metal. Conventional borax-based fluxes are unsuitable when brazing aluminium and aluminium alloys, but there are proprietary fluxes, basically mixtures of the alkali metal chlorides and fluorides. For example, a standard aluminium brazing flux contains chlorides of potassium, sodium and lithium. Extreme care must be taken when aluminium brazing as there is only a small margin of temperature which permits the joint to be made without the parent metal melting and collapsing.

7.8 Fusion welding

In the soldering and brazing processes described so far, the joints are formed by a thin film of metal having a lower melting point than the parent metals. In fusion welding any additional metal (filler metal) added to the joint has a similar composition, strength and melting temperature as the parent metal. Figure 7.23 shows the principle of joining two pieces of metal by fusion welding, where not only the filler metal is melted but also the edges of the components being joined. The molten metals fuse together and, when solid, form a homogeneous joint whose strength is equal to the parent metal.

Fig. 7.23 *Fusion welding: (a) before welding – a single V-butt requires extra metal; (b) after welding – the edges of the V are melted and fused with the molten filler metal*

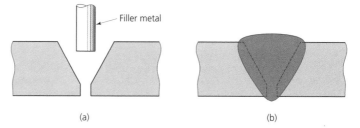

(a)   (b)

7.8.1 *Oxyacetylene welding*

The heat source for oxyacetylene welding is a mixture of oxygen and acetylene gases burning to produce a flame whose temperature can reach 3250 °C, this is above the melting point of most metals. Figure 7.24 shows a typical set of gas welding equipment. Since the gases are stored in the cylinders under very high pressures and form highly flammable and even explosive mixtures, the equipment must be handled with very great care. This equipment must be used only by persons who have been fully trained in the operating and safety procedures recommended by the Health and Safety Executive and the equipment suppliers.

Fig. 7.24 *Oxyacetylene welding equipment*

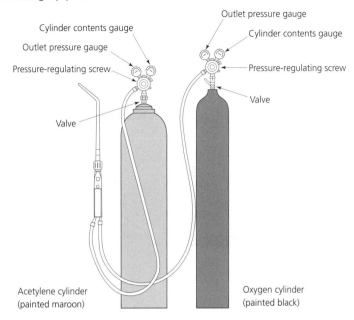

7.8.2 *Metal-arc welding*

Metal-arc welding is a fusion welding process where the energy required to melt the edges of the components and the filler material is provided by an electric arc. An arc is the name given to the prolonged spark struck between two electrodes. In this process the filler rod forms one electrode and the work forms the other electrode. The filler rod is coated with a flux that shields the joint from attack by atmospheric oxygen at the very high temperatures involved, and it stabilises the arc so that alternating current can be used (the average arc temperature is about 6000 °C). Figure 7.25 shows the general arrangement of a manual metal-arc welding installation.

Fig. 7.25 *Manual metal-arc welding equipment*

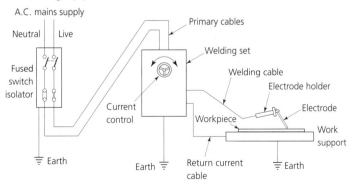

A transformer is used to reduce the mains voltage to a safe low-voltage high-current supply suitable for welding. As with gas welding equipment, arc welding equipment must not be used by unskilled persons except under the closest supervision during training. Appropriate protective clothing must be worn, especially goggles or visors with the correct filter glasses. The gas and arc welding processes are compared in Fig. 7.26.

Fig. 7.26 *Comparison of oxyacetylene and metal-arc welding: (a) oxyacetylene welding; (b) metal-arc welding*

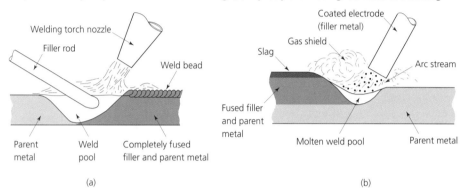

7.9 Resistance welding

The resistance welding process depends upon pressure as well as temperature to achieve the joining of the parent metals; fusion does not occur. The parent metal is raised to just below its melting point and the weld is completed by the application of high local pressure as in forge welding. The principle of this process is shown in Fig. 7.27. The temperature of the components to be joined is raised locally by the passage of a high current at low voltage through the components; resistance to its passage causes intense local heating. The cycle of events is always controlled automatically to ensure uniform quality:

- *Squeeze time* The time between the first application of pressure and the moment when the welding current begins to flow.
- *Weld time* The time for which the welding current flows whilst the pressure is maintained.
- *Hold time* The time for which the joint is maintained under pressure after the current has ceased to flow.
- *Off time* The time for which the electrodes are open to allow loading and unloading of the work.

No additional filler metal is required to produce a resistance weld. Unlike fusion welding, where the joint metal is in the as-cast condition, the metal in a resistance weld is in the wrought condition. Therefore resistance welds are surprisingly strong.

Fig. 7.27 *Principles of resistance heating: (a) electric spot welding machine (schematic); (b) spot welding; (c) seam welding; (d) projection welding*

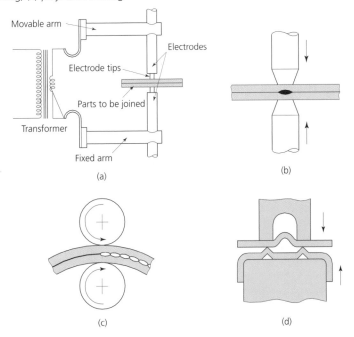

(a)

(b)

(c)

(d)

7.9.1 *Spot welding*

Spot welding is shown in Fig. 7.27(a) and (b). It is the most common of the resistance welding processes and is much quicker and neater than riveting, but it can only be applied to sheet metal. The joint is produced by making a row of individual spot welds side by side at regular intervals. These joints are not fluid-tight and they require sealing to prevent leakage or corrosion. Apart from ensuring the joint surfaces are clean, no special preparation is required. Spot welding is widely used for joining the body panels of motor vehicles.

7.9.2 *Seam welding*

The components to be joined are clamped between revolving circular electrodes, as shown in Fig. 7.27(c). The current is applied in pulses to create a seam of overlapping spot welds. Seam welding can be used to manufacture fluid-tight containers and fuel tanks.

7.9.3 *Projection welding*

In projection welding, the electrodes act as locations for holding the parts to be joined. The joint is designed so that projections are preformed on one of the parts, as shown in Fig. 7.27(d). Projection welding enables the welding pressure and heated welding zone to be localised at predetermined points.

7.9.4 *Butt-welding*

The resistance welds described so far are lap joints between sheet-metal components. Butt-welding (Fig. 7.28) is used for connecting more solid sections. For example, the expensive high-speed steel bodies of the larger twist drills can be resistance butt-welded onto low-cost shanks of medium-carbon steel. And the solid high-speed steel tips of some lathe tools are treated in a similar way.

To ensure a sound weld, the temperature is raised slightly above the value suggested by theory. Any molten metal which may occur at the joint interface is displaced by the pressure in a shower of sparks until layers of metal are created at the correct welding temperature and a weld is achieved.

Fig. 7.28 *Resistance butt-welding*

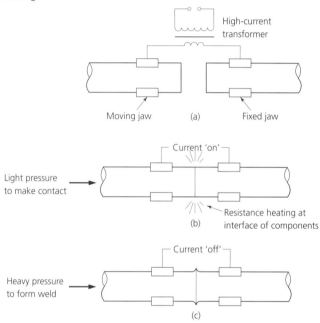

7.10 Adhesive bonding

Adhesive bonding, with modern synthetic adhesives, is widely used throughout industry for joining a wide range of materials in a variety of combinations. The strength of adhesive-bonded joints is now so high that adhesive bonding has replaced welding and riveting for some quite highly stressed applications. Adhesives can be used for joining metals to metals, metals to non-metals, and non-metals to non-metals with equal facility.

Figure 7.29(a) shows a typical adhesive-bonded joint and explains the terminology used to describe its features. The strength of the joint depends upon two factors:

- *Adhesion* The ability of the bonding material (the adhesive) to stick (adhere) to the materials being joined (the adherends). The two ways in which this bond can occur are shown in Fig. 7.29(b).
- *Cohesion* The ability of the adhesive to resist the applied forces within itself. Two ways in which cohesive failure can occur are shown in Fig. 7.29(c) and (d). Adhesive failure is shown in Fig. 7.29(e).

Fig. 7.29 *Adhesive bonding: (a) elements of the bonded joint; (b) types of bond; (c) cohesive failure of the adherend, adhesive too strong; (d) cohesive failure of the adhesive, adhesive too weak; (e) adhesive failure, inadequate preparation of the joint faces has produced a poor bond*

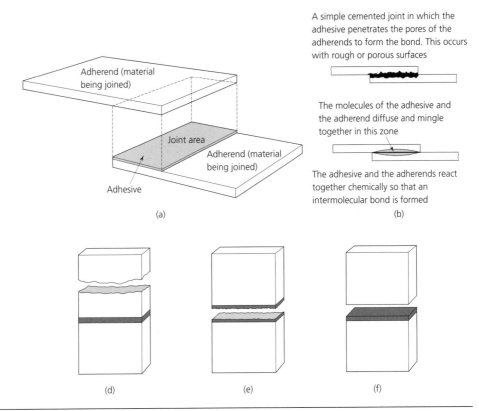

A simple cemented joint in which the adhesive penetrates the pores of the adherends to form the bond. This occurs with rough or porous surfaces

The molecules of the adhesive and the adherend diffuse and mingle together in this zone

The adhesive and the adherends react together chemically so that an intermolecular bond is formed

Adherend (material being joined)

Joint area

Adherend (material being joined)

Adhesive

(a)

(b)

(d)

(e)

(f)

Table 7.4 lists some of the more important advantages and limitations of adhesive bonding as compared with the mechanical and thermal jointing processes discussed earlier.

Table 7.4 *Advantages and limitations of bonded joints*	
Advantages	Limitations
1. The ability to join dissimilar materials, and materials of widely different thicknesses	1. The process is more complex than for mechanical and thermal bonding; it includes surface preparation, temperature and humidity control of the working atmosphere, ventilation and other precautions against health problems caused by the adhesives and their solvents. The assembly must be jigged up for some time whilst setting (curing) takes place
2. The ability to join components of difficult shape that would restrict the application of welding or riveting equipment	
3. Smooth finish to the joint which will be free from voids and protrusions such as weld beads, rivet and bolt heads, etc.	
4. Uniform distribution of stress over entire area of joint. This reduces the chances of the joint failing in fatigue	2. Inspection of the joint is difficult
5. Elastic properties of many adhesives allow for flexibility in the joint and give it vibration-damping characteristics	3. Joint design is more critical than for many mechanical and therm processes
6. The ability to electrically insulate the adherends and prevent corrosion due to galvanic action between dissimilar metals	4. Incompatibility with the adherends; the adhesive itself may corrode the materials it is joining
7. The joint will be sealed against moisture and gases	5. Degradation of the joint when subject to high and low temperatures, chemical atmospheres, etc.
8. Heat-sensitive materials can be joined	6. Creep under sustained loads

No matter how effective the adhesive and how careful its application, the joint will be a failure if it is not correctly designed and executed. It is bad practice to apply adhesive to a joint which was originally proportioned for bolting, riveting or welding. The joint must be proportioned to exploit the properties of adhesives. Most adhesive-bonded joints are strong in tension and shear (providing the joint area is adequate) but weak in cleavage and peel; these terms are explained in Fig. 7.30. The success of an adhesive-bonded joint depends upon the joint faces being carefully prepared so they are physically and chemically clean, free from grease and, where appropriate, roughened to form a key by wire brushing, vapour or shot blasting or chemical etching. The adhesive must 'wet' the joint surfaces thoroughly, otherwise voids will occur and the effective joint area will be less than the designed area. This will weaken the joint considerably.

7.10.1 *Thermoplastic adhesives*

Thermoplastic adhesives are based upon synthetic materials such as polystyrene, polyamides, vinyl and acrylic polymers and cellulose derivatives. They are also based upon

Fig. 7.30 *Stressing of bonded joints: (a) tension; (b) cleavage; (c) shear; (d) peel*

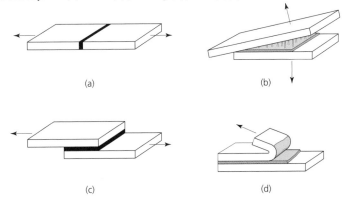

(a)

(b)

(c)

(d)

naturally occurring materials such as resin, shellac, mineral waxes and rubber. Thermoplastic adhesives are not as strong as thermosetting plastic adhesives but, being more flexible, they are more suitable for joining non-rigid materials.

- *Heat-activated* The adhesive is softened when heated and, when sufficiently fluid, is spread over the whole joint area. The materials to be joined are brought into contact and the adhesive adheres to them. When cooled to room temperature the adhesive sets and a bond is achieved.
- *Solvent-activated* The adhesive is softened by a suitable solvent and a bond is achieved by the solvent evaporating. Because evaporation is essential to the setting of the adhesive, a sound bond is almost impossible to achieve at the centre of a large joint area, particularly when joining non-absorbent and non-porous materials.
- *Impact adhesives* These are solvent-activated adhesives which are spread separately on the two joint faces then left to dry by evaporation. When dry, the treated joint faces are brought together, whereupon they instantly bond by intermolecular attraction. This enables non-absorbent and non-porous materials to be joined, even when there is a large joint area.

7.10.2 *Thermosetting adhesives*

Thermosetting adhesives depend upon heat to cause a non-reversible chemical reaction (curing) that makes them become set. Once cured they cannot be softened again by reheating. This makes the strength of the joint less temperature sensitive than when using thermoplastic adhesives.

The heat required to cure the adhesive can be applied externally by heating the assembly and adhesive in an autoclave; this method is used to cure phenolic resins. Or the heat can be generated internally by adding a chemical hardener; this method is used to cure epoxy resins. Since the setting process is a chemical reaction, independent of evaporation, the area of the joint can be as large as necessary to give the appropriate strength.

Thermosetting adhesives are very strong and can even be used for making structural joints in high-strength materials such as metals. The body shells of cars and stressed members of aircraft increasingly depend on thermosetting adhesives for their joints in place of spot welding and riveting. The stresses are more uniformly distributed and the joints are sealed against corrosion. And the relatively small temperature rise that occurs on curing does not affect the crystallographic structure of the metal. Thermosetting adhesives tend to be rigid when cured, so they are not suitable for joining flexible (non-rigid) materials.

When working with adhesives, care must be taken as the solvents and their fumes are both toxic and flammable. The working area must be declared a no-smoking zone.

7.11 The manipulation and fabrication of sheet plastic

7.11.1 *Heat bending*

Simple, straight bends in thermoplastic materials follow the techniques used for sheet metalworking. The only difference is that the plastic material needs to be heated before bending, and bending must take place whilst the plastic sheet is still hot. For this reason the bending jig must be faced with materials having a low thermal conductivity, e.g. wood or Tufnol.

Electric strip heaters may be used to ensure that heating is localised along the line of the bend. This makes the plastic sheet easier to handle. To avoid loss of shape and degradation of the plastic, rapid cooling is required immediately after bending is complete. Before bending thick sheet, it should be heated on both sides because the plastic materials have low thermal conductivity.

Other heat bending techniques such as vacuum forming, blow forming and pressing are outlined in Fig. 7.31. In all these examples the thermoplastic sheet is preheated before forming.

Fig. 7.31 *Forming of plastic sheet materials: (a) vacuum forming; (b) blow forming; (c) simple pressing*

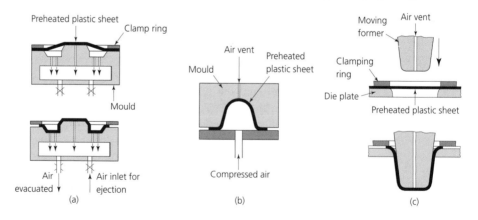

7.11.2 *Heat welding*

Heat welding can join only thermoplastic materials since it is only thermoplastics that soften upon heating. The temperature of an oxyfuel gas welding torch is much too high for plastic welding and would destroy the material being joined. The low thermal conductivity and softening temperatures of thermoplastic materials dictate the use of a low welding temperature so the heat can penetrate the body of the plastic before the surface degrades.

Heat is normally applied to the joint by a welding gun. Air or nitrogen gas is heated in the gun and directed as a jet of hot gas into the weld zone. Figure 7.32 shows the principle of an electrically heated welding gun. The easiest plastics to weld are polyvinyl chloride (PVC) and polyethylene (PE) as they have a wide softening range. The basic technique is to apply a jet of heated air or nitrogen that softens the edges of the parent plastic sheets. A filler rod, made of the same material as the plastic being welded, is added into the joint in much the same way as when welding metals, except that the joint edges and filler rod are softened but not melted. Some degradation inevitably occurs, so the strength of the joint is slightly below that of the surrounding material.

Fig. 7.32 *A plastic welding gun using electrically heated hot air*

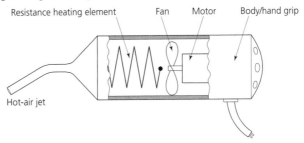

Figure 7.33 shows examples of edge preparation for a range of joints in thermoplastic materials. A small root gap should be provided, no feather edges should be left, and the weld bead (reinforcement) should not be removed as it can increase the joint strength by up to 20 per cent. The technique of welding sheet plastic is shown in Fig. 7.34.

Fig. 7.33 *Edge preparation: (a) single-V butt; (b) double-V butt; (c) corner weld; (d) edge weld; (e) fillet weld; (f) lap weld; (g) how to avoid a feather edge; (g) reinforcement*

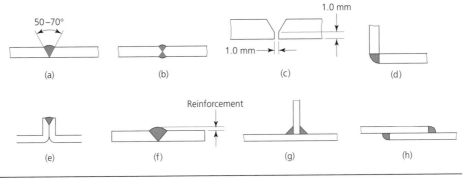

Fig. 7.34 *Welding sheet plastic*

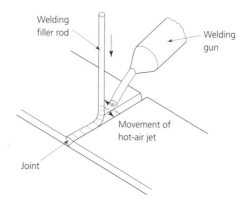

7.11.3 *Solvent welding*

Solvent welding can only be applied to thermoplastic materials since cured thermosets cannot be softened by solvents. When solvent welding, the edges or surfaces being joined are softened by a suitable solvent cement instead of by heat. The surfaces are pressed together after application of the solvent until evaporation is complete. The solvent has to be chosen to suit the material being joined. Often the solvent contains a small quantity of the parent material already dissolved in it to give the cement *gap-filling* properties. Volatile solvents are required to ensure rapid evaporation; many of these solvents give off flammable and toxic fumes, so great care is required in their use.

7.12 Miscellaneous plastic welding techniques

7.12.1 *Friction welding*

When two surfaces are rubbed together without a lubricant, the friction between them converts mechanical energy into heat energy at the interface. This effect is exploited in friction welding (also called spin welding). The components to be joined are pressed lightly together, one component rotates and one remains stationary. To avoid using excessively high speeds when the diameters are small, both components can be rotated but in opposite directions. As soon as the welding temperature has been reached, rotation ceases and the axial pressure is increased to complete the weld. The principle of friction welding is shown in Fig. 7.35.

Friction welding has one major drawback, the heating effect is not uniform across the joint face. This is because the surface speed diminishes towards the centre of the rod. At the very centre the surface speed is zero and there is no heating, so friction welding is best suited to tubular components (Fig. 7.36).

Fig. 7.35 *Principles of friction welding*

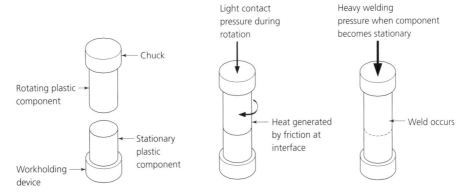

Fig. 7.36 *Types of friction weld: (a) centre-relieved; (b) spigot and register; (c) tongue and groove, for cylindrical components*

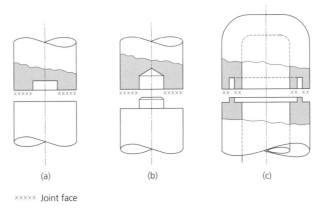

(a)   (b)   (c)

××××× Joint face

7.12.2 *Resistance (hot wire) heating*

Resistance welding is suitable for solid rods since the heating effect can be virtually uniform. A mat of resistance wire is placed in the joint (Fig. 7.37), the two components are brought together and an electric current is passed through the mat. As soon as the joint faces have softened, the current is disconnected and the welding pressure is applied. The resistance mat remains permanently in the joint and the ends of the wires are cut off flush.

7.12.3 *Induction heating*

Instead of a resistance mat, induction heating uses a ring of metal foil. The plastic components, complete with the foil, are passed through an induction coil. The induction coil acts as the primary winding of a transformer and the foil disc acts as the secondary winding of a transformer. An electric current is induced in the foil disc and it becomes hot, softening the plastic at the joint face. Pressure is applied to complete the joint (Fig. 7.38). The advantage of induction heating is that no connections have to be made and no trimming is required after the joint is complete; it lends itself to quantity production.

Fig. 7.37 *Resistance welding: (a) stage 1; (b) stage 2; (c) stage 3*

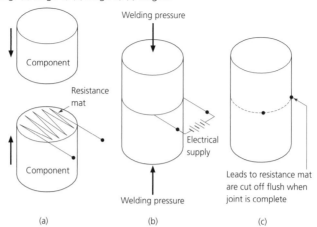

(a)   (b)   (c)

Fig. 7.38 *Induction welding: the electric current induced in the foil ring causes it to heat up; the hot ring softens the plastic joint face*

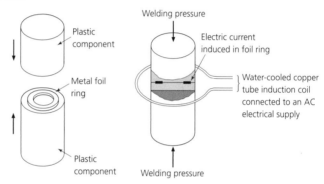

7.12.4 *Dielectric welding*

Dielectric welding exploits the insulating properties of plastic materials. The plastic to be joined forms the dielectric of a capacitor by being placed between two electrodes. The electrodes are connected to a high-frequency alternating current generator operating at about 30 MHz. The dielectric losses that occur in some thermoplastics at this frequency cause internal heating in the zone between the electrodes. The joint is completed by applying a pressure once the welding temperature has been reached. Dielectric welding overcomes the degrading effects of externally heating plastic materials in the presence of atmospheric oxygen. The principle of this process is shown in Fig. 7.39. The following plastic materials cannot be welded by dielectric heating:

- Polyethylene (PE).
- Polypropylene (PP).
- Polycarbonate (PC).
- Polytetrafluoroethylene (PTFE, Teflon).
- Polystyrene (PS).

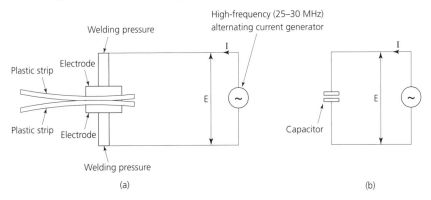

(a)

(b)

7.12.5 *Ultrasonic welding*

This process uses acoustic energy at a frequency of 20 kHz or higher, beyond the range of normal human hearing, hence the name ultrasound or ultrasonic. Figure 7.40 shows the principle of an ultrasonic welding tool. *Magnetostriction* is the property of the core material to expand and contract in sympathy with the electromagnetic field produced in the solenoid (coil) by the high-frequency alternating current.

To concentrate the heating effect of the high-speed vibration, the weld zone must be prepared so that the joint contains conical *energy directors* (Fig. 7.41). These directors localise the heat and limit the required energy. The small volume of softened plastic spreads out under pressure to complete the joint. The generator is switched off and the pressure is held on for about 1 s for the joint to set. This process can be applied to most thermoplastics, except for the vinyls and the cellulosics.

Besides welded joints, ultrasonics can also join plastics to non-plastics using the process of *staking*, as shown in Fig. 7.42(a). It can also be used for inserting metal parts into plastic components, as shown in Fig. 7.42(b).

Fig. 7.40 *Ultrasonic plastic welding tool*

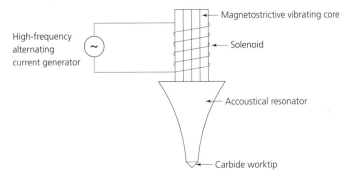

Fig. 7.41 *Weld preparation: (a) butt-weld, prepared; (b) butt-weld, joint complete; (c) corner joint; (d) combined tongue-and-groove joint/spigot-and-register joint*

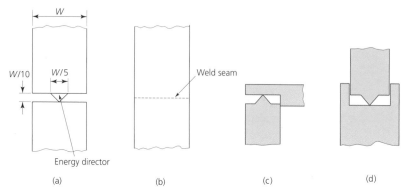

(a)   (b)   (c)   (d)

Fig. 7.42 *Further applications of ultrasonics: (a) ultrasonic staking; (b) metal insert fixing*

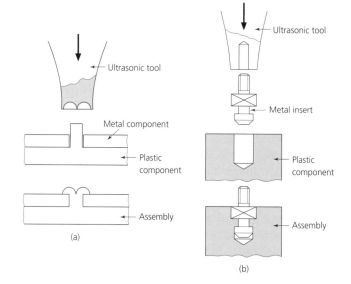

(a)   (b)

7.1 With the aid of sketches, describe suitable applications for the following forming equipment:
(a) folding machine
(b) press brake
(c) bending rolls

7.2 With the aid of sketches, describe **three** essential steps to achieve a satisfactory joint when using screwed fastenings to secure rolled-steel sections (RSJ or BSB).

7.3 Figure 7.43 shows a section through a typical riveted joint using a snap head rivet. Calculate the hole diameter and the shank length of the rivet to ensure a satisfactory joint.

Fig. 7.43 *Exercise 7.3*

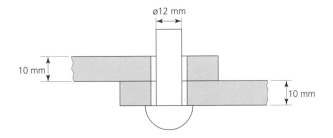

7.4 (a) When soft soldering, describe what precautions must be taken to ensure a satisfactory joint.
(b) Explain why a flux must be used when soft soldering. Explain the difference between a *passive* flux and an *active* flux. Under what circumstances each would be used?

7.5 (a) Describe the essential difference between hard soldering (brazing) and soft soldering, and describe the four basic conditions upon which a successful brazed joint depends.
(b) With the aid of sketches, explain how a collar may be brazed onto the end of a shaft using preformed spelter and electric induction heating.

7.6 With the aid of sketches, describe the essential principles of the following joining processes:
(a) oxyacetylene welding
(b) manual metal-arc welding
(c) spot welding

7.7 Describe in detail the safety precautions which must be taken when:
(a) oxyacetylene welding
(b) manual metal-arc welding

7.8 (a) List the advantages and limitations of adhesive bonding for metal components compared with soldering, brazing and welding.
   (b) With the aid of sketches, explain how bonded joints may fail.
   (c) Describe the precautions which must be taken to ensure a satisfactory bonded joint.

7.9 With the aid of sketches, describe how plastic materials may be joined by:
   (a) heat welding
   (b) solvent welding
   (c) adhesive bonding
   (d) ultrasonic welding

7.10 With the aid of sketches, describe how sheet plastic materials may be fabricated by heat bending, vacuum forming, blow forming and pressing techniques.

8 Heat treatment processes

The topic areas covered in this chapter are:

- Annealing of plain carbon steels.
- Annealing of non-ferrous metals.
- Solution and precipitation treatment of aluminium alloys.
- Normalising.
- Quench (through) hardening.
- Case-hardening.
- Tempering.
- Heat treatment furnaces.
- Temperature measurement.
- Quenching and quenching media.

8.1 Heat treatment

Heat treatment processes are used to modify the properties of materials by controlled heating and cooling cycles. The process chosen depends upon the material specification and the properties required. The temperature to which the material is heated depends upon certain critical temperatures at which chemical and physical changes take place within the material in the solid condition, and also the processing (e.g. cold rolling) the material has received before heat treatment.

8.2 Recrystallisation

During the cold-working processes, described earlier in this book, the grain of the metal becomes distorted. This distortion of the grain causes the metal to become work hardened and lacking in ductility. To prevent the metal cracking, its grain structure and ductility have to be wholly or partially restored. Internal stresses exist within the distorted crystals, and if the temperature of the metal is raised sufficiently, seed crystals will form along the grain boundaries at points of high internal stress. The minimum temperature at which this occurs is called the *temperature of recrystallisation*. The more severe the cold working, the greater the internal stress in the metal and the lower the temperature of recrystallisation. Continued heating at this temperature, or above, causes the seed crystals to grow as they feed off the work-hardened crystals. Eventually the distorted, work-hardened crystals are completely consumed and the structure once more consists of ductile and malleable equiaxed crystals.

8.3 Annealing: plain carbon steels

All annealing processes associated with plain carbon steels are concerned with rendering the steel soft and ductile so that it can be cold worked or machined easily. Annealing is achieved by heating the steel to the appropriate temperature for its carbon content and previous processing (Fig. 8.1(a)) then cooling it very slowly. This slow cooling is usually achieved by turning off the furnace and allowing the furnace and the work to cool down together. The annealing temperatures shown in Fig. 8.1 are related to the upper and lower critical temperatures of the steel section of the iron–carbon phase equilibrium diagram, as indicated by the lines ABC and DBE. (see *Engineering Materials*, Volume 1). Let's now consider the three basic annealing processes applicable to plain carbon steels.

8.3.1 *Stress-relief annealing*

Stress-relief annealing is also known as subcritical annealing, process annealing and interstage annealing. Carried out at subcritical temperatures (Fig. 8.1(a)), it only applies to lower-carbon steels that can be cold-worked. This is because stress-relief annealing is a recrystallisation process, so the cold working must be severe enough to trigger recrystallisation at subcritical temperatures when the steel is annealed. Stress-relief annealing aims to restore the grain structure of the steel before it became work hardened. This allows further processing to take place and also prevents the work-hardened material from cracking and failing in service.

8.3.2 *Spheroidising annealing*

When steels containing more than 0.5 per cent carbon are heated to just below the lower critical temperature (Fig. 8.1(a)), the iron carbide in the grain of the metal tends to ball up; this is called *aspheroidisation* (Fig. 8.1(d)). No chemical changes occur in the steel; aspheroidisation is simply a surface tension effect. Spheroidising annealing is usually applied to previously quench-hardened steels (Section 8.5) which already have a fine grain structure. This produces fine globules of iron carbide (cementite) after aspheroidisation. After treatment the steel can be cold worked and it will also machine freely to a good surface finish. Furthermore, steel which has been subjected to spheroidising annealing will quench harden more uniformly and with less chance of cracking and distortion. As with any other annealing process, heating is followed by very slow cooling.

Fig. 8.1 *Annealing: (a) annealing temperatures; (b) lamellar pearlite; (c) pearlite begins to ball up; (d) aspheroidisation of the pearlitic cementite (balling up) complete*

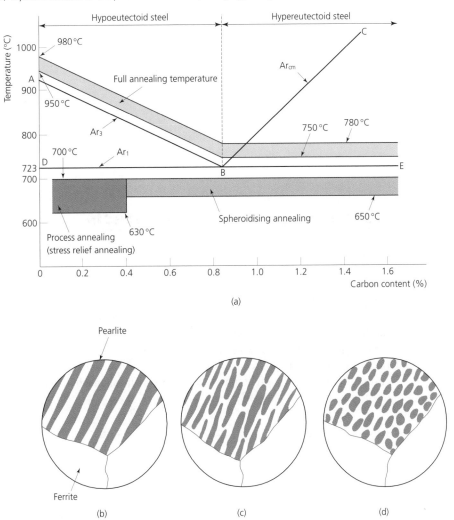

8.3.3 *Full annealing*

Steel which has been cast or forged will have cooled from very high temperatures. This leads to a coarse grain structure with a mesh-like appearance and poor mechanical properties. This mesh-like structure is referred to as a Widmanstätten structure (Fig. 8.2). To render the steel usable, it has to be reheated to approximately 50 °C above the upper critical temperature for hypoeutectoid steels, and to approximately 50 °C for hypereutectoid steels. Very slow cooling is required after heating. It is usual to leave the steel in the furnace, turn off the furnace, close the dampers to eliminate draughts and let the steel and the furnace cool down together.

Fig. 8.2 *Widmanstätten structure*

8.4 Normalising: plain carbon steels

Annealing processes, especially full annealing, are intended not only to soften the steel but to give it maximum ductility for subsequent flow-forming operations. This can lead to a poor surface finish as the metal tends to tear when machined. The normalising temperatures for steels are shown in Fig. 8.3. Cooling is much quicker for normalising than for annealing; this produces a finer grain structure in the steel, which improves its machining properties at the expense of its ductility. It is usual to remove the work from the furnace after heating, allowing it to cool down in freely circulating air, but away from draughts. Normalising is frequently carried out after rough machining of large forgings and castings. This removes any residual stresses before finish machining and avoids subsequent distortion of the finished work.

Fig. 8.3 *Normalising temperatures for plain carbon steels*

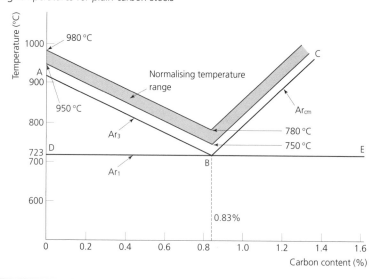

8.5 Hardening: plain carbon steels

If components made from plain carbon steels with a carbon content greater than 0.5 per cent are heated to the same temperature as for full annealing (Fig. 8.1), but instead of being cooled slowly they are quenched (cooled quickly) in water or oil, they will become hard and brittle. To reduce the brittleness, the components have to be *tempered*. This involves reheating them to the temperatures listed in Table 8.1 and quenching them again. Although they are very much less brittle after tempering, there is some slight loss of hardness.

Table 8.1 *Tempering temperatures*

Colour*	Equivalent temperature (°C)	Application
Very light straw	220	Scrapers, lathe tools for brass
Light straw	225	Turning tools, steel-engraving tools
Pale straw	230	Hammer faces, light lathe tools
Straw	235	Razors, paper cutters, steel plane blades
Dark straw	240	Milling cutters, drills, wood-engraving tools
Dark yellow	245	Boring cutters, reamers, steel-cutting chisels
Very dark yellow	250	Taps, screw-cutting dies, rock drills
Yellow–brown	255	Chasers, penknives, hardwood-cutting tools
Yellowish brown	260	Punches and dies, shear blades, snaps
Reddish brown	265	Wood-boring tools, stone-cutting tools
Brown–purple	270	Twist drills
Light purple	275	Axes, hot setts, surgical instruments
Full purple	280	cold chisels and setts
Dark purple	285	Cold chisels for cast iron
Very dark purple	290	Cold chisels for iron, needles
Full blue	295	Circular and band saws for metals, screwdrivers
Dark blue	300	Spiral springs, wood saws

*This is the colour of the oxide film that forms on heating a polished surface.

For a steel to quench harden, it has to be cooled more rapidly than its critical cooling rate. In a thick component, heat will be trapped in the centre of the component, which will cool more slowly than the outer layers in contact with the quenching bath. In practice the centre of the component may not achieve its critical cooling rate. This leads to a variation in hardness across the section of the component (Fig. 8.4), known as the *mass effect*. Plain carbon steels have a high critical cooling rate, so thick sections cannot be fully hardened throughout. However, a 3 per cent nickel steel containing only 0.3 per cent carbon will harden uniformly across its section because it has a relatively low critical cooling rate; this steel has good *hardenability*. Hardenability can be defined as the ease with which hardness is attained throughout the mass of the metal.

Fig. 8.4 *Mass effect (hardenability)*

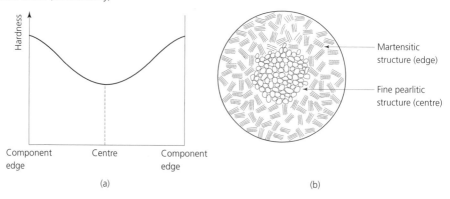

(a)

(b)

Martensitic structure (edge)

Fine pearlitic structure (centre)

Lack of uniformity in structure and hardness for steels with a poor hardenability can seriously affect their mechanical properties, so a maximum diameter or *ruling section* is specified for which the stated mechanical properties can be achieved under normal heat treatment conditions. One reason why steels contain alloying elements, such as nickel and chromium, is to reduce the mass effect and to increase the ruling section for which the required properties can be achieved.

SELF-ASSESSMENT TASK 8.1

1. You want to increase the hardness of components being made from 0.8 per cent carbon steel. Which of these options should you choose? Give the reason for your choice.

    (a) increase the temperature from which the steel is being quenched.
    (b) increase the rate of cooling.

2. Describe what is meant by the mass effect. How does it affect the design of components that are to be quench hardened throughout?

3. You need to alter a high-carbon steel cutting tool that is already in the hardened state.

    (a) Describe how you would soften the tool so that you can alter it.
    (b) Describe how you can reharden and temper it after alteration.

8.6 Case-hardening: plain carbon steels

Sometimes components require a hard, wear-resistant case and a strong tough core. These composite properties are achieved by case-hardening. The component is made from a steel that is easily processed and will provide the properties required in the core. The case-hardening process is as follows.

8.6.1 *Carburising*

The component is heated in contact with a carburising medium so that it will absorb additional carbon into its outer layers. Various materials with high carbon contents are used in the carburising process.

Solid media

Solid media may be bone charcoal or charred leather, together with an energiser such as sodium and barium carbonates. The carbonates make up to 40 per cent of the total composition. The components to be carburised are packed into fabricated steel or cast-iron boxes along with the solid carburising media in granular form. An airtight lid is sealed in place with fireclay and the boxes are heated to between 900 °C and 950 °C for up to 5 hours depending upon the depth of case required. The case depth should not exceed 2 mm as beyond that it will tend to flake off.

Fused salts

Fused salts may be sodium cyanide together with sodium carbonate and varying amounts of sodium and barium chloride. Since sodium cyanide is a deadly poison and since it represents 20–50 per cent of the furnace content, stringent safety precautions must be taken in its use. The components are suspended in the fused salts in a gas or electrically heated salt bath furnace (Section 8.11). Large components are suspended individually from a bar lying across the pot or crucible. Copper wire must not be used as it will be dissolved by the cyanide and the component will drop to the bottom of the pot. Small work is suspended in baskets made from a non-reactive, heat-resistant metal such as Inconel. The advantages of using fused salts are:

- Loading is quicker, easier and cheaper.
- Heating and carburisation are more uniform with less chance of distortion.
- The components can be hardened by quenching in water straight out of the cyanide without further heat treatment.

The cyanide 'cracks' off the surface of the components, leaving them clean. Since the quenching bath becomes contaminated with dangerous chemicals, it must be removed by a specialist waste treatment firm, not just poured down the drain.

Gaseous media

Gaseous media are increasingly used. They are based on the hydrocarbon gas methane (natural gas). Methane is usually enriched with the vapours from heated oil that has been cracked by heating it in the presence of a catalyst such as hot platinum. Gas carburising is carried out in batch and continuous furnaces. The components are heated up to between 900 °C and 950 °C in an atmosphere of enriched methane. The gaseous atmosphere is cleaned of moisture and carbon dioxide before being allowed into the furnace to avoid corrosion of the work. This atmosphere is highly flammable, and great care must be taken to avoid the build-up of an explosive mixture. Gas carburising is very clean and is used for the mass production of cases up to 1 mm deep.

8.6.2 *Heat treatment*

The carburising process does not harden the steel, it merely adds carbon to the outer layers of the metal and leaves it in a fully annealed condition with a coarse grain structure. Therefore additional heat treatment is required to harden the case and refine the grain of the case and the core. The temperatures associated with this additional heat treatment are shown in Fig. 8.5. Before carburising, the steel from which the components were made had a uniform carbon content of 0.1 per cent. After carburising, the diffusion of carbon through the component is not constant. The carbon content of the case is about 1.0 per cent and the carbon content of the core is about 0.3 per cent.

- *Refining the core* The metal is heated to a temperature of approximately 870 °C, just above the upper critical temperature for a 0.3 per cent steel (temperature 1 in Fig. 8.5). Quenching the carburised steel from this temperature will produce a fine-grained, tough core. At 0.3 per cent the carbon content of the core is too low to harden.
- *Hardening and refining the case* The core-refining temperature is sufficiently high to cause grain growth in the case, which has a carbon content of 1.0 per cent. Therefore the component is reheated to the hardening temperature for the case, which is approximately 760 °C (temperature 2). This temperature is too low to cause grain growth in the core, providing the steel is heated quickly through the range 650–760 °C and is not soaked at the hardening temperature. The component will now have a hard, fine-grained case and a tough, fine-grained core.
- *Tempering* Tempering at 200 °C is advisable to relieve any quenching stresses which may be present.

This procedure is used to give ideal results; however, in the commercial interests of speed and economy the process is often simplified to a single heating and quenching (temperature 1). This can be satisfactory where the work is lightly stressed or where alloy steels are used which harden from lower temperatures and have less critical grain growth characteristics.

Fig. 8.5 *Case-hardening temperatures*

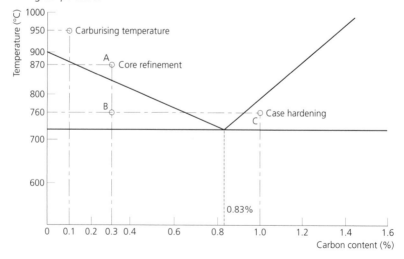

8.7 Localised case-hardening: plain carbon steels

It is not always desirable to case-harden a component all over. For example, it is undesirable to case-harden screw threads. Not only would they be extremely brittle, but any distortion occurring during heat treatment could only be corrected by expensive thread grinding. Various means are available for avoiding the local infusion of carbon during the carburising process:

- Heavily copper plating those areas to be left soft. This is suitable for pack carburising and gas carburising, but it cannot be used with salt bath carburising as the cyanide dissolves the copper.
- Encasing the areas to be left soft with fireclay.
- Leaving surplus metal on (Fig. 8.6). This is machined off between carburising and hardening, and the infused carbon is removed with the swarf. Although expensive, it is the surest way of leaving soft areas.

Fig. 8.6 *Localised case-hardening: surplus metal is left on the blank during carburising; additional carbon is then removed during screw cutting so the thread remains soft after heat treatment*

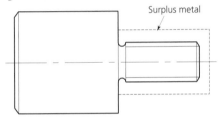

Surplus metal

Localised case-hardening can also be achieved in medium- and high-carbon steels and some cast irons by rapid local heating and quenching. Figure 8.7(a) shows the principle of flame hardening (the Shorter process). A carriage moves over the component so that the surface is rapidly heated by an oxyacetylene flame to the hardening temperature. The same carriage carries the water quenching spray. Thus the surface of the workpiece is heated and quenched before the core can rise to the hardening temperature. This process is widely used for the surface hardening of machine-tool slideways.

Figure 8.7(b) shows how the same effect can be produced by electromagnetic induction. The higher the frequency of the alternating current, the shallower the case. This is because the higher the frequency, the more superficial the eddy currents, and it is the eddy currents that cause the heating. The heating coil is often made of perforated tube with fine sprayholes so that it can be used for both heating and quenching. And the coil can be formed to suit the contour of the component being hardened, often used for the surface hardening of gear teeth.

SELF-ASSESSMENT TASK 8.2

1. Compare and contrast quench (through) hardening and case-hardening

2. Describe a process for hardening the slideways of machine tools.

Fig. 8.7 *Surface hardening: (a) flame hardening (Shorter process); (b) induction hardening*

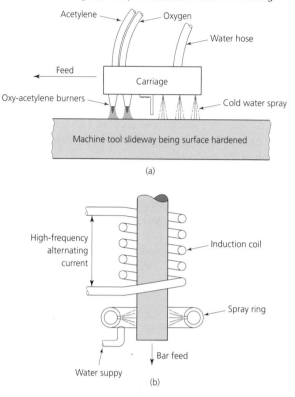

8.8 Annealing: non-ferrous metals and alloys

Unalloyed non-ferrous metals such as pure copper and pure aluminium, together with many non-ferrous alloys such as the brasses and tin bronzes, can only be hardened by severe cold working. That is, they can only be work hardened, and they do not respond to quench hardening treatments. However, they do respond to annealing processes and this is the only way in which they can be stress-relieved and softened after cold working.

8.8.1 *Recovery*

The minimum treatment for a cold-worked metal is simple stress relief. This treatment is performed at quite low temperatures. There is no recrystalisation and no change in the grain structure as a whole, but the individual atoms can move to equilibrium positions within the crystal lattice. This effect is called *recovery*. Treatment at the recovery temperature does not adversely affect the increased hardness and strength obtained from cold working, and it may even enhance them. High-ductility brass consisting of 70 per cent copper and 30 per cent zinc is particularly susceptible to cracking after severe cold working (season cracking). Stress-relief annealing at the recovery temperature of 200 °C not only prevents the metal cracking but also improves its mechanical properties (Fig. 8.8).

Fig. 8.8 *Effect of heat treatment on cold-worked 70/30 brass*

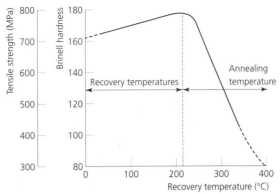

8.8.2 *Annealing*

Where a cold-worked material is to undergo further cold forming, it must first be annealed to restore its ductility. In most non-ferrous metals and alloys this is achieved by annealing at the recrystallisation temperature. The recrystallisation temperature is substantially higher than the recovery temperature; it will depend upon the composition of the alloy and the severity of any cold working before the annealing.

Alloys tend to have a higher recrystallisation temperature than pure metals. Unlike ferrous metals, non-ferrous metals and alloys can be quenched after recovery treatment or annealing since they do not quench harden. This speeds up the process, prevents excessive grain growth and improves the surface finish since the rapid contraction of the metal strips any oxide film (scale) from the surface.

Recrystallisation annealing is equally applicable to low-carbon steels and was described as subcritical annealing in Section 8.3. For steels the cooling has to be very slow. The term *subcritical annealing*, as used for low-carbon steels, is not appropriate to non-ferrous metals and alloys since their heat treatment temperatures are not directly related to the critical temperatures of their phase equilibrium diagrams. The more common terms associated with non-ferrous metals and alloys are *process annealing* and *interstage annealing*. These names are derived from the fact that annealing is carried out after cold working or between the stages of cold working.

8.9 Solution and precipitation treatment

These heat treatment processes apply to a number of non-ferrous alloys, but especially to those aluminium alloys containing copper.

8.9.1 *Solution treatment*

Although this is a process by which the alloy is softened and made more ductile ready for cold working, it is not a recrystallisation process. The alloy is raised to a temperature of

500 °C and quenched in water. The hard intermetallic compounds in the structure become a *solid solution* of copper in aluminium; this is what gives the process its name. All solid solutions tend to be soft and ductile, suitable for cold-working.

8.9.2 *Precipitation treatment*

The solid solution produced by solution treatment is a *supersaturated* solution and is therefore unstable at room temperature. On standing, particles of the copper–aluminium intermetallic $CuAl_2$ precipitate out and become finely dispersed through the whole component. These hard particles interfere with the slip planes in the grains of the metal and reduce the ability of the metal to deform, rendering it harder and more brittle. In this condition it cannot be cold worked as it tends to crack. This hardening process is called *precipitation age hardening* and is a form of particle hardening. In some alloys, such as *Duralumin*, it is so pronounced that hardening occurs naturally at room temperature. This is called *natural ageing*. Components made from these alloys, e.g. rivets, need to be solution treated immediately before processing and kept under cool (refrigerated) conditions to delay the onset of precipitation until they are required. The precipitation can be accelerated by reheating the solution-treated alloy in a furnace for several hours at 165 °C. This is known as *artificial ageing*.

SELF-ASSESSMENT TASK 8.3

1. Since brass alloys cannot be quench hardened, explain why they sometimes have to be annealed and how is this done.

2. Explain why 70/30 (α) brass alloys need to be heat treated to relieve process stresses after severe cold working. Name the process.

3. Explain briefly what is meant by solution treatment and precipitation age hardening. To which group of non-ferrous alloys do these processes refer.

8.10 Requirements of heat treatment furnaces

The successful heat treatment of metals depends upon carefully controlled heating and cooling, so a heat treatment furnace must satisfy several requirements.

8.10.1 *Uniform heating of the charge*

The components to be heat treated may be placed into the furnace singly or in batches, or passed through the furnace on some form of conveyor. However the furnace is loaded, the components are known as the *charge*. Uniform heating of the charge is necessary to prevent cracking and distortion of the component due to unequal expansion. Uniform heating is also necessary to ensure that chemical and physical changes which occur in the metal during heat treatment are uniformly distributed throughout the charge.

8.10.2 *Accurate temperature control*

Since the temperatures involved in heat treatment processes are critical, not only must heat treatment furnaces be capable of operating over a wide range of temperatures, they must be easily and accurately adjustable to the required temperature.

8.10.3 *Temperature stability*

Not only should it be accurately adjustable, but once it has been set, the furnace should remain at the required temperature. This can be achieved by ensuring that the mass of the heated furnace lining (refractory) is very much greater than the mass of the charge. Also, the furnace should have some form of automatic temperature control to measure the temperature of the furnace and control the level of energy input.

8.10.4 *Atmosphere control*

If the charge is heated in the presence of air, the surface of the metal becomes heavily oxidised, and if the oxide film becomes extremely thick, it is called a *scale*. Besides that, steels may have the carbon content of their surface layers reduced by oxidation of the carbon; this is called *decarburisation*. Both oxidation and decarburisation can be avoided by controlling the atmosphere of the furnace; air is displaced from the furnace chamber by an inert gas (a gas which does not react with the metal). Alternatively the charge may be immersed in molten salts heated to the process temperature. These salts may be chemically inert or they may react with the surface of the work in a controlled manner, as in casehardening.

The simplest form of atmosphere control is where the combustion products of the burnt fuel replace the air in the furnace chamber. However, this is not wholly satisfactory as it may produce the following contaminant:

- Sulphur from the fuel.
- Residual nitrogen from the combustion air.
- Any residual oxygen.
- Water vapour formed when the fuel is burned in air.
- Corrosive acid formed when carbon dioxide reacts with water vapour.

A more expensive solution is to keep the products of combustion out of the furnace chamber (muffle) and replace the air in the chamber with a mixture of inert and carefully controlled gases, specially generated for atmosphere control.

8.10.5 *Economical use of fuel*

Fuel costs are the most important element in the operating costs of a furnace. Energy costs are continually rising over the long term and the economical use of fuel is essential, both commercially and ecologically. To this end it is better to run a furnace continually on a shift basis than to repeatedly heat it up and cool it down. The energy required to keep heating up a furnace from the cold is much greater than the energy required to keep it operating. And the repeated expansion and contraction of the furnace lining, when used intermittently, leads to its cracking and early failure. Instead of doing their own heat treatment on an occasional, jobbing basis, it is more economical for small and medium-sized workshops to contract out their heat treatment to specialist firms that run their furnaces continuously.

8.10.6 *Low maintenance costs*

In order to keep maintenance to a minimum and preserve the refractory lining as long as possible, the furnace should not be used intermittently, nor should it be used beyond its maximum operating temperature, even for a short time. Operating at excessively high temperature rapidly reduces the life of the heating elements in electric muffle furnaces and is wasteful of energy. Excessively high temperatures can also degrade the refractory lining and increase the frequency of its replacement. This is costly in materials, labour and lost production time.

8.11 Types of heat treatment furnace

8.11.1 *Semi-muffle furnace*

The semi-muffle furnace (Fig. 8.9) is made in a variety of sizes from small benchtop furnaces for toolrooms and laboratories up to very large furnaces for the batch treatment of large components. They may be gas-fired or oil-fired, and the flame does not play onto the charge directly but passes under the hearth to provide bottom heat. Bottom heat is provided by conduction and radiation from the hearth; supplementary heating is provided by the circulation of flue gases and by radiation from the furnace crown. The combustion air is carefully controlled so that no excess oxygen is present in the flue gases to cause oxidation. The flue gases are finally drawn off each side of the furnace door so that air entering at this point is immediately swept up the flue. Full atmosphere control is not possible with this type of furnace.

Fig. 8.9 *Gas-heated semi-muffle furnace: the furnace arch focuses radiated heat on the component and circulates flue gases to promote uniform heating*

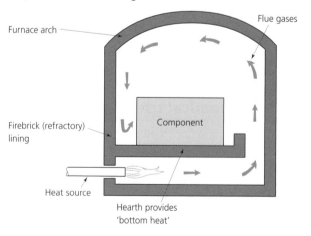

8.11.2 *Gas-heated muffle furnace*

Figure 8.10 shows a gas-heated muffle furnace. The muffle chamber completely separates the charge from the combustion chamber. The conditions in the two chambers can be controlled independently to ensure combustion is most economical in the combustion chamber and the appropriate treatment atmosphere exists in the muffle chamber.

Fig. 8.10 *Gas-heated muffle furnace*

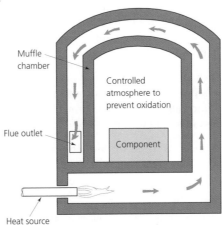

8.11.3 *Electric resistance muffle furnace*

Figure 8.11(a) shows a typical electric resistance muffle furnace. Since the electric resistance heating elements can operate independently of the atmosphere in which they are placed, they may be installed directly into the muffle chamber. Figure 8.11(b) shows the construction of a typical electric resistance heating element. This type of furnace is available in a wide range of sizes. Full atmosphere control can be provided, and the use of electric resistance heating allows automatic control systems to be fitted.

Fig. 8.11 *Electrically heated muffle furnace: (a) electric resistance muffle furnace; (b) electric furnace heating element*

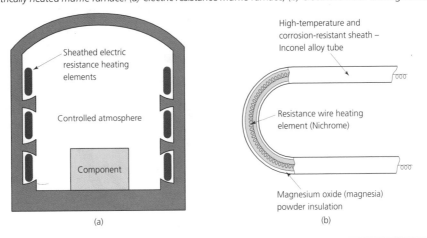

8.11.4 *Gas-heated salt bath furnace*

Figure 8.12 shows a typical gas-fired salt bath furnace. Points to note are:

- *Tangential firing* The flame does not play directly onto the pot.
- *Top heat* In the interests of safety, the salts must be melted from the top downwards. If heated from the bottom, the expanding salts could erupt red-hot through the solid crust, like a miniature volcano, creating a very dangerous situation.
- *Preheating* All work must be preheated to ensure it is *thoroughly dry*. Failure to do this will risk an explosion, which could throw the molten salts out of the pot.
- *Protective clothing* A face visor and other protective items must be worn when using salt bath furnaces.

The choice of salts depends upon the process being carried out; all reputable manufacturers provide advice on suitable salts for various applications. To prevent loss through oxidation and fuming, it is normal to use an economiser in the form of mica flakes floated on the surface of the salts.

- *Nitrate-based salts* Used for low-temperature applications such as tempering and for the solution and precipitation treatment of light alloys. Great care must be taken in their use for, if overheated, they can cause a serious explosion.

Fig. 8.12 *Gas-heated salt bath furnace: firing is tangential so the flame does not play directly onto the pot*

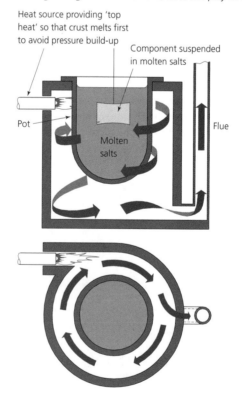

- *Chloride-based salts* Suitable for all applications above about 750 °C, e.g. the annealing, normalising, and quench hardening of plain carbon and alloy steels.
- *Cyanide-based salts* Used for case-hardening small low-carbon steel and alloy steel components. These salts are exceptionally and fatally poisonous, so special care must be taken in their use.

8.11.5 *Electrically heated salt bath furnace*

Figure 8.13 shows an electrically heated salt bath furnace; electrodes are immersed in the salts. The resistance of the salts to the passage of an electric current converts electrical energy into heat energy. Since this conversion occurs within the salts, the furnace has a high efficiency. Electrical heating is readily adapted to automatic control in situations where the temperature must be highly stable.

Fig. 8.13 *Electrically heated salt bath furnace*

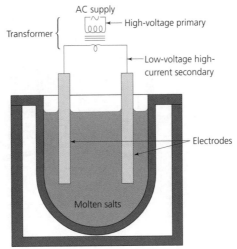

Advantages
- Uniformity of heating as the charge is enveloped in molten salt at the treatment temperature.
- Accurate temperature control.
- High temperature stability if the mass of the salts is substantially greater than the mass of the charge.
- No expensive atmosphere control as the charge is immersed in the molten salt.

Limitations
- Only economical in fuel utilisation if run on a continuous basis.
- Regular maintenance required.
- Salt baths use toxic chemicals and present an explosion hazard; a highly trained workforce is required.
- Only relatively small components can be accommodated.

8.11.6 *Hot-air tempering furnace*

Figure 8.14 shows the principle of a hot-air tempering furnace. The air is heated by an electric resistance element and circulated by a fan. The air is recirculated to conserve heat and save energy. Tempering is carried out at low temperatures, so this arrangement is quite efficient. Atmosphere control is not possible, but it is seldom required at these low temperatures. There will be slight discoloration of the metal surface. Small components tend to nest together and restrict the air circulation. Agitation and/or rotation of an inclined or horizontal work basket is sometimes used to overcome this problem.

Fig. 8.14 *Electrically heated hot-air tempering furnace: the arrows indicate the path of the hot air*

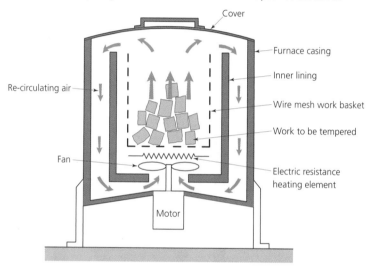

8.12 Temperature measurement

8.12.1 *Thermocouple pyrometer*

This is the most widely used temperature measuring device and is also used as the basis for automatic control systems. Figure 8.15(a) shows the principle of the thermocouple pyrometer. Heating the junction of two dissimilar metal wires, such as iron and copper, will cause an electrical potential to appear across their ends; the size of the potential depends on the composition of the wires and on the temperature differential between the junction and the remote ends of the wires. If the wires forming the junction are part of a closed electric circuit, the generated potential will cause a small current to flow. In a circuit of constant resistance, this current will depend on the magnitude of the potential. In the circuit of Fig. 8.15(a) the magnitude of the current will be indicated by the galvanometer. The connections between the dissimilar wires and the galvanometer form the cold junction.

Fig. 8.15 *Thermocouple pyrometer: (a) principle of operation; (b) pyrometer circuit – the flexible 'compensating' leads are made from an alloy that only forms a cold junction at the black box; the indicating instrument is calibrated in degrees Celsius; (c) thermocouple probe*

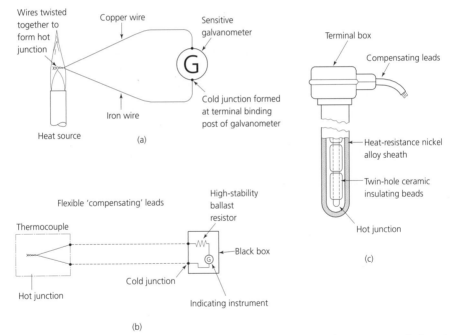

Figure 8.15(b) shows the circuit for a practical thermocouple pyrometer; it has the following components:

- *Thermocouple probe (hot junction)* This consists of a junction of two wires of dissimilar metals contained within a tube of refractory metal or porcelain to protect the wires from chemical or physical damage. Porcelain beads are threaded over the dissimilar wire of the junction to insulate them and to locate them within the sheath, as shown in Fig. 8.15(c).

- *Indicating instrument* This is a sensitive milliammeter calibrated in degrees of temperature so that direct readings can be taken. When the furnace is cold, the milliammeter should read ambient temperature or there will be a zero error. Since it forms the cold junction, it should be mounted in a cool position unaffected by the heat of the furnace.
- *Ballast or 'swamp' resistor* This is contained within the case of the indicating instrument. It is made from manganin wire whose resistance is unaffected by temperature change. The ohmic value of this resistor is made very large compared with the resistance of the instrument and the resistance of the external circuit. This ensures the readings are accurate, since its large value 'swamps' any resistance variations in the rest of the circuit due to changes in temperature.
- *Compensating leads* These are used to connect the thermocouple probe to the indicating instrument. Made of a special alloy, they form a cold junction at the indicating instrument but have no effect when connected to the thermocouple probe terminals. To avoid changes in calibration, the compensating leads must not be changed in length, nor must alternative conductors be used. The thermocouple, compensating leads and the indicating instrument must always be kept together as a set.

8.12.2 *Radiation pyrometer*

The thermocouple pyrometer measures the temperature of the furnace atmosphere. This is not necessarily the temperature of the charge. Figure 8.16 shows a typical radiation pyrometer. Aimed at the furnace charge using the telescopic sight, it measures the temperature of the charge rather than the temperature of the furnace. The radiation pyrometer is based on the same principle as the thermocouple pyrometer, but it uses a parabolic mirror to focus the radiant heat from the heated body onto the hot junction. This type of pyrometer is used in the following circumstances:

- Where the temperature of the charge needs to be measured rather than the temperature of the furnace atmosphere.
- Where the temperature of the furnace is so high that the normal thermocouple probe would be damaged.
- Where the heated body is inaccessible.

Fig. 8.16 *The radiation pyrometer: the parabolic reflector concentrates heat rays at the hot junction of the thermocouple*

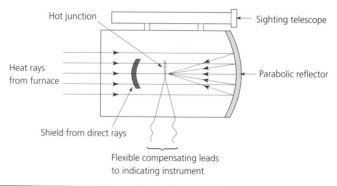

8.13 Quenching media

The choice of quenching medium depends upon the material being quenched and the rate of cooling to be achieved. Steels are quench hardened using the slowest possible rate of cooling that lies above the critical cooling rate; this is to avoid cracking and distortion.

8.13.1 *Water*

Water is widely used as it is cheap, readily available and has a high specific heat capacity. Heated water gives off no noxious fumes and is non-flammable. It can be used for quenching non-ferrous metals, case-hardening plain carbon steels and quench hardening medium-carbon steels, as long as it exceeds the critical cooling rate.

8.13.2 *Quenching oil*

Quenching oil is specially formulated; do not use lubricating or cutting oils. The cooling rate with oil is much lower and will only impart toughness to medium-carbon steels. However, it will quench harden high-carbon steels; it is widely used for quenching alloy steels as the critical cooling rate for alloy steels is very much lower than for plain carbon steels. Quenching in oil is less likely to cause cracking as the cooling rate is lower than for water.

8.13.3 *Air blast*

Air blast can be used to quench harden small components made from steels containing a very high percentage of alloying elements, e.g. some high-speed steels.

8.14 The quenching bath

Where water or oil baths are used, the volume of the quenching bath should be sufficient to avoid undue increase in temperature when the heated work is immersed in it. If the bath is used continuously, the quenching medium should be circulated through a cooling system. The work should be constantly agitated in the bath to prevent a blanket of steam or oil vapour forming around it; any blanket would insulate the work from the quenching medium. There should be a fume hood and extractor over the bath to draw off steam produced by water or fumes produced by oil. Where oil is used as the quenching medium, an airtight lid should be available so that the quenching bath can be sealed off instantly in the event of the oil overheating and igniting.

As soon as the hot workpiece is plunged into the water or the oil, it becomes surrounded by a blanket of vaporised quenching medium. This reduces the rate of cooling since vapours have a lower thermal conductivity than liquids. The work must be agitated as soon as it enters the quenching bath to disperse the vapour as quickly as it forms and to keep the work in contact with the liquid. Agitation of the quenching medium also helps to keep the quenching bath at a uniform temperature, which prevents local overheating.

Care must be taken to minimise distortion during quenching, so it is usual to dip long thin components vertically into the quenching bath. Figures 8.17 and 8.18 show how cracking and distortion may be caused by incorrect design and quenching.

Fig. 8.17 *Designing to avoid cracks during quench hardening: (a) incorrect; (b) correct*

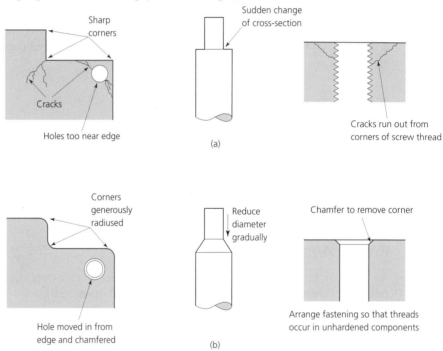

SELF-ASSESSMENT TASK 8.6

State the precautions that must be taken when using oil as a quenching agent.

Fig. 8.18 *Distortion during quench hardening: (a) an unbalanced shape before hardening (top) will become distorted after hardening; (b) on first contact with the quenching oil, the lower surface of the component cools more rapidly than the upper surface, so the component bends; (c) long slender components should be dipped into the quenching bath end-on*

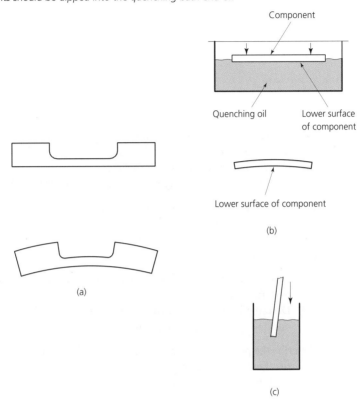

EXERCISES

8.1 Describe what is meant by recrystallisation. How can recrystallisation be exploited when softening work-hardened non-ferrous metals.

8.2 (a) With reference to plain carbon steels, describe:
  (i) subcritical annealing
  (ii) spheroidising annealing
  (iii) full annealing
  (iv) normalising
  (b) State an appropriate application for each of the processes in part (a); give a reason for each choice.

8.3 Describe in detail how a component made from a plain carbon steel containing 1.0 per cent carbon may be quench hardened. Pay particular attention to the hardening temperature, the rate of cooling and the precautions which must be taken to avoid grain growth, distortion and cracking.

8.4 Describe in detail how the component shown in Fig. 8.19 may be case hardened whilst leaving the thread and hexagon head soft.

Fig. 8.19 *Exercise 8.4*

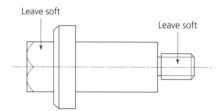

Leave soft

Leave soft

8.5 With reference to Duralumin aluminium alloy, describe how:
(a) it may be softened by solution treatment
(b) natural age hardening may be delayed
(c) it may be precipitation age hardened by an artificial ageing process

8.6 List the essential requirements of a heat treatment furnace and explain how these requirements are met in:
(a) an electrically heated muffle furnace
(b) a gas-heated salt bath furnace

8.7 (a) Describe in detail the general precautions which must be taken when using salt bath furnaces.
(b) Describe in detail the particular precautions which must be taken when using nitrate-based salts for tempering and cyanide-based salts for carburising.

8.8 Describe the principles of the thermocouple and how they are applied to:
(a) the thermocouple pyrometer
(b) the radiation pyrometer

8.9 Another method of checking the temperature of a furnace is to use Seger cones. Find out what they are and how they compare with a pyrometer for temperature control.

8.10 In order to prevent cracking and distortion, discuss the precautions that should be taken at the design stage of a component to be quench hardened.

9 Finishing processes

> The topic areas covered in this chapter are:
>
> - Chemical treatments.
> - Physical treatments.
> - Metallic coatings.
> - Plastic coatings.
> - Paint systems.

9.1 Preparatory treatments: chemical

Satisfactory finishing always depends on the careful surface preparation of the base material. Most component surfaces are contaminated with one or more of the following:

- Oxide films resulting from reaction of the base material with atmospheric oxygen and moisture, e.g. the rusting of steel.
- Metal salt deposits, such as sulphates and carbonates of the base material, caused by acid rain attack, e.g. the patina on copper.
- Soils in the form of grease, dust and dirt together with swarf, grinding wheel dross and polishing dross from machining and finishing processes.
- Previous protective films, e.g. paint films which need to be stripped to provide a sound base for replacement.

Any of these contaminants must be removed before finishing treatments are carried out.

9.1.1 *Acid Pickling*

Unless steels have been very thoroughly protected rust, will usually be present. Unfortunately rust will continue to spread under any decorative or protective coating and will eventually destroy the base metal. It will also lift the protective coating, forming bubbles underneath. This bubbling effect is often the first sign of 'body rot' in a motor vehicle. Acid pickling in hydrochloric or sulphuric acid is used to remove rust and scale.

The acid cannot distinguish between the oxide and the metal; if the oxide scale is thick, the acid will often attack the metal in preference to the oxide. This produces uneven pickling and pitting of the metal surface. Inhibitor chemicals are usually added to the acid to prevent any attack on the base metal. Although high, the cost of the inhibitor is more than recovered by greatly extending the useful life of the acid. Untreated acid has to be discarded when between 6 and 10 per cent iron is present. It is essential to wash and neutralise the pickled metal before storage, and treat it with a corrosion inhibitor such as lanolin or oil. The discarded acid must not be poured down public drains, it must be removed by a specialist waste disposal firm for neutralisation.

9.1.2 *Degreasing*

Greases and oils prevent the wetting of a surface to be treated; they must be removed before any pretreatment or finishing processes is applied.

Solvent degreasing

Trichloroethylene or perchloroethylene are still widely used in vapour degreasing plants. The solvent is boiled and the vapours are condensed on cooling coils, the condensed liquid cascading down over the suspended components. Oil and grease removal is effective, but inorganic soils are only removed by the washing action of the condensed liquid.

Kerosene

Kerosene (paraffin) will dissolve many oils and greases. Nowadays kerosene is blended with oil-soluble surface active agents and becomes emulsifiable. These systems have the advantage over vapour degreasing that soils and residues can be rinsed cleanly away from the metal surfaces by the detergent and flushing action of the liquid. And there is no toxicity factor to be considered, as is the case with chlorinated hydrocarbons, which also attack the ozone layer of the upper atmosphere.

Alkali cleaning

Alkali cleaning may be preferable where degreasing is to be followed by electroplating; this is because residual solvent from solvent degreasing will lead to poor adhesion of the plating. Also, alkalis do not have the toxicity and flammability of the chlorinated hydrocarbons and kerosene. Alkali detergents range from washing soda and caustic soda to sophisticated blends of silicates, phosphates, carbonates and surface active agents. Phosphates and silicates are valuable detergents in their own right, but they become even more effective when a surface active agent is added. The *surfactant* performs a dual function. Firstly, it lowers the surface tension of the liquids, so they wet and penetrate the soils more efficiently. And secondly, it enhances the emulsification of oils and greases.

Alkali solutions are used at temperatures of about 80–90 °C. The components need to be rinsed adequately after alkali cleaning to avoid carry-over into the plating baths, where the presence of alkalis would be highly undesirable. Aluminium and zinc suffer from alkali attack, so alkali cleaning must not be used with aluminium- and zinc-based alloys unless a suitable buffered mixture is used.

SELF-ASSESSMENT TASK 9.1

Investigate and report on the precautions that should be taken when using the following substances to clean components preparatory to finishing:

(a) acids (pickling)
(b) trichloroethylene or perchloroethylene
(c) kerosene (paraffin)
(d) alkali solutions

9.2 Preparatory treatments: physical

9.2.1 *Wire brushing*

Wire brushing with a rapidly rotating, coarse wire brush is used to dislodge loose debris and soils from structural steelwork before painting or repainting. The surface left by brushing provides a key for the first coat of paint. Fine wire brushing is also used as a decorative finish on sheet aluminium.

9.2.2 *Shot and vapour blasting*

Fine particles are blasted against the metal surface at high velocity using compressed air. This is used to descale structural steelwork on site, as well as descaling smaller components under factory conditions. It is also used for cleaning and descaling sand castings. Although more expensive and labour-intensive than acid pickling, particle blasting enhances the mechanical properties of materials, whereas acid pickling can detract from the mechanical properties of materials (Section 9.9).

9.2.3 *Flame descaling*

Flame descaling depends upon the difference in expansion between the scale and the base metal when subjected to local heating. It is used for cleaning heavily rusted structural steelwork before maintenance painting. The steel surface is heated with an oxyfuel gas torch fitted with a specially designed nozzle that gives a broad fan-shaped high-intensity flame. Compared with the relatively cool steel beneath, the scale or rust expands rapidly and flakes off. Any entrapped moisture turns to steam and assists the stripping action. For best results, the prepared steelwork should be painted with a primer whilst the metal is still warm (45 °C), and the primer should contain a corrosion inhibitor.

9.2.4 *Abrasive finishing*

Abrasive finishing covers a variety of manual and automatic processes similar to grinding. For rapid surface removal, finely powdered aluminium oxide (emery) or silicon carbide is used. Grinding uses a rigid abrasive wheel and the geometry of the process is closely controlled. But in abrasive finishing the abrasive particles may be glued directly to a flexible leather polishing wheel called a *stitched basil*, or an abrasive belt may be used as in *linishing* (Fig. 9.1(a)) or *back-stand grinding* (Fig. 9.1(b)). The greater area of abrasive means that belts have the advantages of lasting longer and cutting cooler than wheels. The work is usually hand-held.

Fig. 9.1 *Abrasive belt finishing: (a) linishing; (b) back-stand grinding*

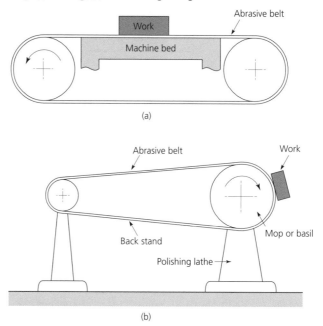

9.2.5 *Polishing*

Although it is not fully understood how a lustrous polished surface is produced, there are currently two schools of thought. One is that polishing is an extension of the grinding principle but using very much finer abrasive particles. The other is that polishing produces very high surface temperature in the metal, causing the topmost layer to melt and become smeared out by the polishing action to form a smooth *Beilby layer*. Whichever theory is correct, polishing mops are used in practice with grease-based polishing compounds to achieve smooth and glossy polished surfaces with a high lustre.

Polishing mops are made from linen, calico, cotton and sisal cloth. They are mounted on the spindles of polishing lathes and polishing takes place on the circumference of the rapidly spinning mops. There are three basic types of mop:

- *Loose fold mops* Soft and flexible, they are used for the final polishing of contoured surfaces and are usually made from discs of calico cloth.
- *Stitched mops* Discs of cotton cloth are spirally stitched to secure them and increase their rigidity. These mops are used for general and heavy-duty manual grease mopping.
- *Bias type mops* Used for back-stand contact wheels, which need to be more flexible than stitched leather basils. However, they are mainly used on automatic polishing machines.

Polishing compounds are available in bar form for manual application and in liquid form for automatic application. The compound consists of the abrasive particles suspended in a grease base that causes them to adhere to the mop. When the particles become dulled, their temperature rises, the grease melts locally, the dulled grains are released and fresh grains are exposed. The choice of abrasive depends on the material being polished:

- *Hydrated silica* (Tripoli) is used for non-ferrous metals and thermosetting plastics.
- *Aluminium oxide* (flour of emery) is used for ferrous metals.
- *Chromium oxide* is used for stainless steel and for the final buffing of chrome-plated products.

Because plated finishes are very thin, abrasive finishing and polishing must take place before plating. Only the lightest buffing should be necessary after plating to give the final lustre to the work.

9.2.6 *Barrelling*

Barrelling or tumbling is a common method of polishing small components that are inconvenient to handle individually (Fig. 9.2). It involves rotating a 'barrel' containing the components, the fine abrasives (polished stones and ceramics) and water. The whole box is sealed and rotates slowly so that tumbling takes place. If it were rotated too quickly, rather like a centrifuge, the components would be thrown against the sides of the barrel, where they would remain and no tumbling action would occur. The polishing action is a combination of abrasive polishing and burnishing. It is most effective for polishing lumpy components rather than flat components.

Fig. 9.2 *A barrel polishing system*

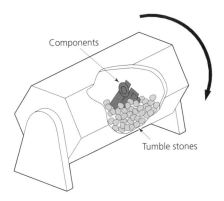

Components

Tumble stones

Briefly describe the following preparatory processes and state the purpose for which they would be used:

(a) wire brushing
(b) shot or vapour blasting
(c) flame descaling
(d) abrasive finishing
(e) barreling

9.3 Protective coatings

To retard the onset of corrosion and the eventual destruction of metallic components, assemblies and structures, various protective coatings are applied to insulate the base metal from the environment. The various mechanisms by which corrosion can occur are discussed in *Engineering Materials*, Volume 1. The coatings applied to metals may be protective, decorative or a combination of both. This section deals only with coatings which are protective.

9.3.1 *Galvanising*

Galvanising is the coating of ferrous metals (usually low-carbon steels) with zinc. It may be performed on sheet material or on finished goods. There are two processes. The sheet or finished goods may be cleaned, fluxed and dipped into molten zinc (hot-dip galvanising) or the zinc coating may be deposited on the base metal electrolytically (electrolytic galvanising). The hot-dip process provides a thicker coating of zinc and better protection; it is also gap-filling as it seals the joints of vessels designed to hold liquids. Electrolytic galvanising is not gap-filling, but the thickness of the coating and the amount of zinc can be more closely controlled. Electrolytic galvanising is widely used as a pretreatment before paint finishing ferrous metal components.

9.3.2 *Sherardising*

Sherardising, calorising and chromising are the three cementation processes in common use. Sherardising applies a coating of zinc to the components that require protection. Cleaned and dried components are placed into a rotating barrel along with zinc powder at a temperature of about 370 °C, just below the melting point of zinc. The time to produce the coating depends on the thickness required, but is usually about 12 hours. The zinc bonds to the surface of the ferrous workpiece by diffusion and forms a hard even layer of iron–zinc intermetallic compounds. The slight roughness of the surface provides an excellent key for subsequent painting.

9.3.3 *Calorising*

Calorising is similar to sherardising except that the work is heated in aluminium powder to between 850 °C and 1000 °C. It is used to protect steel components from high-temperature oxidation (as in car exhaust systems) rather than against ambient-temperature atmospheric corrosion. It also provides better protection at higher levels of humidity than sherardising.

9.3.4 *Chromising*

Chromising is similar to calorising except that the work is heated in chromium powder, in an atmosphere of hydrogen, to between 1300 °C and 1400 °C. Because of the high temperatures involved, this is an expensive process and is only used where very high levels of protection are required. The high temperatures tend to cause grain growth in the steel, which impairs its mechanical properties.

9.3.5 *Chromating*

Magnesium alloys rely upon the formation of a coherent oxide film to retard the effects of corrosion, but they do not respond to anodising or similar processes (Section 9.4), so chromating has to be used instead. In this process the components made from magnesium alloy are dipped in a solution containing potassium dichromate, together with chromic and phosphoric acids, to form a hard oxide film on the surface. Unfortunately this film is far from decorative and is a dark grey or even black. The oxidised surface is usually sealed with a coat of zinc chromate paint followed by a decorative paint of the required colour.

9.3.6 *Phosphating*

Phosphating describes several processes where the surface of the base metal is chemically converted to complex metal phosphates, oxides and chromates. These processes originally went under the trade names of Parkerising, Bonderising, Granodising and Walterising. More recently they have been standardised under BS 3189. Nowadays phosphating is virtually standard practice before the application of paint films. Conversion coatings are also used to improve the lubricity of bearings, and as a pretreatment for metals being formed by such processes as wire and tube drawing, deep drawing and cold forging, where extreme-pressure lubrication is required. As well as being applied to ferrous metal components, phosphate conversion processes can also be applied to zinc, zinc alloys, zinc-coated steels, aluminium and aluminium alloys.

9.3.7 *Metal spraying*

Metal spraying is used for a variety of purposes. The most important is to apply corrosion-resistant coatings to ferrous metal structures that are too large to be protected in any other way. Other purposes include building up worn shafts and depositing surfaces that are wear resistant. The process consists of spraying molten particles of metal onto a prepared surface. This surface should be free from grease and it should be slightly roughened to provide a key for the coating.

The metal coating is fed into a spray gun in the form of wire or powder. Heat is provided by means of an oxyfuel gas flame, an electric arc or a plasma arc inside the gun. The plasma-arc technique is reserved for metals with a high melting point. Compressed air is used to spray the molten metal onto the work. Here are some typical applications:

- *Zinc* Used to rustproof plain carbon steels. The sprayed zinc coating is slightly porous, so it needs to be sealed by painting. Common in marine and structural applications, zinc is the most widely used deposit metal.
- *Aluminium* Used to rustproof plain carbon steels subjected to moist environments. Aluminium prevents scaling (oxidation) at high temperatures. It is increasingly used to protect car exhaust systems.
- *Copper* Used to coat and recoat printing rolls for high-quality colour reproduction. Also used to provide high-conductivity coatings where electroplated deposits are not thick enough to carry the required current. Also used for decorative coatings that have a high corrosion resistance.
- *Brass* Used for corrosion-resistant decorative coatings; cheaper than copper.
- *Bronze* Used to build up bearing surfaces. The porosity of the spray-deposited metal helps the retention of lubricants and improves lubrication.
- *Plain carbon steels* Rebuilding worn bearing surfaces on shafts.
- *Hard-facing materials* Materials such as Stellite can be deposited on plain carbon steels and low-alloy steels to improve their wear resistance.
- *Stainless steel (18/8)* Extremely corrosion-resistant coatings capable of operating at very high temperatures, e.g. in chemical plant. Also used for architectural decoration.

9.3.8 *Cladding*

A composite billet is made from the base metal and the coating (Fig. 9.3(a)). As the billet is reduced in thickness by rolling or drawing (Fig. 9.3(b)), the thicknesses of the base and the coating are reduced proportionally. A typical application of this process is the cladding of aluminium alloy with corrosion-resistant, but weaker, high-purity aluminium (Alclad).

Fig. 9.3 *Cladding: (a) section through a clad metal composite; (b) proportional reduction of clad billet*

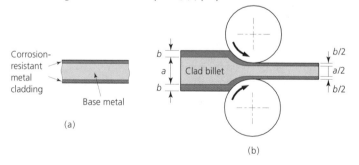

9.4 Metallic decorative coatings

Decorative coatings usually provide some protection as well as enhancing the appearance of the product.

9.4.1 *Anodising*

To prevent corrosion, aluminium and its alloys rely upon the formation of a self-healing, homogeneous, transparent oxide film on the surface of the metal. The process of anodising artificially builds up a thick, adherent layer of aluminium oxide that resists atmospheric corrosion, both indoors and outdoors even when subjected to the pollution from urban environments.

Components to be anodised are first cleaned and degreased by the use of chemical solvents, then they are etched, wire brushed or polished, depending upon the surface texture required. The work is made the anode (positive electrode) of an electrolytic cell and a direct electric current is passed through it. This is the reverse of electroplating, where the work is the cathode (negative electrode) of the cell. The electrolyte is a dilute acid, chosen to produce the required finish. Finally, the surface is sealed, usually with boiling water or steam, to improve the corrosion resistance of the coating and to minimise the absorptive properties of the oxide film.

9.4.2 *Electroplating*

The components to be plated are made the cathodes (negative electrodes) of an electrolytic cell (Fig. 9.4). The electrolyte and the anode (positive electrode) will depend upon the metal being deposited. Briefly, when a direct electric current is passed through the cell, the electrolyte becomes ionised. Metal ions are deposited upon the cathodes (the work) and the strength of the electrolyte is depleted. If the anode is soluble, metal ions are dissolved from the anode to bring the electrolyte back up to strength. If the anode is insoluble, as in chromium plating, it only completes the circuit and makes no contribution to maintaining the strength of the electrolyte.

Fig. 9.4 *Electroplating*

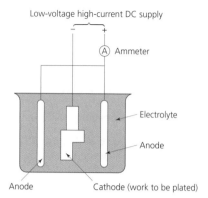

Low-voltage high-current DC supply

(A) Ammeter

Electrolyte

Anode

Anode

Cathode (work to be plated)

Typical electrodeposits are:

- *Cadmium* This is a white metal used to protect steel from atmospheric attack and is frequently used in the electrical and aircraft industries. Its low contact resistance reduces the risk of electrochemical corrosion where plated steel components and aluminium components are assembled together. Cadmium is toxic and should not be used where there is a possibility of contact with foodstuffs and drinking water. It is resistant to alkali attack, so it is preferred to zinc for marine purposes.
- *Copper* The rich colour of this metal enables it to be used for decorative purposes. It can be used directly with a polished finish or an oxidised finish, when it is usually lacquered to prevent the formation of a patina that would spoil its lustre and appearance. Alternatively it can be used as a base before nickel plating. For example, zinc-base die castings are usually copper plated before nickel plating to prevent attack on the zinc by the nickel plating electrolyte.
- *Chromium (decorative and protective)* The brilliant blue-white colour of chromium coupled with its resistance to tarnishing makes it an ideal finishing deposit over a nickel base. Only a thin film is required and this usually ranges from 0.000 25 to 0.000 75 mm in thickness. The chromium itself is not polished, but is applied over a brightly polished nickel base when a bright finish is required, or over a matt or semi-matt nickel base when a satin-chrome finish is required.
- *Chromium (hard)* For engineering applications a thick layer of chromium is built up directly onto the component that is to be protected or reclaimed. This is a specialised process and differs from flash-chrome decorative plating in the techniques used. Deposits up to 0.4 mm thick can be built up where a component is to be sized by grinding after plating. The plated surface is resistant to abrasion and has antifriction properties, so it is frequently used to protect the working faces of gauges and to build up and reclaim worn gauges.
- *Nickel* This is widely used as an initial deposit before chromium plating. It is easier to process than chromium and gives a good corrosion-resistant finish. It can be polished and has good surface-levelling properties; it will build up in the hollows of the surface and improve the appearance of poorly polished and unpolished surfaces. Hence it is widely used for the barrel plating of small unpolished components. Unfortunately nickel has a yellowish tinge to its colour and tends to tarnish, so it is frequently given a

flash coating of chromium. Chromium is a much more expensive metal but it has a more pleasing appearance and resists tarnishing.

- *Silver and gold* These precious metals are highly corrosion resistant and are good electrical conductors. Gold, in particular, is widely used to protect the edge connectors of printed circuit boards used in electronic equipment. Silver is used for plating switch contacts.
- *Tin* This is used as a corrosion-resistant deposit on steel and non-ferrous metals. It is non-toxic and is widely used on food-handling equipment. Tin plate is produced in continuous strip plants where thin low-carbon steel strip is coated electrolytically with bright tin. Printed circuit boards and terminal tags are frequently tin plated to facilitate soldering.
- *Zinc* See electrolytic galvanising (Section 9.3).

SELF-ASSESSMENT TASK 9.4

1. Explain the essential difference between anodising and electroplating. What are these processes used for?

2. Explain the essential difference between decorative chromium plating and hard chromium plating.

9.5 Plastic coatings: decorative and protective

Plastic coatings can be functional as well as being corrosion resistant and decorative. The wide range of plastic coatings provides the designer with a means of achieving:

- Abrasion resistance.
- Cushion coating (up to 6 mm thick).
- Electrical and thermal insulation.
- Flexibility over a wide range of temperatures.
- Non-stick properties.
- Permanent protection against weathering and atmospheric pollution subject to the inclusion of antioxidants and ultraviolet filter dyes.
- Reduction in maintenance costs.
- Resistance to corrosion by a wide range of chemicals.
- The covering and sealing of mechanical joints, welds and porous castings.

There are many ways of applying plastic coatings to metal components, but two of the most widely used are fluidised bed dipping and liquid plastisol dipping.

9.5.1 *Fluidised bed dipping*

This technique is widely used with all fluidisable powders except for epoxy resins. Figure 9.5 shows a section through a fluidised bed dipping bath. The powder has a particle size of 60 to 200 mesh and is supported on a bed of air passing through the porous ceramic tile bed

Fig. 9.5 *Fluidised bed dipping*

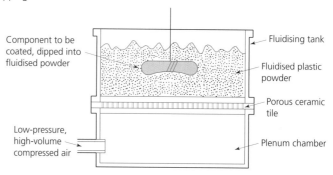

Component to be coated, dipped into fluidised powder

Fluidising tank

Fluidised plastic powder

Porous ceramic tile

Low-pressure, high-volume compressed air

Plenum chamber

from the plenum chamber. The air lifts the powder, causing it continuously to bubble up and drop back. The effect is similar in appearance to boiling water, and the diffused powder offers little resistance to the immersed preheated work. The plastic powder adheres to the heated work to form a homogeneous skin.

9.5.2 *Liquid plastisol dipping*

This process is limited to the use of a PVC plastisol. A plastisol is a resin powder suspended in a plasticiser; it contains no dangerous solvent. The process is similar to fluidised bed dipping, except the preheated work is dipped into a thixotropic (non-drip) liquid plastic instead of a powder.

9.6 Paint coatings: decorative and protective

Painting is widely used for the protection and decoration of metallic components and structures. It is the easiest and cheapest coating to apply with any degree of permanence. As with all applied coatings, its success depends upon careful preparation of the work surface and pretreatment.

Paints consist of a *binder* which sets to provide the protective film; a *pigment* to provide the required colour and sometimes also to act as a *corrosion inhibitor*, and a *solvent* (thinner) which does not form part of the final paint film but gives the paint the required consistency during application and controls the drying process. A complete paint system consists of:

- A primer acts as a corrosion inhibitor and provides a key for the subsequent coats.
- A putty or filler is used to fill surface blemishes in castings.
- Undercoats are used to build up the thickness of the paint film and to give body to the final colour. Flatting down is essential between each undercoat and before the topcoat to provide a smooth surface where a high gloss is required (e.g. car bodywork).
- A topcoat or finishing coat is not only decorative, it seals the previous coats against moisture and atmospheric attack, and provides most of the corrosion and abrasion resistance.

Let's now consider the four main groups of paints.

9.6.1 *Group 1*

In this group, atmospheric oxygen reacts with the binder, causing it to polymerise into a solid film. This reaction is speeded up by *forced drying* at 70 °C. Paints that dry by oxidation include the traditional linseed oil paints, the oleoresinous paints and the modern general-purpose air-drying paints based on oil-modified alkyd resins.

9.6.2 *Group 2*

These paints are based on amino-alkyd resins which do not cure (set) at room temperature and have to be stoved at 110–150 °C. When set, they are tougher and more resistant to abrasion than air-drying paints. Paints in this group are used for car bodies.

9.6.3 *Group 3*

In this group, polymerisation is caused by the addition of an activator or hardener. Since this is stored separately and only added to the paint immediately before use, the paints are known as two-pack paints. Polymerisation commences as soon as the hardener is added to the paint. At first this is slow, but as the paint is spread, a solvent starts to evaporate, increasing the concentration of the hardener. This increase in concentration leads to rapid polymerisation and the paint is soon touch-dry. However, it does not attain its full mechanical properties and resistance to damage until after a few days.

   Paints in this category are based upon polyester, polyurethane and epoxy resins. The tendency nowadays is to use one-can paints. The hardener is added at the time of manufacture, but at too low a concentration for polymerisation to begin.

   As in the previous example, once the paint is spread, a volatile solvent evaporates and increases the concentration of the hardener to a level at which polymerisation can occur.

9.6.4 *Group 4*

The fourth group consists of the lacquers. A lacquer is a paint which dries by simple evaporation of the volatile solvent. There is no chemical hardening or polymerisation. The name comes from lac, the sticky, resinous secretion of the tiny lac insect *Laccifer lacca*, which was dissolved in a spirit before a pigment or dye was added to give the required colour. Nowadays synthetic resins are used for lacquers and the most suitable are cellulose products and acrylics.

9.7 Paint application

9.7.1 *Brushing*

Brushing is the simplest and most traditional method of applying paints to the work surface. Unfortunately it requires considerable skill, it is labour-intensive and therefore costly, and the quality of the paint film is difficult to control.

9.7.2 *Spraying*

Spraying is one of the most versatile methods of coating surfaces with paint. Originally introduced for finishing mass-produced cars, it is now used for large panels, structural steelwork and small components.

Conventional spraying uses compressed air to atomise the paint and to project it onto the surface of the component. It is a quick and fairly simple process that requires relatively low-cost equipment. It is versatile and can accommodate frequent colour changes. In the hands of a skilled operator, it can give consistently high standards of finish, but a lot of paint and solvent is wasted due to overspray and bounce. Conventional spraying can be easily robotised to reduce labour costs and improve quality.

Airless spraying does not use compressed air to atomise the paint, but pumps it under pressure through a fine jet. This creates less overspray and bounce, reducing the amount of hazardous spray dust compared with conventional spraying. Therefore airless spraying is safer to use away from the spray booth and is widely adopted for site work and maintenance.

9.7.3 *Dipping*

In the dipping process, work to be coated is lowered into a paint bath. The work is then removed and any surplus paint is allowed to drain off. The coated work is then passed through a stove to set and dry the paint. The whole process is usually automated and the work passes through the various stages suspended from a conveyor. Dip painting is highly productive and the labour costs are low, but close control is required for consistent quality.

9.8 Hazards of painting

The hazards of industrial painting fall into two main categories:

- Explosion and fire hazards resulting from the use of flammable solvents and the creation of flammable dust particles as the spray mist dries in the atmosphere.
- Toxic and irritant effects due to the inhalation of paint mist (wholly or partially solidified) and solvent fumes.

These hazards are particularly related to spraying and stoving processes, but the storage of paints and solvents on a large scale also presents special problems. The local Health and Safety Executive (HSE) inspector and the fire authorities should be consulted before painting on an industrial scale.

It is essential when spray painting to provide an efficient means of extraction to remove the excess spray mist and solvent fumes. Spray booths serve the double function of removing the spray mist and fumes from the working area then treating the exhausted air so it is cleansed before being released back into the atmosphere. This prevents it from becoming an environmental nuisance or hazard. The operator must wear full protective clothing and a respirator. Health problems are largely overcome by the increasing use of industrial robots for spray painting under factory conditions.

The main hazard associated with stoving ovens results from the use of unsuitable paints having volatile and flammable solvents, and from the accumulation of explosive dusts and gases in the fume extraction ducts. Stoving ovens and their extractors should have pressure release vents so that any explosion is carried upwards and away from the working area.

All electrical equipment associated with paint-spraying booths has to be to Buxton Approved Standards for flame- and explosion-proof fittings.

SELF-ASSESSMENT TASK 9.5

1. Explain the essential differences between plastic cladding and painting.

2. You have just made some steel-plate security shutters for your windows. The shutters are to be fitted to the outside of the building and need to be protected from the weather as well as having visual appeal.

    (a) Describe a suitable paint system for this application and give reasons for your choice.
    (b) Describe any preparatory treatment the steel would require and say how the paint would be applied.

9.9 Effects of finishing on material properties

All finishing processes have some effect upon the properties of the material being treated. This is a complex subject but some of the more important points are now briefly introduced.

9.9.1 *Mechanical finishing*

Processes such as machining and polishing change the surface of the material mechanically. Positive rake machining leaves the surface in tension, which reduces the fatigue performance of the material; negative rake machining leaves the surface in compression, which improves the fatigue performance of the material. Many machining processes, especially grinding, cause local heating and thermal stressing of the surface layers of the material. This can lead to surface cracking and early failure of the material through fatigue failure. Polishing using a grease-based compound tends to leave the surface of the work in compression and, providing the work is not overheated, this can improve the fatigue performance of the material. Polishing also removes surface discontinuities (incipient cracks) from which fatigue cracks can spread.

9.9.2 *Thermal effects*

Many finishing processes are carried out above ambient temperatures; this means they can impair the mechanical properties. For example, low-carbon steels become brittle when heated to 200 °C for any period of time, yet many finishing processes are carried out at around this temperature. Similarly, aluminium alloys are particularly susceptible to processing at temperatures between 100 °C and 150 °C, yet this is the temperature for force drying paints.

9.9.3 *Chemical effects*

Hydrogen is released in many finishing processes and this tends to be absorbed by metals, particularly high-strength alloy steels, causing a marked deterioration in the mechanical properties of the steel. The hydrogen absorbed by the steel can be driven off by heating the work to 200 °C. But prolonged heating at this temperature can cause embrittlement in low-carbon steels. Many deposited coatings seal the surface of the metal and prevent the hydrogen gas molecules from escaping.

Chemical etching and polishing also reduce the fatigue strength of materials, so particle blasting and mechanical polishing are preferable for highly stressed components. Even a light vapour blast after chemical treatment is frequently all that is required to restore the mechanical properties of the metal.

9.9.4 *Surface alloying effects*

The deposition of metals on metallic surfaces invariably leads to chemical interaction at the interface. This usually results in some loss of fatigue strength. However, the protection from corrosive fatigue often far outweighs the lowering of the mechanical fatigue performance.

9.9.5 *Shot and vapour blasting*

Shot and vapour blasting processes are often used for cleaning up the surfaces of castings and forgings. They release surface stresses and reduce the risk of surface cracks that lead to fatigue failure. They tend to leave the metal surface in a state of compression, which not only improves the mechanical properties of the metal but also retards the onset of corrosion.

EXERCISES

9.1 Describe one chemical and two physical preparatory treatments which may have to be carried out before the application of decorative or corrosion-resistant finishes to metal components.

9.2 Describe suitable applications for the following finishing processes, giving reasons for your choice:
(a) galvanising (hot-dip)
(b) sherardising
(c) chromating
(d) phosphating

9.3 (a) Describe the process of metal spraying.
(b) Describe **three** typical applications of metal spraying and give reasons for your choice.

9.4 Describe the principle of anodising and explain how it protects aluminium and aluminium alloy components.

9.5 (a) Describe the principle of electroplating.
(b) Explain the following:
(i) zinc alloy die castings should be copper plated before nickel plating
(ii) steel components are usually nickel plated before chromium plating

9.6 Discuss the advantages and limitations of plastic coating metal components compared with electroplating.

9.7 Describe a complete paint system and explain the need for the various coats which have to be applied.

9.8 Compare and contrast the processes of spraying and dipping as methods of applying paint films to metal components and fabrications.

9.9 Describe the hazards associated with industrial painting processes and explain how they may be overcome.

9.10 Describe briefly how finishing processes affect material properties in terms of:
(a) mechanical finishing
(b) thermal effects and chemical effects
(c) surface alloying effects

10 Quality control

The topic areas covered in this chapter are:

- Standards of length.
- Measurement of length.
- Measurement of angles.
- Straightness, flatness and squareness.
- Surface texture.
- Gauging.
- Quality control.
- Quality assurance.

10.1 Quality

Quality is the fitness of a product for achieving its designated purpose. This does not imply high cost and high precision. A wheelbarrow does not have to be manufactured to the precision of a machine tool, but providing it can be pushed easily, does not corrode away or rot, and it carries the required load, then it is of suitable quality.

To ensure that the quality, or fitness for purpose, of a component or assembly has been achieved, measurements must be taken to see if the design and manufacturing standards have been satisfied.

10.2 Standards of length

For practical measurements there is a hierarchy of working standards. For example, in most engineering companies, inspection grade slip gauges would be used to check and measure work produced using workshop grade slip gauges, micrometers, verniers, etc. The inspection grade slip gauges could themselves be calibrated against a laser standard at the National Physical Laboratory (NPL) and the company would receive a calibration chart. A large company might well have its own laser standard. The calibrated slip gauges could then be used to check and calibrate the micrometers, verniers and other instruments and gauges in day-to-day use in the workshop.

It is usually assumed that any measuring device is 10 times more accurate than the component feature it is measuring. Table 10.1 lists some typical linear measuring instruments and the accuracy of measurement which can be achieved by them. The accuracy given is the reading accuracy. The measuring accuracy actually achieved will always be less than or equal to the reading accuracy and will depend upon the skill with which the measurement is made.

Table 10.1 *Workshop standards of length*

Name	Range (mm)	Reading accuracy (mm)
Steel rule	150 to 1000	0.5
Vernier caliper	0/150 to 0/2000	0.02 or 0.01
Micrometer caliper	0/25 to 1775/1800	0.01
Slip gauges	1.0025 to 327	0.0025
	(105-piece set)	

All linear measurements are comparative measurements. That is, the component feature being measure is compared directly or indirectly with a standard of length, be it a rule or a set of slip gauges. However, these working standards are themselves based upon the International Standard Metre and the International Yard, which is defined as 0.9144 metres. From 31 January 1964 this became the British legal standard. The International Standard Metre is now defined as the length of the path travelled by light in a vacuum in $1/299\,792\,458$ second. This can be realised in practice through the use of an iodine-stabilised helium–neon laser. The reproducibility is 3 parts in 10^{11}, comparable to measuring the Earth's mean circumference to within one millimetre.

10.3 Measurement of length

10.3.1 *Engineer's rule*

A steel rule is frequently used for making quick measurements of limited accuracy. Figure 10.1 shows various ways of using a rule. To reduce sighting errors and increase the accuracy of the measurement, the datum end of the rule should be aligned with the edge of a component using an abutment as shown. Another way of transferring measurements to or from a rule so as to reduce sighting errors is to use calipers (Fig. 10.2). The use of calipers requires considerable skill in achieving the correct feel.

10.3.2 *Micrometer caliper*

This is one of the most familiar precision measuring devices used in the workshop and the inspection room. Figure 10.3 shows the construction of a typical micrometer and names and describes the more important parts. The operation of a micrometer depends upon the principle that the distance moved by a nut along a screw is proportional to the number of revolutions made by the nut.

Figure 10.4 shows the scales for a metric micrometer with a screw which has a lead of 0.5 mm. The illustrated reading is as follows:

	mm
9 whole millimetres	9.00
1 half millimetre	0.50
48 hundredths of a mm	0.48
Total reading	9.98

Fig. 10.1 *Use of a steel rule: (a) measuring the distance between two scribed lines; (b) measuring the distance between two faces using a hook rule; (c) measuring the distance between two faces using a steel rule and abutment*

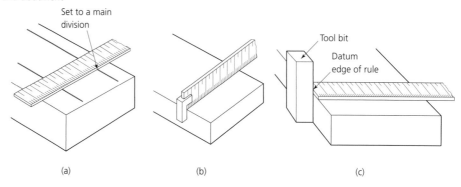

Set to a main division

Tool bit

Datum edge of rule

(a)   (b)   (c)

Fig. 10.2 *Construction and use of calipers*

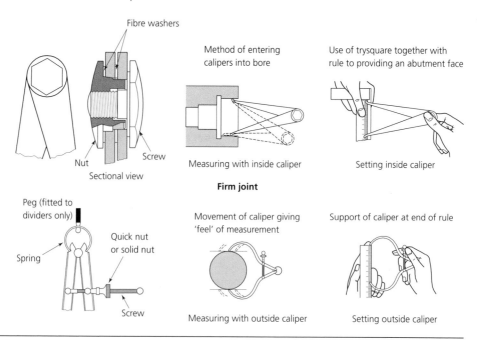

Fibre washers

Method of entering calipers into bore

Use of trysquare together with rule to providing an abutment face

Nut

Screw

Sectional view

Measuring with inside caliper

Setting inside caliper

Firm joint

Peg (fitted to dividers only)

Movement of caliper giving 'feel' of measurement

Support of caliper at end of rule

Quick nut or solid nut

Spring

Screw

Measuring with outside caliper

Setting outside caliper

Fig. 10.3 *Construction of a micrometer caliper*

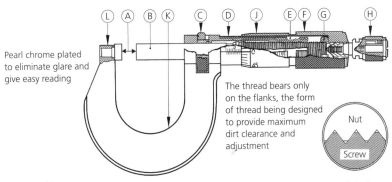

Pearl chrome plated to eliminate glare and give easy reading

The thread bears only on the flanks, the form of thread being designed to provide maximum dirt clearance and adjustment

Nut

Screw

A *Spindle and anvil faces* – Glass hard and optically flat, also available with tungsten carbide faces
B *Spindle* – Thread ground and made from alloy steel, hardened throughout, and stabilised
C *Locknut* – Effective at any position. Spindle retained in perfect alignment
D *Barrel* – Adjustable for zero setting. Accurately divided and clearly marked, pearl chrome plated
E *Main nut* – Length of thread ensures long working life
F *Screw adjusting nut* – For effective adjustment of main nut
G *Thimble adjusting nut* – Controls position of thimble
H *Ratchet* – Ensures a constant measuring pressure
I *Thimble* – Accurately divided and every graduation clearly numbered
J *Steel frame* – Drop forged
K *Anvil end* – Cutaway frame facilitates usage in narrow slots

Fig. 10.4 *Micrometer scales (metric)*

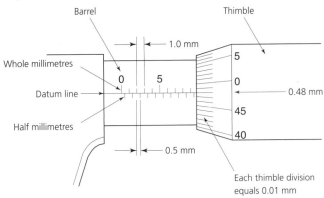

Barrel

Thimble

1.0 mm

Whole millimetres

Datum line

Half millimetres

0.48 mm

0.5 mm

Each thimble division equals 0.01 mm

10.3.3 *Internal micrometer*

An internal micrometer is shown in Fig. 10.5(a). Its range is 50 mm to 210 mm. For any one rod, the scale range is 20 mm. It suffers from severe practical limitations; for instance, it cannot be adjusted readily once it is in a hole or slot, which adversely affects the feel that can be obtained. And it cannot be used for small holes and slots. For cylindrical bores, the cylinder gauge in Fig. 10.5(b) is a more satisfactory instrument. This employs a micrometer-controlled wedge to expand three equispaced anvils until they touch the wall of the bore.

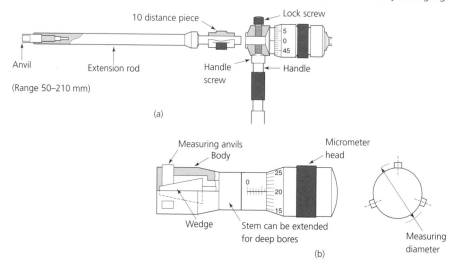

(a)

(b)

10.3.4 *Depth micrometer*

A depth micrometer is shown in Fig. 10.6. Notice that it consists of a micrometer measuring-head together with a number of extension rods. The desired rod can be easily inserted by removing the thimble cap. When the cap is replaced, the rod is held firmly against the positive datum face. The rods are marked with their respective measuring range and are perpendicular to the base at any setting. The measuring faces of the base and rods are hardened. Note that the scales of a depth micrometer give reverse readings when compared with the scales of a micrometer caliper or inside micrometer. Care has to be taken when using a depth micrometer as it is easy for the measuring force on the rod to lift the base off the work.

Fig. 10.6 *Micrometer depth gauge: the measuring faces of the base and rods are hardened; the depth gauge reading is the reverse of ordinary readings*

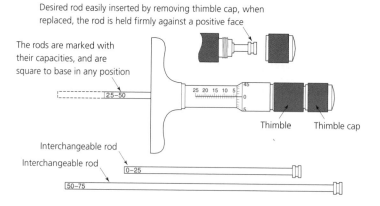

10.3.5 *Vernier caliper*

Figure 10.7(a) shows a vernier caliper. Unlike the micrometer, the same instrument can take inside and outside measurements, as shown in Fig. 10.7(b). The vernier caliper reads from zero to the full length of its beam scale, whereas the micrometer only reads over a range of 25 mm. Unfortunately it is difficult to use a vernier caliper as accurately as a micrometer for the following reasons:

- It is difficult to obtain a correct feel due its size and weight.
- The scales are difficult to read even with the aid of a magnifying glass.

Figure 10.7(c) shows a typical 50-division vernier scale as used on a metric instrument. The zero mark of the vernier scale is just beyond the 32 on the main scale, so the measurement is slightly in excess of 32 mm. The first pair of vernier and main scale graduations to come into line with each other is at 11 on the vernier scale. For the scales shown, since each vernier division is 0.02 mm, the total reading is as follows:

	mm
32 whole millimetres	32.00
11 vernier divisions of 0.02 mm each	0.22
Total reading	32.22

Fig. 10.7 *Vernier caliper: (a) design; (b) use of internal and external jaws; (c) the scale*

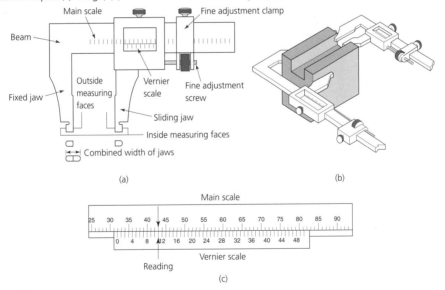

10.3.6 *Vernier height gauge*

In the vernier height gauge of Fig. 10.8(a) the fixed jaw becomes the base of the instrument. This base is the datum from which the measurements and settings can be made. The reading obtained from the main and vernier scales represents the distance from the underside of the base to the upper side of the moving jaw (lower surface of the scribing blade).

The height gauge can be used for a number of applications in the workshop and for inspection. Figure 10.8(b) shows the height gauge being used for accurate marking-out, and Fig. 10.8(c) shows how it can be used to check the centre distance and the height of hole centres from a datum surface. If the scribing blade is used for the application shown in Fig. 10.8(c), the accuracy is limited to the skill of the operator. It is very difficult to obtain a satisfactory feel, due to the mass of the instrument and the friction between its base and the surface plate (datum surface) on which it stands. A *datum* is a point, line, edge or surface which forms a common basis from which a number of measurements can be taken.

As an aid to accuracy, a dial test indicator (DTI) can be used to ensure a constant measuring pressure (feel), as shown in Fig. 10.8(c). In this case the readings obtained from the scales are not absolute distances from the base, and the centre distances and heights from the base of the component are obtained by subtraction of the vernier readings, as shown on the diagram. The setting of the DTI must not be disturbed between readings, and the correct measuring pressure will have been achieved when the vernier scale is adjusted so the DTI reads zero when the stylus is in contact with the surface being checked. In this example the DTI is being used as a *fiducial indicator* to remove errors of feel by ensuring a constant measuring pressure.

Fig. 10.8 *Vernier height gauge: (a) design; (b) marking-out; (c) measuring the height of a surface; errors of 'feel' are eliminated by adjusting the vernier height gauge until the dial test indicator reads zero for each measurement*

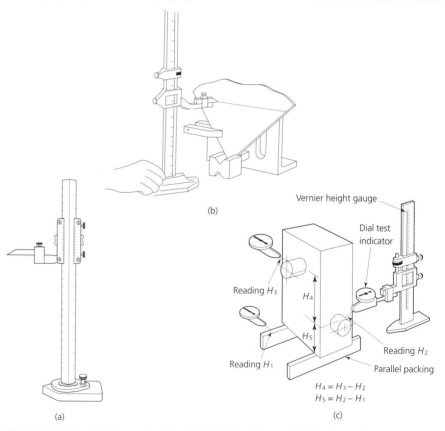

Electronic measuring devices with digital read-outs are now used increasingly; a digital caliper and a digital micrometer are shown in Fig. 10.9. Their read-outs are easier to interpret than the conventional scales and they can be switched between inch and metric units at the touch of a button. Unfortunately, electronic instruments are still more expensive than their conventional counterparts and they are also heavier; the extra weight can give problems of feel. Electronic instruments are more bulky and can give false readings when their internal batteries are running low. The batteries always seem to run down when a replacement is not immediately available.

Fig. 10.9 *Digital measuring equipment: (a) micrometer; (b) caliper*

(a)

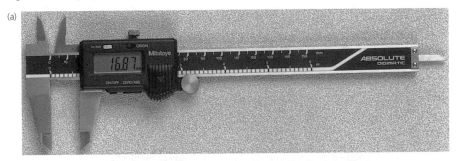

(b)

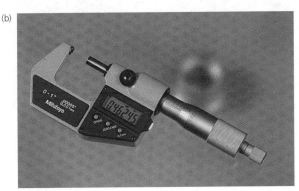

SELF-ASSESSMENT TASK 10.1

1. Sketch micrometer scales (similar to Fig. 10.4) to show the following readings:

   (a) 6.60 mm
   (b) 5.12 mm
   (c) 0.88 mm
   (d) 12.50 mm

2. Sketch vernier caliper scales (similar to Fig. 10.7) to show a reading of 55.62 mm:

   (a) with a 25-division vernier scale
   (b) with a 50-division vernier scale

10.3.7 *Dial Test Indicator (DTI)*

A dial test indicator (DTI) measures the displacement of its plunger or stylus and displays it using a rotating pointer and calibrated dial. Figure 10.10 shows the two most popular types of DTI that are normally used.

The *plunger type* instrument shown in Fig. 10.10(a) relies upon a rack and pinion followed by a geartrain to magnify the displacement of the plunger and rotate the pointer. This type of instrument has a long plunger movement and is fitted with a secondary scale to count the number of revolutions of the main pointer. Various dial markings and magnifications are available.

The *lever type* instrument shown in Fig. 10.10(b) relies upon a lever-and-scroll system of magnification. It has only a limited range of stylus movement, little more than one revolution of the pointer. It is more compact than the plunger type and is widely used for machine setting and inspection, where the position of the dial makes it convenient to read. Typical applications of DTIs are described throughout this book.

Fig. 10.10 *Types of dial test indicator (DTI): (a) plunger; (b) lever*

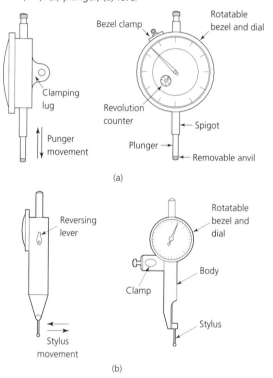

(a)

(b)

10.3.8 *Slip gauges*

Slip gauges are blocks of high-carbon steel which have been hardened and stabilised by heat treatment. They are ground and lapped to very high standards of accuracy and finish (Table 10.2). The lapped measuring surfaces must satisfy the following requirements:

Table 10.2 Accuracy of slip or block gauge

Maximum permissible errors is 1 μm

Size of gauge (mm)		Grade 2*			Grade 1†			Grade 0‡		
Over	Up to and including	Flatness	Parallelism	Gauge length	Flatness	Parallelism	Gauge length	Flatness	Parallelism	Gauge length
–	20	2.5	3.5	+5.0 / −2.5	1.5	2.0	+2.0 / −1.5	1.0	1.0	±1.0
20	60	2.5	3.5	+8.0 / −5.0	1.5	2.0	+3.0 / −2.0	1.0	1.0	±1.5
60	80	2.5	3.5	+12.0 / −7.5	1.5	2.5	+5.0 / −2.5	1.0	1.5	±2.0
80	100	2.5	3.5	+14.0 / −10.0	1.5	2.5	+6.0 / −3.0	1.0	1.5	±2.5

*General workshop applications.
†Precision workshop (toolroom) applications.
‡Inspection.
Note: Two grades are not listed in the table, grade calibration and grade 00. Grade calibration is a reference standard for testing grades 2, 1, 0. Grade 00 is a reference standard used only by slip gauge manufacturers to test all other grades.

- The faces are parallel.
- The faces are flat.
- The face separation is accurate.
- The finish is sufficiently high. A good surface finish means that two slip gauges may be wrung together so they adhere to each other by molecular attraction.

Table 10.3 lists the sizes of the individual slip gauges in a standard 78-piece metric set. The set also contains two protector slips, which are 2.50 mm thick and made from wear-resistant steel or tungsten carbide. They may be added to the end of the stack to protect the other gauge blocks from wear. The required dimension is obtained by wringing together various slip gauges. To obtain maximum accuracy, the following rules must be observed:

- Use the minimum number of blocks.
- Wipe the measuring faces clean.
- *Wring* the individual blocks together.
- *Slide* the blocks apart immediately after use.
- Clean and return the blocks to their case immediately they have been separated.

If slip gauges are left wrung together for too long, they will form a cold weld and their measuring surfaces will be damaged. Therefore the gauge blocks must be separated, cleaned, wiped with petroleum jelly (Vaseline) and returned to their case immediately after use.

Table 10.3 *Slip gauges in a standard 78-piece set*

Range (mm)	Steps (mm)	Pieces
1.01 to 1.49	0.01	49
0.50 to 9.50	0.50	19
10.00 to 50.00	10.00	5
75.00 and 100.00	–	2
1.0025	–	1
1.005	–	1
1.0075	–	1

Figure 10.11 shows how a stack of slip gauges is built up to give a dimension of 39.9725 mm using the sizes listed in Table 10.3. The first slip selected is always the one giving the right-hand digit of the required dimension and the other slips are selected in sequence. If protector slips (2 × 2.5 mm) were to be used with the stack in Fig. 10.11, block C would be reduced to 2.50 mm.

10.3.9 *Comparator*

All measurement is comparative. However, in engineering, the term *comparative measurement* is reserved for the technique shown in Fig. 10.12. Here a plunger type DTI is used to determine the difference between the component size and a length standard by *comparison*. In this example a stack of slip gauges is used as the standard of measurement.

Fig. 10.11 *Building up slip gauges*

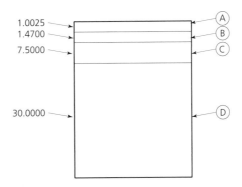

Fig. 10.12 *Comparative measurement: (a) dial gauge mounted on a simple comparator stand; (b) setting and using the comparator*

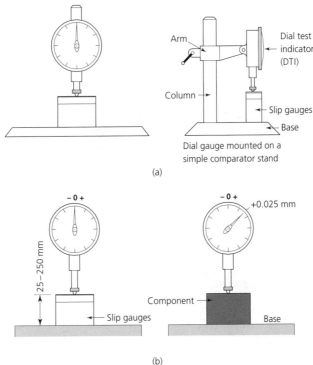

The accuracy of a DTI decreases as the displacement of the plunger gets greater, so the stack of slip gauges should be made as near as possible to the required dimension, making the plunger movement as small as possible.

Figure 10.12(b) shows that the DTI is set to read zero when slip gauges totalling 25.250 mm are placed under the plunger. When the workpiece is exchanged for the slip gauges, the DTI indicates a reading of +0.025 mm. Thus the workpiece dimension is $25.250 + 0.025 = 25.275$ mm. More accurate measurements may be made using pneumatic, electronic and optical comparators, considered in *Manufacturing Technology*, Volume 2.

10.3.10 *Cosine error*

The measuring head of a comparator, be it a DTI or a more sophisticated device, only indicates the displacement of the measuring stylus. Whether or not this is the same as the error in the workpiece will depend upon how the measuring head is set. The most common error of setting is *cosine error*; this occurs when the axis of the measuring stylus is inclined to the datum surface from which the measurement is made (Fig. 10.13).

Fig. 10.13 *Cosine error*

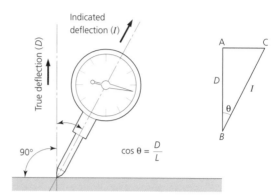

The true deflection D can only be measured along AB, which is perpendicular to the datum surface. When the DTI is inclined along BC, it indicates the stylus deflection I, and the indicated reading is greater than the actual error value:

$$I = D/\cos\theta$$

where D = true deflection
  I = indicated deflection

Thus, for the DTI to indicate the actual error, it must be set *perpendicular to the datum surface*.

SELF-ASSESSMENT TASK 10.2

1. Using the slip gauges listed in Table 10.3, select suitable gauge blocks to build up the dimension 57.6825 mm:

   (a) without protector slips
   (b) with protector slips

2. A DTI and slip gauges are used to make a comparative measurement of a component. The DTI is set to zero over a stack of slip gauges of height 30.375 mm. The DTI shows a reading of −0.032 mm over the component. State the actual component size.

3. A DTI shows a reading of 0.5 mm when inclined at 15° to the perpendicular.

   (a) Calculate the reading on a DTI perpendicular to the component surface.
   (b) What is the numerical value of the cosine error for the inclined DTI.

10.4 Measurement of angles

10.4.1 *Protractors*

Simple bevel protractors of the type shown in Fig. 10.14(a) have a reading accuracy of only ±0.5°. However, by the addition of a circular vernier scale, as shown in Fig. 10.14(b), the reading accuracy can be greatly increased. Figure 10.14(c) shows the scales greatly enlarged. The main scale is graduated in degrees of arc and the vernier scale has 12 divisions each side of zero. These divisions mark out the range 0 to 60 minutes of arc, so each division equals $1/12 \times 60 = 5$ minutes of arc. Thus the reading of a vernier bevel protractor is the sum of two quantities:

- The largest whole-degree value on the main scale indicated by the vernier zero graduation.
- The reading on the vernier scale in line with a main scale graduation.

Thus the reading of the scales shown in Fig. 10.14 is:

17 'whole' degrees	17°
Fifth vernier division (25 mark) is in line with a main scale graduation	0° 25′
Total reading	17° 25′

Fig. 10.14 *Protractors: (a) plain bevel; (b) universal bevel (vernier protractor): A = blade; B = stock; C = graduated dial; D = vernier scale; (c) vernier scales*

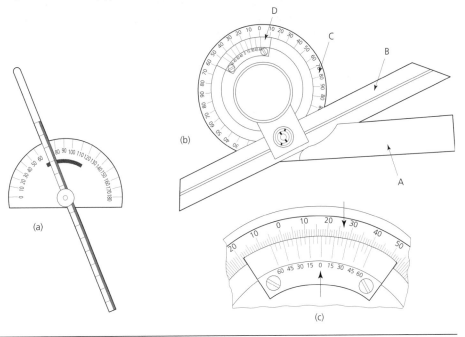

10.4.2 *Sine bar*

There are no standards for angles. But all angles can be defined in terms of linear dimensions, so ultimately they are referred back to the International Standard Metre. A sine bar uses linear measurements to define angles to a high degree of accuracy (Fig. 10.15). Figure 10.15(b) shows that the sine bar, slip gauges and datum surface form a right-angled triangle. The sine bar itself forms the hypotenuse and the slip gauges form the opposite side to the angle being measured.

Since $\sin \theta = \dfrac{\text{opposite side}}{\text{hypotenuse}}$

then $\sin \theta = \dfrac{\text{height of slip gauges}}{\text{centre distance of roller axes}}$

$= H/L$

The centre distance of the contact roller axes is the nominal length of the sine bar; the nominal length is usually 10 in or 250 mm.

Figure 10.16 shows how the sine bar is used to check small components mounted upon it. The DTI is mounted upon a suitable stand, such as a vernier height gauge, which is rigid and can provide fine adjustment. The DTI is moved over the component at position A in Fig. 10.16 and zeroed. The stand and DTI is then slid along the datum surface to position B, and the DTI reading is noted.

Method 1
The height of the slip gauges is adjusted until the DTI reads zero at each end of the component. The actual angle can then be calculated using trigonometry, as previously described, and any deviation from the specified angle is the error.

Method 2
The sine bar is set to the specified angle. The DTI will then indicate any error as a run of so many hundredths of a millimetre along the length of the component. Providing the DTI was set to zero in the first position, the error or run will be shown as a plus or minus reading at the second position.

Fig. 10.15 *The sine bar: (a) design; (b) principles*

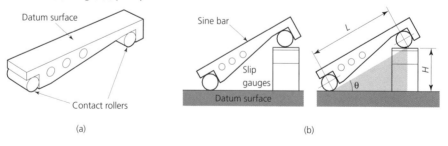

(a)   (b)

Fig. 10.16 *Using the sine bar on small components*

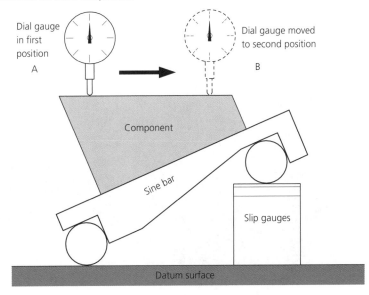

Dial gauge in first position
A

Dial gauge moved to second position
B

Component

Sine bar

Slip gauges

Datum surface

Fig. 10.17 *Types of sine table: (a) simple; (b) compound*

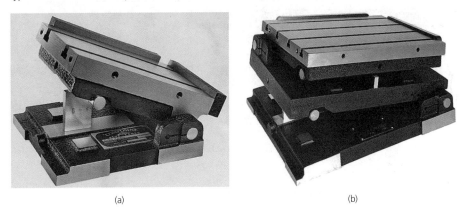

(a)

(b)

A *sine table* (Fig. 10.17) may be used for large work. In effect, it is a sine bar with a wide working surface and a T-slot for clamping the work. *Sine centres* (Fig. 10.18) may be used for cylindrical work which has been turned or ground between centres or which can be mounted on a mandrel. Further applications of the sine bar and alternative methods of precision angular measurements using optical measuring instruments, such as an autocollimator or an angle-dekkor, together with combination angle gauges, will be considered in *Manufacturing Technology*, Volume 2.

Fig. 10.18 *Sine centres*

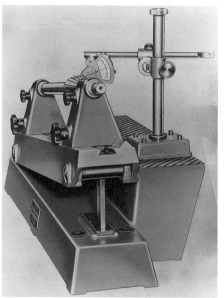

SELF-ASSESSMENT TASK 10.3

1. Sketch vernier protractor scales to show a reading of 30° 15'.

2. Calculate the height of the slip gauge stack needed to give an angle of 20° when using a 250 mm sine bar.

3. What will be the actual angle of a component being checked on the set-up in Question 2 if it shows a run of +0.5 mm in 125 mm at the narrowest end of the component.

10.5 Reference surfaces

Datum surfaces for linear distances and angles have already been introduced. But they are equally important when determining straightness, flatness and squareness. Recognising that an accurate reference surface was essential for precision measurement, as long ago as 1830 Sir Joseph Whitworth devised the three-plate technique of surface generation (Fig. 10.19).

The three-plate technique is based upon the principle that it is impossible to devise any combination of mating three surfaces together unless they are flat. Plates A and B can be scraped so that they mate together without being flat. This out-of-flatness is shown up as soon as they are compared with plate C. It is only by working the plates together so that they are fully interchangeable that three perfectly flat surfaces are generated.

Fig. 10.19 *Generation of flat surfaces*

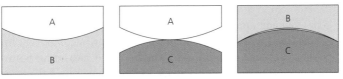

Fig. 10.20 *Surface plate*

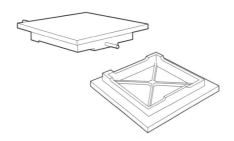

Reference surfaces in engineering workshops and inspection rooms consist of surface plates and surface tables. The most accurate reference surface is the toolmaker's flat. This is a disc of hardened and stabilised steel some 200 mm diameter and with finely lapped surfaces. These surfaces should be finished so that slip gauges may be wrung to them, and when viewed through an optical flat should show straight interference fringes.

Surface plates and surface tables are made from cast iron and their undersurfaces are heavily ribbed for strength and rigidity (Fig. 10.20). The working surfaces are planed for workshop use and hand scraped for inspection and reference use. They are stabilised by weathering or by heat treatment between rough machining and finishing. Plate glass and granite are also used for reference surfaces as they are extremely stable. They also have the advantage that, if accidentally scratched, they do not throw up a burr. Unfortunately they do not have the antifriction properties of cast iron and they have a less pleasant feel in use.

10.6 Straightness and straight edges

Steel straight edges are made up to 2 m in length and may be rectangular in section or have a bevelled edge as shown in Fig. 10.21(a). Cast-iron straight edges are made from close-grained cast iron. After rough machining they are stabilised by weathering or heat treatment and then finish-machined and scraped to a reference surface. They are made up to 3 m long and are widely used for checking machine-tool slideways.

Cast-iron straight edges can be heavily ribbed and bow-shaped (camel-backed) to prevent distortion, as shown in Fig. 10.21(b). When not in use they should be placed upon their feet, which are provided at the points of minimum deflection. Some confusion often occurs between the Airey points and the points of minimum deflection.

The *Airey points* (Fig. 10.21(c)) are used for supporting length standards such as combination length bars so that the measuring ends are perpendicular and parallel, and are drawn up into this position by the sag along the unsupported length of the bar.

The *points of minimum deflection* are shown in Fig. 10.21(d). Slight sagging of the ends of the bar help to draw up the unsupported middle length of the bar. Thus all deflections are kept to a minimum, but the ends are not perpendicular or parallel.

Just as it is not possible to make or measure a component to an exact size, neither is it possible to make it to an exact shape. When measuring both straightness and flatness, reference is made to an imaginary, perfect plane called the *mean true plane*. Figure 10.22 shows that it is a perfectly flat plane situated relative to an actual surface so that the plus errors and minus errors are equally balanced above and below the plane.

In fairly coarse measuring situations, the deviation of a surface from the mean true plane can be determined approximately (Fig. 10.23). The irregular gap between the straight edge and the surface being tested can be determined by the use of a feeler gauge.

Fig. 10.21 *Straight edges: (a) steel straight edge; (b) cast-iron straight edge; (c) Airey points – the sag S_1 that occurs in a bar supported at the Airey points is so arranged that it pulls the ends of the bar up square with the measuring plane; (d) points of minimum deflection – for minimum deflection some sag S_2 at the ends of the bar is permissible to reduce the sag S_3 at the centre; when $L_2 = 0.544L_1$, S_2 and S_3 are at a minimum and considerably less than S_1*

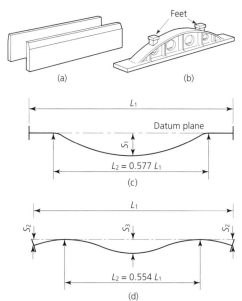

Fig. 10.22 *Mean true plane*

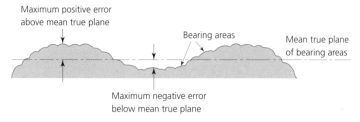

Fig. 10.23 *Use of a straight edge: feeler gauge*

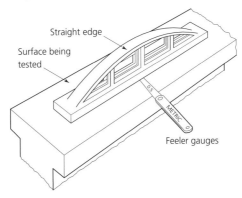

Alternatively the cast-iron straight edge may be used as a narrow surface plate. Engineer's blue, smeared on the straight edge, is transferred on contact to the high spots of the surface being tested. These high spots are hand scraped until the pattern shows uniform bearing surfaces along the length being tested (Section 10.8).

Small components can be tested for straightness with a precision straight edge and a light box (Fig. 10.24). The light box provides a source of uniform illumination that is independent of the viewing angle. If there is a gap between the straight edge and the component, a strip of light will be visible. With practice the error of straightness can be estimated from the quality of the light:

- White light indicates a gap greater than 0.002 mm.
- Tinted light indicates a gap between 0.001 mm and 0.002 mm.

This technique can also be used with precision squares. Figure 10.25(a) shows a simple technique to calibrate a straight edge and to test components when the straight edge has been calibrated. It uses the principle of a wedge to magnify any errors which are present. Since some of the errors may be attributed to the datum surface upon which the tests are

Fig. 10.24 *Light box: any error of flatness in the component will leave a gap through which light will be visible*

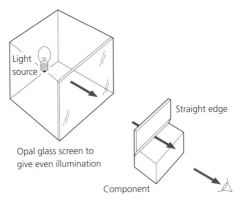

Fig. 10.25 *Wedge method: (a) using a straight edge; (b) results of the straightness test*

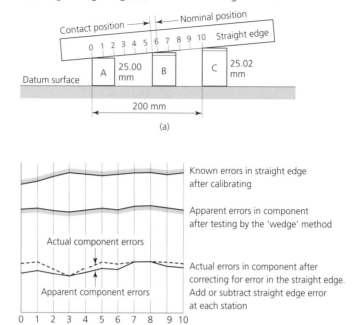

being made, only a grade A inspection or reference surface is suitable. First the straight edge is calibrated as follows:

- The straight edge is supported on slip gauges at the points A and C. These points are chosen to be as near as possible to the points of minimum deflection (explained in Fig. 10.21) and the slip gauges are chosen to give a rise of 0.02 mm.
- The distance on the straight edge between the points of support is divided into 10 equal parts. Thus the rise should be 0.002 mm per division.
- Knowing the height at A and the increase in height at each successive point along the straight edge, slip gauges can be used to test the height at each point, and any deviation is noted on a calibration chart. For instance, at point 6 the slip gauges B should equal 25.00 mm + (6 × 0.002 mm) = 25.012 mm. However, a slip gauge stack of this magnitude makes contact with the straight edge between points 5 and 6, showing there is a hollow at this point.
- The slip gauges are now built up to the actual height at point 6. The difference between the actual height and 12.012 mm is the error at this point and is recorded on the calibration chart.

Once the straight edge has been calibrated against a reference surface, it can then be used to test other surfaces by the same technique. The apparent errors for the surface under test are plotted graphically along with the known errors in the calibrated straight edge, as shown in Fig. 10.25(b). The two sets of errors are then added algebraically to obtain the actual errors in the surface.

10.7 Flatness

Testing large areas for flatness requires sophisticated equipment and techniques. However, for surfaces of limited accuracy and size, the following technique can be used.

The surface to be checked is marked out into a series of parallel bands (Fig. 10.26); use a soft lead pencil, so as not to damage the surface. It is then checked using a bevel-edge straight edge and feeler gauges. Errors should be plotted for each band as for testing straightness. If the surface is large enough, the wedge technique used for testing straightness can be used for each band lengthways and crossways. These results can then be fed into a computer set up for surface modelling and a magnified image of the surface can be produced. Care is necessary when checking for surface flatness with a straight edge. Figure 10.26 shows that, although the surface is winding, it will appear flat if measured parallel to the sides. Therefore additional checks should be made across the diagonals to detect errors of flatness due to winding.

Small surfaces can be checked against a reference surface. The surface to be checked is first carefully cleaned and any burrs are removed with an oilstone. Engineer's blue is then smeared over this surface. The blued surface is inverted over a reference surface (inspection grade surface plate) and gently moved in a series of small circles, as shown in Fig. 10.27(a).

Fig. 10.26 *Testing for flatness using a straight edge*

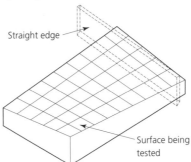

Fig. 10.27 *Testing for flatness using a reference surface: (a) the component is moved across the blued surface plate in small circles; (b) engineer's blue is transferred to any high spots on the component*

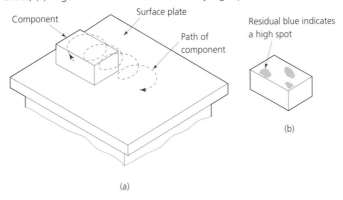

This will transfer the blue to the high spots of the scraped reference surface lightly and uniformly. Any attempt to apply blue directly to the reference surface usually produces a poor distribution, leading to a false interpretation of flatness.

The surface to be checked is now cleaned of blue and once more brought into contact with the reference surface, using the same rotary motion. Residual blue will now be transferred back to any high spots on the surface under test, as shown in Fig. 10.27(b). These high spots may be broken down by hand scraping, and the checking process repeated. The sequence of checking and scraping is repeated until the surface under test has a uniform pattern of small, closely spaced high spots over its whole area.

10.8 Squareness

As well as straightness and flatness, the engineer is concerned with perpendicularity or squareness. Although sophisticated optical equipment and techniques are available for work of the highest precision, small components can be checked using a standard try square. Some examples of try squares to BS 939 are shown in Fig. 10.28:

- Figure 10.28(a) shows a grade B try square with precision ground, hardened and tempered blade.
- Figure 10.28(b) shows a grade AA (reference) try square for sizes up to 300 mm blade length. Both the stock and the blade are hardened and precision ground. The bevel edges are left glass hard.

Fig. 10.28 *Engineer's try square*

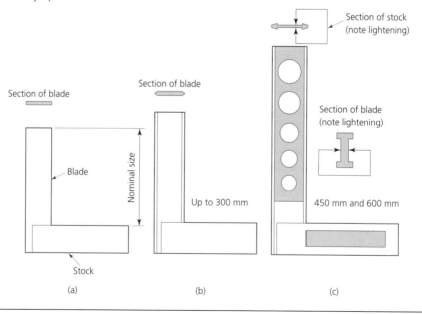

- Figure 10.28(c) shows a grade AA (reference) try square for sizes from 450 mm to 600 mm in length. The stock and blade are both hardened and precision ground. The bevel edges of the blade are left glass hard. The stock and the blade have been designed in conjunction with the National Physical Laboratory (NPL) to provide maximum rigidity with minimum weight.

Although not a British Standard specification, many inspection rooms favour a cylinder square (Fig. 10.29) when working from the datum surface of a surface plate or table. The cylinder square is usually made from through-hardened or case-hardened steel; the body and the base are ground at the same setting on a precision cylindrical grinding machine. The advantages claimed for this type of square are its stability when standing on a datum surface, and the fact that it makes line contact with a flat surface.

Figure 10.30 shows applications of a try square for testing perpendicular surfaces. In Fig. 10.30(a) the stock is placed against edge AB of the component and slid gently downwards until the blade comes into contact with edge BC. Any lack of squareness between edges AB and BC will allow light to be seen between BC and the blade of the try square.

Fig. 10.29 *Cylinder square*

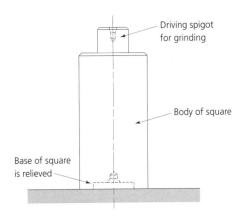

Fig. 10.30 *Use of a try square*

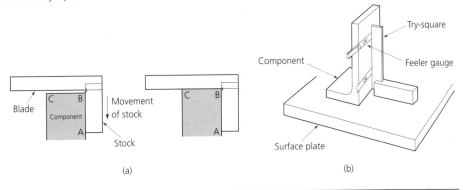

(a)

(b)

It is not always convenient to hold a large component and try square up to the light, and Fig. 10.30(b) shows an alternative technique. A surface plate is used as a datum and the squareness of the component is checked with a feeler gauge. Alternatively, a light box may be used for a visual check of any gap between the component and the blade of the try square.

A more accurate method is to use a squareness comparator (Fig. 10.31). The DTI is set to zero with the comparator in contact with a known squareness standard such as a grade AA try square, as shown in Fig. 10.31(a). When the comparator is brought into contact with the component being checked, as shown in Fig. 10.31(b), any error is indicated as a plus or minus reading on the DTI. Since the centre distance between the fixed contact and the DTI plunger is known, the angular error may be calculated using trigonometry.

Fig. 10.31 *Squareness comparator: (a) squareness comparator zeroed against a known square; (b) squareness error shows up as a + or − reading on the DTI*

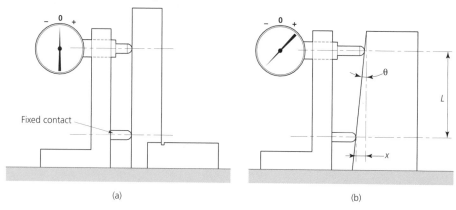

(a)   (b)

SELF-ASSESSMENT TASK 10.4

1. List the equipment used for checking flatness and straightness under workshop conditions and give an example where each would be used.

2. Compare the advantages and limitations of a try square and a cylinder square for checking perpendicularity.

3. Calculate the error of perpendicularity if the DTI of a squareness comparator shows an error of 0.15 mm over a length of 150 mm.

10.9 Surface texture

The marked improvement in the reliability of engineering mechanisms in recent years is largely due to a better understanding of surface texture (finish) and new techniques for measuring it. It is no use specifying close dimensional tolerances to a process whose inherent surface roughness lies outside that tolerance, as shown in Fig. 10.32(a). The

Fig. 10.32 *Surface characteristics: (a) mismatch between process and limits of size; (b) process suitable for limits of size; (c) surface with low rate of wear characteristics; (d) surface with high rate of wear characteristics*

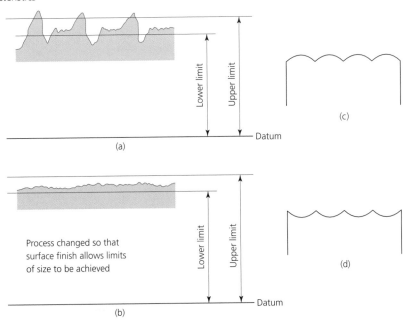

tolerance and the surface finish of the manufacturing process must be matched, as shown in Fig. 10.32(b).

Inversely proportional to the rate of wear, the life of any mating surfaces, such as shafts and bearings, will depend on their surface finishes. A rough surface with large peaks and valleys will have less contact area and will wear more quickly than a smoother surface. Even two surfaces having the same roughness index can have different wear characteristics. Figure 10.32 shows two surfaces having the same roughness index – their peaks have the same height, their valleys have the same depth, and their peaks and valleys have the same spacing. But under the same conditions of service it is obvious that the surface in Fig. 10.32(c) will wear less quickly than the surface in Fig. 10.32(d).

BS 1134 discusses fully the assessment of surface texture, but some fundamental principles will now be considered. Figure 10.33 will help to explain the following definitions as applied to surface texture assessment:

- *Real surface* The actual physical surface separating the component from the surrounding space.
- *Effective surface (measured surface)* The close representation of a real surface obtained by instrumental means.
- *Effective profile (measured profile)* The contour which results from the intersection of the effective surface by a plane conventionally defined with respect to the geometrical surface.
- *Irregularities* The peaks and valleys of a real surface.
- *Spacing* The average distance between the dominant peaks on the effective profile.

Fig. 10.33 *Surface texture terminology*

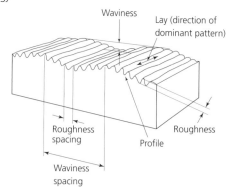

- *Surface texture* Irregularities with regular or irregular spacing which tend to form a pattern or texture on the surface. The texture itself may contain roughness and waviness components.
- *Roughness* Irregularities in the surface texture that are inherent in the production process, but not waviness and errors of form.
- *Waviness* The component of surface texture upon which the roughness is superimposed. Waviness may result from such factors as machine or work deflections, vibrations, chatter, and heat treatment or warping strains.
- *Lay* The direction of the predominant surface pattern, ordinarily determined by the production method used. Surface measurement is usually carried out at right angles to the lay.
- *Sampling length* The profile length selected for an individual measurement of surface texture.
- *Reference line* The line chosen by convention to serve for quantitative evaluation of the roughness of the effective profile. The preferred methods of grading surface texture currently employed are the arithmetic mean deviation (R_a), formerly known as the centre line average (CLA) height index, and the peak-to-valley height index (R_z).

10.9.1 *Determining* R$_a$

R_a is the universally recognised parameter of roughness; it is the arithmetic mean of the departures of the profile from the mean line. It is rare to quote R_a from a single sample on the surface. Several samples are normally chosen and R_a is obtained for each one. These R_a values are then averaged to give an overall value for the surface.

When determining R_a from graphical recordings of surface texture, it is necessary to determine the centre line of the sample. This can be done electronically or manually, as shown in Fig. 10.34(a):

- Draw a line AB so that it grazes the deepest valley and lies parallel to the general course of the record over the sampling length L.
- Measure the area P (shaded) with a planimeter; thus the mean height $H_m = P/L$.

Fig. 10.34 *Determination of R_a values*

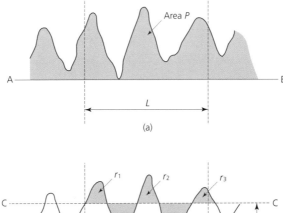

(a)

(b)

- The centre line CC can now be drawn parallel to the line AB at a height H_m above it. Figure 10.34(b) shows that R_a can now be obtained as follows:

$$R_a = \frac{\sum \text{areas } r + \sum \text{areas } s}{L} \times \frac{1000}{V_m}$$

where R_a = arithmetical mean deviation in microns (0.001 mm)
L = sampling length in millimetres
r = areas of peaks in square millimetres
s = areas of valleys in square millimetres
V_m = vertical magnification
$\sum$ = a mathematical symbol meaning 'the sum of'

This gives R_a for one sample along the entire traversing length. By repeating the process for several samples and taking the average, an overall value can be obtained for the entire traversing length. Table 10.4 relates this overall value to the roughness grade number and Table 10.5 lists the roughness grade numbers for some typical manufacturing processes.

10.9.2 *Determining* R$_z$

R_z is also known as the ISO 10-point height parameter. It is measured over a single sampling length and is itself an average of several positive and negative peak values; it is a useful parameter when only a short length of surface is available for assessment. Figure 10.35 shows how R_z is determined graphically.

Table 10.4 *Relation between R_a and roughness grade number*

R_a value (mm)	Roughness grade number
50	N12
25	N11
12.5	N10
6.3	N9
3.2	N8
1.6	N7
0.8	N6
0.4	N5
0.2	N4
0.1	N3
0.05	N2
0.025	N1
0.0125	–

Table 10.5 *Roughness grade numbers for some typical processes*

Process	Roughness grade number
Casting, forging, hot rolling	N11–N12
Rough turning	N9
Shaping and planing	N7
Milling (HSS cutters)	N6
Drilling	N6–N10
Finish turning	N5–N8
Reaming	N5–N8
Commercial grinding	N5–N8
Finish grinding (toolroom)	N2–N4
Honing and lapping	N1–N6
Diamond turning	N3–N6

Fig. 10.35 *Determination of R_z values*

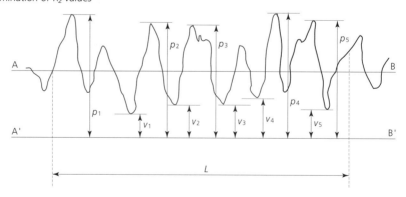

The five highest peaks and the five deepest valleys are conveniently measured from an arbitrary datum line A′ B′ drawn parallel to the centre line AB for a chosen length L, so that:

$$R_z = (p_1 + p_2 + \ldots + p_5) - \tfrac{1}{5}(v_1 + v_2 + \ldots V_5)$$

Figure 10.33 showed that most machined surfaces consist of a number of irregularities superimposed upon each other. If a very short length of surface is tested, only the roughness will register, so R_a will be low. But if a longer length is tested, both the roughness and the waviness will register, so R_a will be higher. Therefore, for comparative purposes, the test length must be standardised. Sample lengths currently in use are 0.25 mm, 0.80 mm and 2.50 mm; they are called cut-off wavelengths and the most frequently used is 0.80 mm.

Various methods of surface texture assessment have been derived, but the most common is electronic magnification as in the Taylor–Hobson Talysurf. The Talysurf uses a device similar to the pick-up on a record-player; the pick-up is drawn over the surface under test and the signals, produced by displacement of the stylus, are amplified and printed on a paper tape.

Figure 10.36 shows a simple, portable set of electronic surface texture assessment equipment. Notice that it consists of a traverse unit and an amplifier/recorder unit. The amplifier contains electronic filters that enable different parameters of the surface texture to be studied. The traverser may also be mounted on a column and base plate to facilitate setting up. Figure 10.37 shows a typical printout for a ground surface. In this example the vertical magnification is ×20,000; the horizontal magnification is ×100, and one division represents one micrometre (1 μm = 0.000 001 m). The R_a value for this example is approximately 0.2 μm.

Fig. 10.36 *Electronic surface texture assessment equipment*

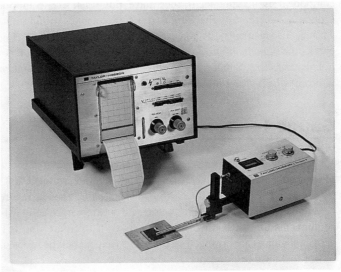

Fig. 10.37 *Typical surface texture traces*

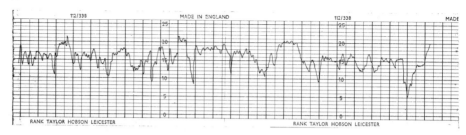

SELF-ASSESSMENT TASK 10.5

1. Explain why it is important to take surface texture into account when setting the tolerance for a dimension.

2. With reference to BS 308, show how surface texture values are added to the standard machining symbol and how they are applied to a typical component drawing.

10.10 Gauging

When we discussed assembly in Chapter 6, we considered the following topics in some detail: accuracy of fit, toleranced dimensions, systems of limits and fits, tolerance grade and interchangeability. The individual measurement of toleranced dimensions is too costly for the inspection of mass-produced components; this is because of the time required for each test and the skill required by each inspector.

For routine inspection, knowledge of the actual size of the component is rarely required. All that needs to be known is whether or not the component dimensions have been produced within the limits of size laid down by the designer. If they have, then the component is acceptable; if they have not, then the component is scrap. The most economical method of checking that a component is within limits is to use a *limit gauge*. This has two elements, the go element and the not-go element (Fig. 10.38). The go element checks the *maximum metal condition* and indicates whether or not a shaft is oversize or a hole is undersize. Similarly, the not-go element checks the *minimum metal condition* and indicates whether or not a shaft is undersize or a hole is oversize.

10.11 Gauge tolerances

Like any other component, gauges cannot be manufactured exactly to size but must have toleranced dimensions. These tolerances must be so arranged that the gauge wears 'better'. You can see from Fig. 10.39 that, at first, the gauge may reject marginally correct components. However, as the go element wears with use, it will accept more and more components until it has worn outside its own tolerance and is itself rejected. Note that gauge tolerances are arranged so the gauge element will be rejected before it starts to accept out-of-tolerance components.

Fig. 10.38 *Gauge elements: (a) plug gauge; (b) caliper gauge*

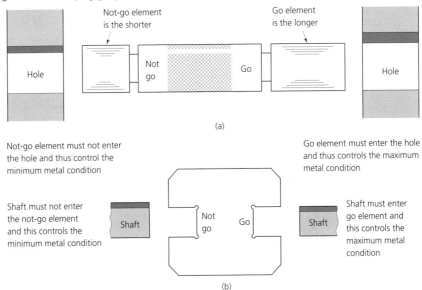

(a)

Not-go element must not enter
the hole and thus control the
minimum metal condition

Go element must enter the hole
and thus controls the maximum
metal condition

Shaft must not enter
the not-go element
and this controls the
minimum metal condition

Shaft must enter
go element and
this controls the
maximum metal
condition

(b)

Fig. 10.39 *Gauge tolerance zones: (a) plug gauge; (b) gap and ring gauges. (T= gauge tolerance; W = wear allowance (go gauge only))*

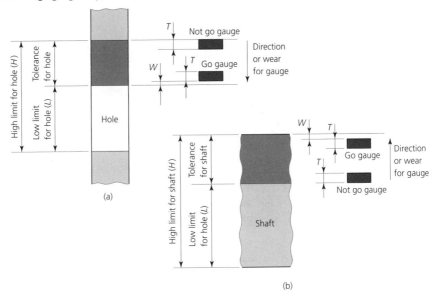

Figure 10.40 shows how the British Standard for limits and tolerances on plain limit gauges are applied to a plug gauge for checking a 75 mm H8 hole. The hole dimensions are determined from tables (Section 6.5) and are shown in Fig. 10.40(a). Notice that the hole tolerance is:

$$75.045\,\text{mm} - 75.00\,\text{mm} = 0.046\,\text{mm}$$

Fig. 10.40 *This example is based on tolerancing a plug gauge suitable for checking a 75 mm H8 hole: (a) to use the BS limit system, first calculate the hole tolerance, 75.046 – 75.00 = 0.046 mm; (b) refer to the appropriate BS 969: 1982 tables for a 0.046 mm work tolerance; (c) apply the limits obtained from the tables; T = tolerance; (d) dimension the plug gauge in accordance with diagram (c)*

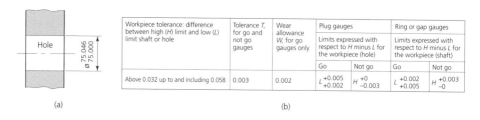

(a)

Workpiece tolerance: difference between high (H) limit and low (L) limit shaft or hole	Tolerance T, for go and not go gauges	Wear allowance W, for go gauges only	Plug gauges		Ring or gap gauges	
			Limits expressed with respect to H minus L for the workpiece (hole)		Limits expressed with respect to H minus L for the workpiece (shaft)	
			Go	Not go	Go	Not go
Above 0.032 up to and including 0.058	0.003	0.002	$L\,^{+0.005}_{+0.002}$ $H\,^{+0}_{-0.003}$		$L\,^{+0.002}_{+0.005}$ $H\,^{+0.003}_{-0}$	

(b)

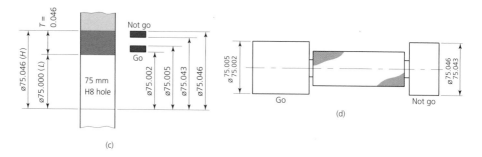

(c)

(d)

The tables for gauge tolerances are entered for a 0.046 mm work tolerance, as shown in Fig. 10.40(b). The remaining columns give the tolerances for the go and not-go gauge elements, the wear allowance (applicable to the go element only) and the upper and lower limits for the go and not-go elements of the plug gauge used to check the hole we are discussing. Figure 10.40(c) shows how the go limits are applied to the lower limit of the hole and the not-go limits are applied to the upper limit of the hole. Therefore:

- Go element
  upper limit = 75.000 mm + 0.005 mm = 75.005 mm
  lower limit = 75.000 mm + 0.002 mm = 75.002 mm

- Not-go element
  upper limit = 75.046 mm + 0.000 mm = 75.046 mm
  lower limit = 75.046 mm + 0.003 mm = 75.043 mm

Figure 10.40(d) shows how these limits of size are applied to the plug gauge itself. Similarly, gap and ring gauges for shafts are toleranced using the prescribed limits in the appropriate British Standard.

10.12 Taylor's principle of gauging

Taylor's principle states that a gauge should be designed so it checks the *geometrical* and the *dimensional* accuracy of a component.

10.12.1 *Not-go element*

Figure 10.41 shows that the not-go element of a gauge should be *form relieved* so it can check each feature of a particular dimension individually. You can see from Fig. 10.41(a) that a full-form not-go element (shaded) would not enter the oval hole and would accept it as correct in size and shape. However, the form-relieved not-go element in Fig. 10.41(b) would detect the ovality by entering in position A but not entering in position B.

Fig. 10.41 *Not-go element gauge design: (a) incorrect; (b) correct*

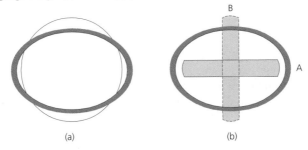

(a)   (b)

10.12.2 *Go element*

The go element of a gauge should be *full form* and it should simultaneously check as many features as possible for a particular dimension. Figure 10.42(a) shows how a full-form go element (shaded) will not enter a hole that is dimensionally correct but has an incorrect geometry (constant-diameter lobed figure). The hole is rejected for incorrect geometry. Had the go element been form relieved, as shown in Fig. 10.42(b), it would have

Fig. 10.42 *Go element gauge design: (a) and (c) correct; (b) and (d) incorrect*

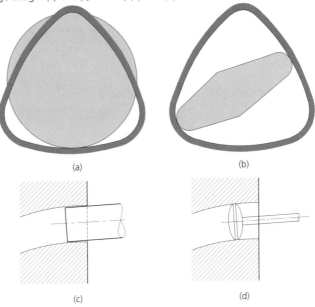

(a)   (b)

(c)   (d)

entered and accepted the out-of-round hole. Therefore the go gauge element must never be form relieved.

Figure 10.42(c) shows how a full-form go element will reject a hole that is dimensionally correct but is distorted axially (not straight). A form-relieved go element would have followed the run-out and accepted the hole, as shown in Fig. 10.42(d). Unfortunately, for practical reasons, it is not always possible to incorporate Taylor's principle into gauge design.

10.13 Plug gauges

Figure 10.43 shows both solid and renewable double-ended plug gauges. The solid type is usually reserved for small diameters, where it is lighter in weight and gives a better feel to the gauging operation. The renewable type is satisfactory for general-purpose medium-range applications. A renewable plug gauge has the following advantages over the solid type:

- The plastic handle does not transmit the heat of the inspector's hand to the gauge elements. It also reduces the weight of the gauge and has a more pleasant feel than metal.
- The gauge elements are kept in a range of stock sizes and only require grinding and lapping to size, reducing manufacturing costs and lead times.
- Since the go element wears out first, it can be replaced without scrapping the whole gauge. The go element of a plug gauge is identified by the fact that it is always longer than the not-go element.

Both solid and renewable gauge elements can be reclaimed by hard chrome plating followed by regrinding and lapping to size. Many new gauges are hard chrome plated to

Fig. 10.43 *Plug gauge types: (a) renewable go and not-go plug gauge; (b) solid go and not-go plug gauge*

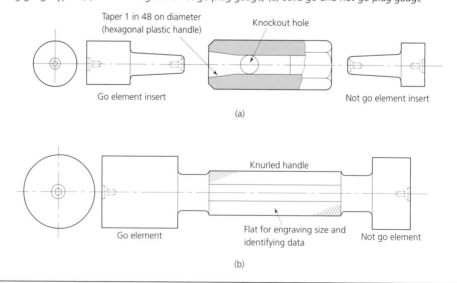

reduce wear. Hard chrome plating must not be confused with decorative chrome plating, which offers little or no mechanical protection.

You can see from Fig. 10.43 that the not-go element is not form relieved for small gauges. This not only speeds up the inspection process and reduce costs, but small holes produced by reaming and similar processes will have correct geometry by the very nature of the process. For larger-diameter holes, single-ended gauges are used to reduce the weight and improve the feel. The go element is a full-form plug gauge, but the not-go element is a bar type gauge (Fig. 10.44). Since a bar type not-go element is by its very nature *form relieved*, it obeys the requirements of Taylor's principle. This is an added advantage, since the processes used to make large holes are more likely to produce holes with geometrical errors than the processes used to make small holes.

Fig. 10.44 *Single-ended bar-type plug gauges: (a) forged steel blank; (b) the no. 6 handle is chamfered at 30° to suit the diameter of the boss*

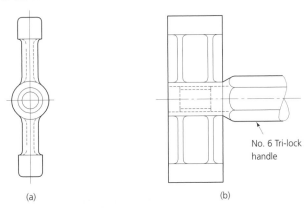

No. 6 Tri-lock handle

(a)   (b)

10.14 Gap and ring gauges

Gap gauges are also known as snap gauges and caliper gauges. Figure 10.45(a) shows examples of single- and double-ended gap gauges; they are made from gauge plate (a ground alloy steel) in the smaller sizes and forgings in the larger sizes. Notice that gap gauges do not follow Taylor's principle so ring gauges are often preferred for cylindrical work (Fig. 10.45(b)). In addition to the solid gap gauges shown so far, adjustable gauges are also available (Fig. 10.46). The advantages of adjustable gap gauges are:

- They can be reset to compensate for wear.
- Unlike the solid gauges, which must be discarded when redundant, adjustable gauges can be reset and reused for each new application.

Against these advantages must be set their higher initial cost and their greater weight, which detracts from their feel. There are many types of limit gauge for checking linear and angular dimensions, and only a few have been considered in this chapter. For the full range of gauge types, refer to specialist texts on metrology and gauging.

Fig. 10.45 *Gap and ring gauges: (a) double-ended gap gauge, ground steel plate stock; (b) solid ring gauge; (c) type A, ground steel plate stock; (d) type B, forged steel blank*

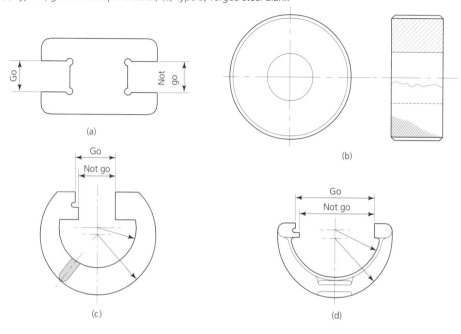

(a)

(b)

(c)

(d)

Fig. 10.46 *Adjustable gap gauges*

1. Compare the advantages and limitations of measurement and gauging when checking the dimensional accuracy of a component.

2. Discuss the difficulties of applying Taylor's principles of gauging.

3. Using the data in Fig. 10.36(b), calculate the dimensions of a plug gauge for checking a hole whose diameter is 70 ± 0.02 mm.

10.15 Monitoring quality

Although the cost of inspection by gauging is very much lower than inspection by measurement, it is still too expensive – and in many cases unnecessary – to gauge every component when using quantity production techniques. In most cases it is possible to use sampling and statistical methods based on the mathematics of probability; that is, the mathematical prediction of the number of rejects in a large batch of components from the inspection of a relatively small sample. *Statistical quality control* (SQC) is a complex subject and only a few basic principles will be considered here. It will be considered in greater depth in *Manufacturing Technology*, Volume 2. Before proceeding here are some relevant definitions.

Control by variables

A *variable* is defined as a characteristic that is appraised in terms of values on a continuous scale. *Control by variables* is the term used when the acceptance or rejection of a batch of components is based upon the inspection of variable characteristics, e.g. the diameters of a batch of turned components. Although the diameters may vary from one component to another, a component will be acceptable if its diameter lies within the prescribed limits, otherwise it will be rejected.

Control by attributes

An *attribute* is defined as a characteristic which is appraised in terms of whether it meets or does not meet a given requirement. *Control by attributes* is the term used when a component property is checked for being right or wrong. A container leaks or it doesn't leak, a nut fits on its bolt or it doesn't fit. A container that leaks is rejected; a container that doesn't leak is accepted. A nut that fits on its bolt is accepted; a nut that doesn't fit is rejected.

10.15.1 *Statistical control*

Acceptance sampling
A random sample is taken from a batch of components. If the sample is satisfactory then the batch is accepted, otherwise the whole batch is rejected. Acceptance sampling is used by the manufacturer during final inspection to reduce the possibility of rejection by the customer. Acceptance sampling is also used by the customer to control the quality of incoming materials.

Process control

Sampling takes place during production to determine whether the machine or tooling requires adjustment. Ideally, corrections are made to the processes before it gets to the point where scrap components are produced. Process control will be considered further in *Manufacturing Technology*, Volume 2. Control by variables and control by attributes are equally applicable to acceptance sampling and process control.

Finally the amount of inspection must be considered. The choice is between total (100 per cent) inspection or inspection of a sample. The larger the sample, the less chance there is of a faulty component being inadvertently accepted, but the greater the cost. For some components, such as critical components in an aircraft, there must be 100 per cent inspection. The safety of the crew and passengers is paramount and no risk can be taken of a faulty component being used. This total inspection is one reason why aircraft are so costly.

10.16 Acceptance sampling

Statistical techniques involving sampling are satisfactory for most items produced on a batch or continuous basis. The problem is to determine the minimum amount of sampling required to ensure acceptable quality.

10.16.1 *Spot checks*

The cheapest way to sample incoming materials is to carry out spot checks. Samples are taken at random from the incoming batch with the hope that the results are a true reflection of the quality of the batch as a whole. This is unscientific and unsatisfactory as the probability of defectives being taken into stock is unknown.

10.16.2 *One hundred percent inspection*

The surest and most expensive method is 100 per cent inspection of all incoming goods and materials. Rarely can the cost of such a system be afforded or warranted. Because of operator fatigue and other unforeseen variables, even 100 per cent inspection cannot guarantee that an occasional defective will not get through the system.

10.16.3 *Statistical sampling*

Statistical sampling falls between the two extremes. The sample size is carefully calculated so that batch rejection occurs if the defectives exceed a prescribed level. Consider a simple example in which the inspector is given the following criteria:

- A random sample of 50 components is taken from each batch of the components.
- The batch is acceptable providing not more than one faulty component is found in each sample, i.e. for 0 or 1 faulty component in the sample then accept the batch, but for 2 or more faulty components in the sample then reject the batch.
- This is specified as $50_{1/2}$ (50 components, accept 1 defective reject 2 defectives).

Figure 10.47 plots the probability of acceptance against the percentage of defective components for $50_{1/2}$. This is called an *operating characteristic* curve. The calculations for plotting this curve are beyond the scope of this chapter. However, notice that when there are no defectives in the batch, the probability of acceptance is unity (one), so the batch will be accepted without question. On the other hand, when there are 100 per cent defectives, the probability of the batch being accepted is nil. The curve represents the intermediate probability of acceptance for various percentages of defective components. This curve applies to any batch size providing the sample does not exceed 20 per cent of the batch.

Fig. 10.47 *Operating characteristics for a sampling system*

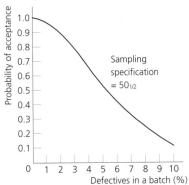

Figure 10.48 shows what happens if a larger sample and acceptance number are specified. The specification $200_{3/4}$ means a batch is accepted if a sample of 200 components contains up to 3 defectives, but it is rejected if the sample contains 4 or more defectives. This specification is more *discriminating*. The curve is steeper, so total rejection (zero probability of acceptance) occurs for a smaller percentage of defectives in the batch. With improved discrimination, a greater number of good batches are accepted and a greater number of bad batches are rejected. Although this is highly desirable, the cost is also greater because of the increased number of samples which have to be checked. Tables and charts are available for drawing up operating characteristics.

Fig. 10.48 *Comparing two characteristic curves*

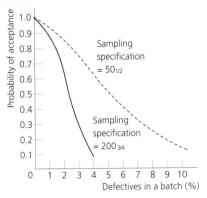

No sampling system is perfect and there is always the risk that a defective component will be taken into stock. The point at which the total percentage of items getting through the system becomes unacceptable is called the *lot tolerance percentage defective* (LTPD). Any batch containing a greater percentage of defective components would be totally rejected. Another parameter is the *acceptable quality level* (AQL). This is the highest percentage of defectives that is acceptable in a batch. Figure 10.49 shows how LTPD and AQL can be superimposed on an operating characteristic. The producer's risk (α) needs to be kept as low as possible to ensure the batch has only an acceptable (or lower) percentage of defectives. The consumer's risk (β) also needs to be kept low to ensure the number of defective components getting through into stock is kept to a minimum and below an acceptable level.

Fig. 10.49 *Application of AQL and LTPD*

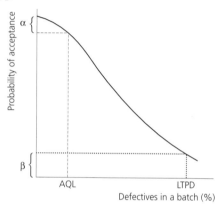

10.17 Quality audit

A quality audit is a systematic and independent examination to determine whether quality activities and related results comply with planned arrangements, whether these arrangements are implemented effectively and whether they are suitable for achieving the prescribed objectives.

This definition shows the similarity between a financial audit and a quality audit. In both instances the audit is carried out by professionally qualified and independent auditors who can act impartially. In both cases not only is the system under scrutiny but also the implementation of the system and the persons using it. And in both instances the audit is used to investigate whether or not prescribed targets are being achieved. Just as it would be most unwise to ignore any weaknesses in the financial structure of a company, so it would be equally unwise to ignore any weaknesses in the quality control structure of a company.

As well as ensuring that existing systems are being properly implemented, quality audits are also used to evaluate the need for improvement or corrective action in existing systems, and to provide a basis for their design. Quality audits should not be confused with surveillance or inspection activities used for the sole purpose of process control and product acceptance. As well as being imposed internally by prudent management, quality audits may also be imposed by external influences. For instance, a company may wish to ensure that a subcontractor can produce and deliver goods to a prescribed quality before placing an order with that subcontractor.

10.18 The cost of quality

10.18.1 *Design quality*

The designer has a very considerable influence on quality. It is at the design stage that attributes such as dimensional, geometrical and surface finish tolerances are set, appropriate British or International Standards are invoked, and materials are chosen. The design must reflect the requirements of the customer, both in performance and reliability. Generally it can be assumed that reliability increases as quality increases. However, care must be taken not to *over-engineer* the product, since increasing the quality and reliability usually increases the material, production and inspection costs.

10.18.2 *Material quality*

The adoption of standard specifications for materials and bought-in components leads to improved product quality and reliability, and reduced waste during production. However, it is essential that materials and components bought from outside sources are sampled and inspected before being issued for manufacture. Traditionally this was done on the purchaser's premises, but modern practice puts the responsibility on the supplier, and most large manufacturers will only purchase from suppliers whose quality control satisfies the purchaser, who specifies the quality standards, decides on the method of testing and conducts regular audits.

10.18.3 *Production quality*

Production quality is sometimes called manufacturing quality. The dimensional accuracy which can be expected from various production processes is summarised in Section 6.6, and the surface finish which can be expected from various production processes is summarised in Table 10.5. The accuracy which can be expected from various measuring devices is summarised in Table 10.1.

The quality of production indicates the degree to which the processing reflects the design requirements. Generally speaking, the greater the precision to which the product has to be manufactured, the greater the manufacturing costs. To some extent, consistency of manufacture has been improved by the adoption of automation and computer-controlled machine tools to remove the chance of human error, resulting from lack of concentration and fatigue (Chapter 5).

10.18.4 *Product quality*

Product quality is the combination of design quality, material quality and production (manufacturing) quality; it is a measure of how well the product's fitness for purpose satisfies the customer's requirements.

10.18.5 *Quality control*

Control costs: prevention

These are the costs associated with activities such as quality management and staff training, intended to inculcate working practices and attitudes to work that ensure only products of a satisfactory quality are produced. Although faulty work can be rejected by rigid inspection, this represents waste, unnecessary expense and reduced profitability. Prevention costs should be less than wastage costs, but in any case, prevention should be preferred over wastage purely on philosophical grounds; wastage should be eliminated at source.

Control costs: appraisal

These are the direct costs of sampling, inspection and testing raw materials, bought-in components, work in progress and the finished product. The aim is to ensure fitness for purpose and to prevent substandard products from reaching the market and damaging the reputation of the manufacturer.

Failure costs

The cost of failure and the resulting waste is inextricably linked with prevention costs. Generally the number of failures varies inversely with the cost of quality appraisal (inspection), which can be expressed graphically as shown in Fig. 10.50.

The cost of failure represents factors such as the cost of spoiled material, the cost of wasted production time, the cost of rectification, the cost of customer compensation, the cost of replacement and the loss of customer loyalty. However, reducing the failure rate increases the cost of quality control. Quality appraisal is expensive, and if the failure rate is to be reduced to zero, it will require total inspection of all material and every component at every stage of manufacture, coupled with extensive testing of the final assembly. This will raise the cost of the product to a level which the market cannot support. This is a good example of prevention being better than cure and is the reason why prevention costs are a worthwhile investment. Ultimately there has to be some compromise between quality and what the customer will pay.

Figure 10.50 shows the relationship between cost of failure and cost of quality control. The *total cost of quality* is the sum of these two cost elements; notice how it reaches a

Fig. 10.50 *Cost of quality*

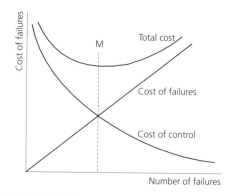

minimum at point M. Point M appears to be an acceptable compromise between quality control costs and failure costs. However, in practice, such a simple decision is not possible and consideration must also be given to factors such as goodwill (customer loyalty) and safety. For example, a zero failure rate has to be attempted for safety-critical components on which human lives depend, e.g. in cars and aircraft.

10.19 Quality assurance

This is a brief introduction to a very wide-ranging subject. Quality assurance and the statistical methods are widely used and are considered in greater depth in *Manufacturing Technology*, Volume 2.

Quality assurance is not a new concept. For many years, in certain areas of industry, customers have demanded certification of quality from their suppliers. These certificates of quality were used to verify that products or services met the specifications demanded by the customer. Some examples are:

- Material specifications where the results of various prescribed tests are recorded on the acceptance certificate, which bears an authorised signature.
- Lifting equipment that includes slings and chains.
- Pressure vessels such as boilers, compressed air receivers and compressed gas bottles.

What is new, however, is that products and services not usually associated with certification have been brought into line by having written evidence that the company has achieved *nationally accredited standards of quality*. These standards are also accepted internationally by keeping the standards of accreditation common amongst all the industrial communities.

In the United Kingdom the British Standard for quality assurance is BS 5750 (ISO 9000). The first standard of quality assurance was published in the United Kingdom as BS 5750: 1979. However, the interlinking of world trade, both within and outside what was then the European Community, led to other countries copying the British approach, but with their own amendments. This international interest in quality standards led to the introduction of ISO 9000: 1987. This combines international expertise with eight years of British experience in operating quality assurance. It has been adopted without alteration by the British Standards Institution, hence the dual numbering. The new ISO Standard also incorporates the European Standard EN 29000.

10.20 The basis for BS 5750/ISO 9000

The definition of quality upon which the standard is based uses the concepts *fitness for purpose* and *safe in use*, and verifies whether the product or service has been designed to meet the customer's needs. The constituent parts of this standard set out the requirements of a *quality-based* system. No special requirements are set out for individual companies or groups of companies; BS 5750 is essentially a practical basis for a quality system or systems that can be used by any company offering products or services anywhere within the United Kingdom or abroad.

The principles of the standard are intended to be applied to *any company of any size*. The basic disciplines are identified with the procedures and criteria specified in order that the product or the service meets with the requirements of the customer. The benefits of BS 5750 are genuine. They include:

- Cost-effectiveness, because a company's procedures become more soundly based and its criteria are more clearly specified.
- Reduction of waste and the necessity for reworking to meet the design specification.
- Customer satisfaction, because quality has been built in and monitored at every stage before delivery.
- A complete record of production at every stage is available to assist product or process improvement.
- In the event of failure in service, the complete history of the failed part is on record, so the source of the fault can be identified and eradicated.

By adopting BS 5750, a company can demonstrate its level of commitment to quality and also its ability to supply goods and services to the defined needs of its customers. For many companies it is merely the formalising and setting down of an existing and effective system in documented form, allowing its validity to be guaranteed by external accreditation.

An agreed standard as the basis of supply can confer major benefits on all parties concerned. A customer can specify detailed and precise requirements, knowing that the supplier's conformance can be accepted, since the quality system has been scrutinised by a third party – *it has been accredited*. However, in the final analysis, successful implementation will depend on the total commitment of the whole management team, especially the person at the very top of the organisation.

10.21 Using BS 5750/ISO 9000

Customers may specify that the quality of goods and services which they are purchasing shall be under the control of a management system complying with BS 5750/ISO 9000. Together with independent third parties, customers may use the standard as an assessment of a supplier's quality management, hence its ability to produce goods and services of a satisfactory quality.

Finally, you may ask whether it is worthwhile becoming involved in what appears to be a rather complex system? Remember that the standard is already used by many major public sector purchasing organisations and accredited third parties. Thus it is in the interests of all suppliers to adopt BS 5750/ISO 9000 in setting up their own quality control systems. Accredited companies can only buy goods and services from other accredited companies. Lack of accreditation can only result in serious loss of business and possibly ultimate closure. Every supplier is someone else's customer, and every customer is someone else's supplier (Fig. 10.51).

Fig. 10.51 *The supply chain: every supplier is someone else's customer; this also applies to departments within a company*

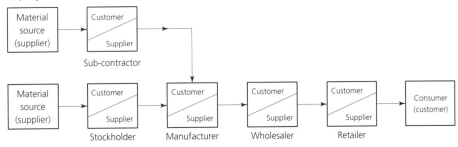

Companies derive several benefits once they have been assessed in relation to the standard and appear in the *Register of Quality Assessed United Kingdom Companies*, published by the Department of Trade and Industry:

- Reduced inspection costs.
- Improved quality.
- Better use of staff and equipment.
- Access to wider markets.

Where a company is engaged in exporting goods or services, there is a direct advantage in possessing mutually recognised certificates which could be required by overseas regulatory bodies.

SELF-ASSESSMENT TASK 10.7

1. Explain briefly why statistical methods are used when controlling the quality of mass-produced parts.

2. Explain briefly what is meant by a quality audit.

3. (a) Explain briefly why the cost of maintaining quality is a justifiable expense.
   (b) State under what conditions a zero failure rate must be aimed at, irrespective of cost considerations.

4. Describe what is meant by quality assurance.

5. To which companies and organisations is BS 5750/ISO 9000 applicable and what is its commercial importance?

EXERCISES

10.1 Define *quality* as applied to manufactured products and explain the importance of quality control in today's markets.

10.2 List the following measuring instruments in descending order of accuracy and state a suitable application for each instrument, giving reasons for your choice: steel rule, slip gauges (inspection grade), laser standard, micrometer caliper.

10.3 (a) Write down the micrometer caliper readings shown in Fig. 10.52.
(b) Explain the difference between reading accuracy and measuring accuracy.

Fig. 10.52 *Exercise 10.3(a)*

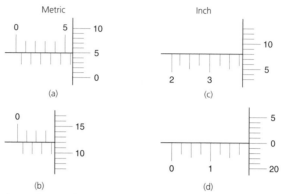

10.4 (a) Write down the vernier caliper readings shown in Fig. 10.53.
(b) List the advantages and limitations of a vernier caliper compared with a micrometer caliper.

Fig. 10.53 *Exercise 10.4(a)*

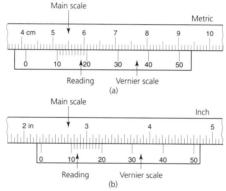

10.5 Figure 10 54 shows a component being checked with a vernier height gauge using a DTI as a fiducial indicator. From the data given, calculate the distance between the hole centres and the centre distance of hole B from the datum surface A.

Fig. 10.54 *Exercise 10.5. (Dimensions in millimetres)*

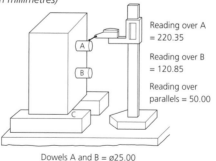

10.6 Figure 10.55 shows a turned component. List the measuring instrument most appropriate for checking each dimension during the machining process, giving reasons for your choice in each instance.

Fig. 10.55 *Exercise 10.6. (Dimensions in millimetres)*

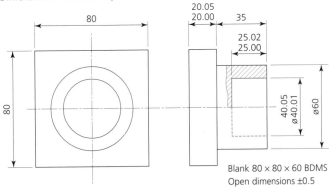

Blank 80 × 80 × 60 BDMS
Open dimensions ±0.5

10.7 With reference to Fig. 10.55 and BS 969, sketch and dimension:
- (a) a double-ended plug gauge for checking the 40 mm bore
- (b) a double-ended caliper gauge for checking the 60 mm diameter

10.8 (a) With the aid of a sketch, explain the principle of a sine bar.
   (b) Calculate the included angle of taper for a taper plug gauge supported between sine centres. The centre distance between the contact rollers is 500 mm and the height of the slip gauge stack is 25.025 mm.

10.9 (a) State the appropriate precautions when assembling and dismantling a stack of slip gauges in order to ensure maximum accuracy and minimum wear.
   (b) List the slip gauges that would be required from a 78-piece set to build up a 25.7925 mm stack:
   - (i) without using the 2.50 mm protector slips
   - (ii) using the 2.50 mm protector slips

10.10 With the aid of sketches, show how a dial test indicator (DTI) may be used for the following applications:
- (a) as a comparator gauge
- (b) to detect lobing (out of roundness)
- (c) to detect eccentricity in a turned component
- (d) to set the fixed jaw of a machine vice parallel to the T-slots in a milling machine table

10.11 (a) Describe two methods for checking the accuracy of a try square.
   (b) Describe how to make a cylinder square that will stand perpendicular to any datum surface on which it may be supported.

10.12 (a) Describe how a straight edge may be used to check flatness over a large area. What precautions should be taken to detect errors of winding?
   (b) Describe how a surface plate may be used to check the surface flatness of a small component.

10.13 (a) Differentiate between roughness, waviness and lay when describing the texture of a machined surface.

(b) The machining symbols in Fig. 10.56 specify R_a for the surface texture. Select a suitable manufacturing process to produce each surface texture and explain how you made your selection.

Fig. 10.56 *Exercise 10.13(b)*

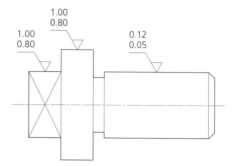

10.14 With the aid of sketches, explain the essential difference between R_a and R_z when specifying surface texture.

10.15 Briefly describe the total cost of quality in terms of control costs and failure costs.

10.16 Briefly explain why it is in the interests of all manufacturing companies to become BS 5750/ISO 9000 accredited.

10.17 Using your own examples, explain the difference between variables and attributes.

10.18 Explain what is meant by the sampling specification $120_{3/4}$.

10.19 Explain the acronyms AQL and LTPD. How are they applied to operating characteristics and what is their significance in acceptance sampling?

10.20 Briefly describe the total cost of quality in terms of control costs and failure costs.

11 Manufacturing relationships

The topic areas covered in this chapter are:

- The need for information flow.
- Marketing and sales.
- Design and development.
- Production planning.
- Estimating and costing.
- Production control.
- Inspection and quality control.
- Maintenance.
- Organisation of a new product.

11.1 The need for information flow

The free flow of information between the departments of an engineering company is essential to the well-being of that company. In a small company, consisting of the proprietor and two or three employees, this is no problem. However, information flow becomes more complex in large companies, and some have failed when the channels of information flow have broken down. In a large company it is impossible for one person to solve all the problems and make all the decisions, so the responsibility for decision making and management has to be devolved. As soon as this happens, the organisation of the company breaks up into specialised departments. Unfortunately, as specialisation increases, the range of interests narrows until ultimately the specialist success of the department becomes an end in itself. This leads to selfish decision making that is independent of its effect on other members of the company. Such a lack of cooperation, and the consequent breakdown in the free flow of information, spells a company's doom.

A widely used method of showing the interrelationships between the various departments in a company is the hierarchical line management tree (Fig. 11.1). Hierarchies of managers are common in companies; the job titles are different but the structure is derived from older institutions such as government offices and the military. This was largely inevitable since, at the time of the Industrial Revolution and the growth of the factory system, the only persons with large-scale organisational experience were soldiers and civil servants.

Fig. 11.1 *Company organisation chart (hierarchical)*

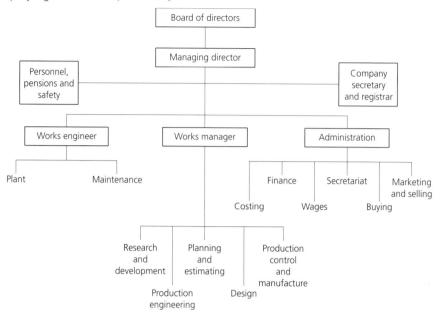

In this system the top manager, or chief executive is responsible to some governing body such as a board of directors. The chief executive transmits the decisions of the board down through the lines of communication to the next level of managers, who are immediately responsible for the implementations of those decisions within their departments and, depending upon the size of the company, this devolvement continues down the line to lower layers of management. Although the primary lines of communication are well established, the secondary lines of interdepartmental communication, essential for cooperation and success, are not so clearly defined; it is all to easy for departmental demarcations to develop, causing a breakdown in the flow of information. A more flexible and less bureaucratic structure is not only desirable, it is often essential for the success of a company.

An alternative method of organisation is shown in Fig. 11.2. Here the policy-forming board of directors forms the hub of the circles. The board's decisions are passed out to the departmental heads along radial lines; feedback follows the same routes. The concentric circles show the levels of management. Thus the radial lines represent the traditional hierarchical system of line management whereas the circles represent interdepartmental information flow.

The circular system of organisation is best suited to problem solving. It is used widely in Japan and firms with Japanese connections. Because of its success it is now increasingly adopted in the United Kingdom. The problem is shown at the hub of the circle (Fig. 11.3) and the radial arrows indicate how many departmental representatives are working on it. The peripheral arrows represents the interchange of ideas between departments. Secondary information flow can take place across the circle as required; they are indicated by the arrow between the shop supervisor and the production engineer.

Fig. 11.2 *Company organisation chart (circular): PM = personnel management; CS&R = company secretary and registrar; WE = works engineer; WM = works manager; AD = administration; PLT = plant; MAIN = maintenance; R&D = research and development; D = design; PL&E = planning and estimating; PE = production engineering; PC = production control; COS = costing; FIN = finance; WAG = wages; SEC = secretariat; BUY = buying; M&S = marketing and selling*

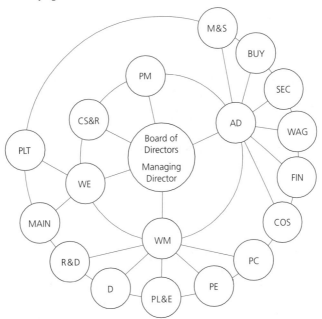

Fig. 11.3 *Round-table problem solving*

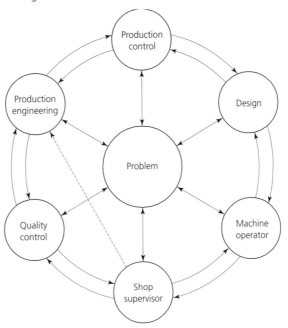

This type of organisation is often used to solve quality problems with the appropriate management and shop-floor workers sorting out the problem on equal terms. Although this circular or round-table approach implies equality between all those present, inevitably one of the group has to act as ringmaster to ensure that everyone receives an equal opportunity for input.

Matrices and flow charts are also appropriate for more complex organisational problems such as the organisation of large-scale projects involving many companies, e.g. the building of the Channel Tunnel.

SELF-ASSESSMENT TASK 11.1

1. Draw an organisation chart for any company or institution with which you are familiar and discuss the routes of information flow.

2. Suggest ways in which the information routes in your example could be improved.

11.2 Marketing and sales

The primary functions of marketing and sales are often confused. *Marketing* is essentially a research operation. The question any company should regularly ask itself is, Are we in the right business? If the answer is no, then it should ask, What business should we be in? For example, a number of prosperous steam locomotive manufacturers went out of business because they had failed to change to alternative products when British Rail started to change over to diesel and electric locomotives. Similarly some household names amongst British machine-tool manufacturers went out of business through failure to update their products for computerised manufacture.

They failed through lack of *market research* and the will and ability of the senior management to understand and embrace the need for change. The organisation of marketing and its relationship with selling varies between those companies selling a standard product and those companies fabricating one-off products to order.

A company making a standard product, or a range of standard products, needs to watch market trends very carefully. Every product has a life cycle. A slowdown in sales growth is a warning signal. It may mean that greater effort in selling is required or that the life of the product is in decline. A positive falling off in sales volume is a very clear signal that, despite the efforts of the sales force, the end of the product life cycle is in sight. Too many companies persist with a product beyond its useful life and even sell it at a loss. Long before this stage has been reached, a successful company will have developed a new product to take the place of the one that has past the peak of its life cycle (Fig. 11.4).

One British company that underwent a successful metamorphosis originally rolled steel strip. Then it rolled steel strip for box strapping. Then it sold complete strapping systems and realised that this was more profitable than rolling steel. Then market research showed that high-tensile plastic strapping was replacing steel. The market research was accurate, the top management recognised the importance of the research

Fig. 11.4 *Product life cycle*

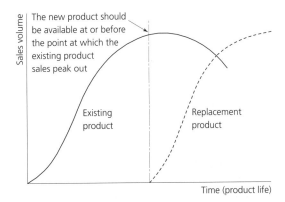

and the complete transition was made from being a roller of steel strip to being a very successful and profitable box-strapping company using largely plastic materials. All because the managing director asked the question, What business should we be in? Fortunately, he came up with the correct answer.

Unfortunately, too many British companies foundered after World War II because, although the market researchers recognised the changing needs, many senior managers found themselves unable to adapt, or they overlooked the significance of rapid changes in technology.

Market research plays a vital role in the development of new products. Apart from providing the data necessary to decide whether a new product or a new market is required, it can help to decide what the new product should be and whether it has a viable market. New products arise in a variety of ways. They can arise from information brought back to the company by its sales staff; they can arise from planned research and development, they can even arise by accident. Stainless steel was discovered by accident when it was noticed that discarded samples of nickel–chromium ferrous alloys had not gone rusty on the scrap heap; they had been thrown out after failing an experiment. However, no company can last long by relying on accidental discoveries.

The responsibilities of the market research department can be summarised as follows:

- Providing the company with up-to-date data of market trends and customer requirements.
- Providing the company with information about the product development, marketing strategies and pricing strategies of its competitors.
- Providing the company with information about the life cycle of its existing products so that new products can be developed before the existing products cease to be marketable and profitable.
- Making a preliminary assessment of ideas for new products.
- Making a detailed study of the viability of a new product before large sums are spent on research and development, capital equipment and tooling.

In addition to these research and advisory functions, the marketing department is also responsible for 'softening up' the ground for the sales force by planning the publicity surrounding the launch of a new product. In contrast to marketing, *selling* has two main functions:

- The procurement of the orders and contracts from customers, upon which the financial prosperity of the company depends.
- The feedback of information on customer requirements, information concerning the effectiveness of promotional material, and customer satisfaction with distribution, servicing and product reliability.

No matter how well researched the market, no matter how well developed the product, without a knowledgeable and enthusiastic sales force in the field, trained in modern selling techniques and capable of conversing with the customers *in their own language*, a company will quickly cease to be profitable and close down.

SELF-ASSESSMENT TASK 11.2

Discuss the essential differences between marketing and selling.

11.3 Design and development

For better or worse, progress is always taking place. The customer is always demanding better-quality products at a lower price. However, these two requirements are often incompatible, so a compromise has to be reached by providing value for money. One customer may prefer to pay more for a handmade car, whereas another customer may prefer to pay less for a mass-produced car. In both cases the customer will expect the car to represent state-of-the-art technology within its class and to reflect current styling trends. In fact, appearance and styling (industrial design) is now not only applied to consumer durables but even to commercial and industrial equipment, which is expected to be pleasing to the eye as well as functional. The achievement of these aims is the function of the design and development department, working closely in collaboration with the marketing department.

11.4 Production planning

The production method for any new product must be carefully planned. Production planners ask questions such as, Will we use the existing plant? If the answer is yes, the production departments and the design team must get together to ensure the existing plant can manufacture it, and manufacture it profitably. This involves consultation with the finance departments.

But if the answer is no, then further questions arise:

- How much can be spent on new plant?
- What new plant is necessary to satisfy the demands of the new product?

- Will the anticipated market and sales volume warrant automation (e.g. CNC machines and robots), and has the company got the appropriate expertise to use it?
- Is a compromise required between the demands of the design team and the practical requirements of production?

Sensible decisions rely on interdepartmental information flow and consultation allowing a working compromise to be reached.

11.5 Estimating and costing

Estimating and costing is closely associated with production planning. In fact, it is often an extension of the planning process. At each planning stage, the production time and the material quantities involved must be estimated and costed so the economic viability of the process can be assessed. The cost of a new product is determined in two stages.

11.5.1 *Estimating*

As the name implies, estimating assesses the time taken to produce a new product based on experience and historical data. There are essentially two techniques:

- Comparison of the new product with a similar product for which actual production times are known. The experience of the estimator is then used to make allowance for any differences.
- The new product is broken down into its constituent elements and the estimated time is synthesised as follows:
  - (a) Productive times can be calculated from the cutting parameters for each machining operation. For example, the time taken to turn a diameter on a component can be calculated as shown in Fig. 11.5. Manual operations are more problematical; estimates can be calculated from tables of *observed times* produced by the work- or time-study engineer.
  - (b) Non-productive times are for operations such as loading and unloading components, removing swarf from the machine, and changing tools. They are also calculated from tables of observed times.

The total floor-to-floor production time for a component is the sum of the productive times and the non-productive times. Figure 11.6 shows a typical turned component and Fig. 11.7 shows the comprehensive planning sheet listing the estimated times for producing the component on a capstan lathe. The times can be much more accurately estimated if the component is produced on a CNC lathe as there are no human elements to take into account.

11.5.2 *Costing*

Once the process times and the material quantities have been estimated, the product can be *costed*. An obvious way to arrive at the *direct labour cost* would be to multiply the cycle

Fig. 11.5 *Calculation of cutting time*

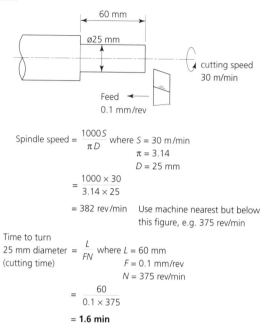

Spindle speed $= \dfrac{1000S}{\pi D}$ where $S = 30$ m/min
$\pi = 3.14$
$D = 25$ mm

$= \dfrac{1000 \times 30}{3.14 \times 25}$

$= 382$ rev/min   Use machine nearest but below
this figure, e.g. 375 rev/min

Time to turn
25 mm diameter $= \dfrac{L}{FN}$ where $L = 60$ mm
(cutting time)
$F = 0.1$ mm/rev
$N = 375$ rev/min

$= \dfrac{60}{0.1 \times 375}$

$= \mathbf{1.6\ min}$

Fig. 11.6 *Typical turned part: dimensions in millimetres; material is free-cutting mild steel*

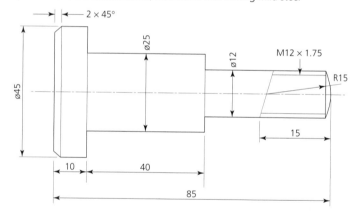

time by the appropriate rate of pay. But this is too easy and once again the experience of the planning engineer is called upon for a variety of reasons:

- The operator will not maintain a constant rate of production but will tire towards the end of the shift.
- Allowance must be made for the operator to leave his or her workstation to attend to personal needs from time to time.
- Allowance must be made for the machine to be adjusted from time to time to compensate for tool wear or accidental tool breakage.

Fig. 11.7 *Comprehensive planning sheet*

A.N. ENGINEERING CO. LTD							Planned by: Checked by:		
DRG NO	MACHINE	MATERIAL	QUANTITY	DATE REQUIRED			SPECIAL FEATURES		
1	2	3	4	5		6	7	8	9
OPERATION NO	OPERATION	TOOLING	TURRET OR X SLIDE	SPEED m/mm	SPEED rev/min	FEED RATE (mm/rev)	LENGTH OF CUT (mm)	CUTTING TIME (min)	NON-PRODUCTIVE TIME (min)
1	Feed to stop	Bar stop	T	–	–	–	–	–	0.100
	Index and lock turret	–	–	–	–	–	–	–	0.100
	Change speed	–	–	–	–	–	–	–	0.085
2	Turn 25 mm diam.	Roller tool box	T	50	600	0.25	75	0.5	–
	Index and lock turret	–	–	–	–	–	–	–	0.100
	Change speed	–	–	–	–	–	–	–	0.085
3	Turn 12 mm diam.	Roller tool box	T	50	1250	0.25	35	0.112	–
	Index and lock turret	–	–	–	–	–	–	–	0.100
4	Radius end	End turning box	T	50	1250	Manual	–	0.08	–
	Index and lock turret	–	–	–	–	–	–	–	0.100
	Change speed	–	–	–	–	–	–	–	0.085
5	Rough thread	S/O die head	T	10	250	1.75	15	0.035	–
6	Withdraw, reset die head, finish	S/O die head	T	10	250	1.75	15	0.035	0.080
	Index to stop and lock	–	–	–	–	–	–	–	0.150
7	Recess	Parting tool	X	30	250	Manual	–	0.200	–
8	Chamfer	RH 45° chamfer tool	X	30	250	Manual	–	0.200	–
9	Part off	Parting tool	X	30	250	Manual	–	0.350	–
10	Break bar end ready for next cycle	LH 45° chamfer tool	X	30	250	Manual	–	0.150	–
	TOTALS							1.562	0.985

Cycle time = 1.562 + 0.985 = 2.547 min per component

These factors are usually taken into account by adding a contingency factor to the estimated cycle time. Finally, other factors such as material costs, overhead expenses and profit margins must be taken into account to arrive at the final selling price (Fig. 11.8). Thus the process of costing and estimating involves cooperation and consultation between the production planning engineers (cycle time), the personnel department (wage rates) and the cost accountants (overheads and profit margins).

Note that costing and estimating is not the same as *cost control*. Cost control is used to ensure that production costs do not get out of step with the original estimates and costing upon which the initial selling price was based. Cost control will be considered in *Manufacturing Technology*, Volume 2.

Fig. 11.8 *Cost structure*

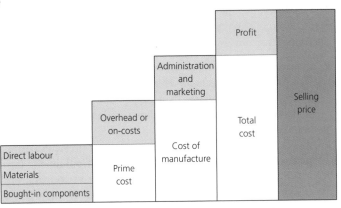

11.6 Production control

Although production planning is largely a technical function, production control is largely an administrative function. Once the product has been accepted for production, the tooling is in place and the materials have been purchased, the production control department takes over. Its responsibilities are to ensure that:

- The right components are produced at the right time to satisfy the customer's delivery requirements.
- The plant is uniformly loaded so that bottle-necks do not occur and expensive capital plant does not stand idle.
- Stocks of work between operations are kept to a minimum.
- Only minimum buffer stocks of finished work are kept to ensure prompt delivery; this is particularly important towards the end of a product's life cycle.

The production control department should work closely with the sales department to ensure that deliveries are correct and on time. It should also work closely with the marketing department to ensure that old stock is run down appropriately when a product is withdrawn, and that stocks of any new product are available at its launch.

11.7 Inspection and quality control

The importance of quality control is considered in Chapter 10. Consumers increasingly expect improved quality at minimum cost in a never-ending search to obtain *value for money*. Quality control is paramount in maintaining the *goodwill* and *customer loyalty* that are vital to a company's success. All successful companies are continually striving to improve quality to internationally acceptable standards. Goodwill takes years to build up and only minutes to lose through lack of attention to detail.

11.8 Maintenance

The maintenance of the plant and premises in good condition is essential if the manufacturing and commercial departments of a company are to maintain a high level of operating efficiency. Maintenance falls into two categories:

- *Routine or preventative maintenance.* The regular servicing of the plant and premises so that major problems are avoided. For example, it is better to paint the window frames of the building regularly than disrupt the work of an office whilst a rotten frame is replaced. A major plant breakdown can cause loss of production, late delivery and a very angry customer.
- *Major repairs or refurbishment.* Even the best preventative maintenance schemes cannot prevent occasional major breakdowns. And from time to time the plant has to undergo major refurbishment. Each requires the closest cooperation between the departments concerned so that disruption is minimised and the lost production is as small as possible.

1. What are the essential differences between estimating and costing?

2. Discuss how design priorities have changed since World War II to make engineering products more marketable.

3. Briefly summarise the objectives of a production control department and explain how it differs from a production engineering department.

4. Summarise the importance of regular preventative maintenance for engineering plant.

11.9 The organisation of a new product

Having considered the various departmental functions, the need for a free flow of information between those departments and some organisational systems to ensure that flow of information, it is now necessary to consider the information itself.

For a number of years, Messrs A. Locksmith Ltd have been manufacturing door locks for cars. They have specialised in locks for the lower-cost end of the volume car market. For some time their design department has been working with their customers to develop a new range of locks that provide greater security against theft. Market research has also established that customers would like a range of similar high-security locks which can be fitted to existing vehicles in place of the original locks. Let us briefly consider the information required to make manufacturing decisions for this project.

11.9.1 *Economic information*

Having revealed the demand for an improved lock, market research must answer a number of fundamental questions in order to justify the expense of a detailed design study:

Quantity
- Does the anticipated demand justify the cost of tooling up to manufacture this range of locks?
- Does the anticipated demand justify the cost of production?
- Will the range cover replacements for all previous cars as well as future production, or will it cover only a limited number of previous cars? Which cars should be included if the range is limited?

Make or buy
- Which parts will be made and which parts will be bought in from specialist firms?
- Has the plant sufficient capacity to make all these replacement locks in addition to its current production?
- If the plant capacity is too small, can some or all of the new production work be contracted to outside suppliers? Can their quality be guaranteed?

Delivery date

What is the launch date? Does the launch need to be at a trade show? Can the launch date be achieved without interfering with other work?

Marketing

Can any inducement be given to the motoring public to buy the new locks? For example, could the cost of fitting be wholly or partially offset by insurance rebates?

11.9.2 *Design information*

Purpose

The design team need to be given a clear design brief that explains the purpose of the new locks. Lack of precise and detailed information can lead to much wasted time and unsatisfactory designs.

Size and shape

Since these proposed locks are to replace existing locks, they will need to be the same size and shape. This information can be obtained from a number of sources, including the company's own records, since it is an established manufacturer of car locks. Other potential sources are British Standards Institution (BSI) publications for the motor trade and the car manufacturers themselves.

Accuracy

The accuracy need only be sufficient to ensure the locks are easy to assemble and will work satisfactorily over their projected life span. Unnecessarily tight tolerances only increase the cost and cause the product to be overengineered.

Materials

The materials need to be closely specified to ensure fitness for purpose. They need to confer adequate properties on the components that are made from them, properties relevant to each component's function. It would be uneconomic to specify a high-duty alloy steel where a plain carbon steel would be adequate.

Treatment

Specify any heat treatment for components subject to wear.

Finish

Specify any decorative or anticorrosion finishes.

11.9.3 *Plant information*

Plant information is required during the planning and design stages to ensure the product can be made without having to resort to the purchase of expensive items of capital equipment.

Range

Is the range of processes adequate to ensure that as many components as possible can be manufactured within the existing plant, thereby minimising any investment in new plant?

Capacity

Can the new product range be manufactured within the capacity of the existing plant or will it have to be subcontracted to outside suppliers?

Availability

Will plant capacity be available when the new product goes into production?

Capability

Is the plant capable of producing components of the required quality in the required quantity?

Batch/flow

Is the new product to be made using batch production or flow production? Batch production will allow the new product to be manufactured between batches of current products. Flow production will require a new production line to be organised.

Maintenance

Will the plant be overutilised? Will the additional loading on the plant interfere with the planned maintenance programme and lead to breakdowns and disruption of production?

11.9.4 *Manufacturing information*

Once the design has been accepted, once any problems concerning plant availability and suitability have been resolved, and once the decision to manufacture has been made, the following information will be required so that production can commence.

Batch size

Batch size will determine the method of manufacture and the sophistication of any special tools required.

Special tooling

As well as the product design, the drawing-office will need to issue drawings for any special tooling such as jigs, fixtures and press tools. Lead time must be allowed for the manufacture of this equipment.

Software

If CNC machine tools and robots are to be used, then part programs will need to be written for the various components.

Quality

Inspection and quality control are an expensive but important element of the manufacturing process. Since the new products chosen for this example are a range of improved locks for the motor industry, it is most likely that the car firms concerned will set their own standards based on the British Standard recommendations for quality control. Since large batches of similar products are being made, inspection costs can be reduced by using automated sampling and gauging along with statistical quality control.

Treatments

Heat treatment, protective and decorative finishes have already been discussed. Treatments are often contracted out, so arrangements will have to be made with suitable subcontractors, and quality control systems will be needed to ensure this outwork is up to standard.

11.9.5 *Industrial engineering*

Method study

This is concerned with finding the best way of doing jobs, particularly manual operations, as quickly and easily as possible. Each selected area of work must be observed by specialist method-study engineers and their observations recorded and analysed. They must then develop the most efficient way of doing the job so as to ensure fast and consistent results with a minimum of operator fatigue and risk of repetitive strain injury. They must install the new method then ensure its use is maintained, in case the workers concerned revert to less efficient methods out of personal preference or habit.

Work measurement

The main objective of work measurement is to achieve *standard times* for jobs. Standard time is the time it takes a trained worker to carry out a job at a specified level of performance. Standard times are used for costing, production planning and to determine staffing levels.

Planning and scheduling

Planning and scheduling are needed to ensure that all the parts arrive at the next stage of production without shortages, causing hold-ups, or the creation of excessive work in progress. Without proper planning and scheduling, the even loading of plant and the smooth flow of production would not be possible and delivery schedules would not be met.

EXERCISES

11.1 Discuss the importance of the free flow of information between the departments of an engineering company and draw up a management structure diagram for any company with which you are familiar.

11.2 Compare and contrast the potential problems with information flow in a very large company and a very small company.

11.3 Differentiate between marketing and sales in a manufacturing company.

11.4 For a familiar engineering company, discuss the functions of the following five departments and the relationships between them:
(a) design and development
(b) production planning and costing
(c) estimating and production control
(d) production engineering and manufacturing
(e) inspection and quality control.

11.5 You have had an idea for a new product. After taking professional advice, you have decided to manufacture it yourself. You have adequate financial backing to set up a small to medium-sized plant. Explain how you would organise the introduction of your new product so as to ensure its success.

12 Health and safety at work

The topic areas covered in this chapter are:

- Health and safety legislation and statutory bodies.
- Accidents.
- Protective clothing.
- Health hazards and personal hygiene.
- Behaviour in workshops.
- Lifting and carrying.
- Hazards associated with tools.
- Electrical hazards.
- Fire hazards, firefighting and fire prevention.
- Welding hazards.
- Workshop layout.

12.1 The Health and Safety at Work, etc., Act

The Health and Safety at Work Act provides a comprehensive, integrated system of law dealing with the health, safety and welfare of workers and the general public as affected by work activity. The Act has six main provisions:

- To overhaul completely and modernise the existing law dealing with safety, health and welfare at work.
- To put *general duties* on employers, ranging from providing and maintaining a safe place of work to consulting on safety matters with the employees.
- To create a Health and Safety Commission.
- To reorganise and unify the various government inspectorates.
- To provide powers and penalties for the enforcement of safety laws.
- To establish new methods of accident prevention and new ways of operating future safety regulations.

The Act now places the responsibility for safe working equally upon:

- The employer.
- The employee.
- The manufacturers and suppliers of goods and equipment.

12.2 The Health and Safety Commission

The Act provides for a full-time, independent chairperson and between six and nine part-time commissioners. The commissioners are made up of three trade union members appointed by the Trades Union Congress (TUC), three management members appointed by the Confederation of British Industry (CBI), two local authority members and one independent member.

The commission has taken over the responsibility previously held by various government departments for control of most occupational safety and health matters. The commission is also responsible for the organisation and functioning of the *Health and Safety Executive*.

12.3 The Health and Safety Executive

This unified inspectorate has amalgamated the Factory Inspectorate, the Mines and Quarries Inspectorate and similar bodies. Since 1975 they have been merged into one body known as the Health and Safety Executive Inspectorate. The inspectors of the HSE have wider powers under the Health and Safety at Work Act than under previous legislation and their duty is to implement the policies of the Health and Safety Commission.

12.4 Enforcement

Should an inspector find a contravention of one of the provisions of the existing Acts or regulations, or a contravention of a provision of the new Act, the inspector can choose from three possible courses of action.

12.4.1 *Prohibition Notice*

If there is a risk of serious personal injury, the inspector can issue a prohibition notice. This immediately stops the activity giving rise to the risk until the remedial action specified in the notice has been taken to the inspector's satisfaction. The prohibition notice can be served on the persons undertaking the dangerous activity, or it can be served on the person in control of the activity at the time the notice is served.

12.4.2 *Improvement notice*

If there is a legal contravention of any of the relevant statutory provisions, the inspector can issue an improvement notice. This notice requires the fault to be remedied within a specified time. It can be served on the person deemed to be contravening the legal provision, or it can be served on any person on whom responsibilities are placed. This responsible person can be an employer, an employee or a supplier of equipment or materials.

12.4.3 *Prosecution*

In addition to serving a prohibition notice or an improvement notice, the inspector can prosecute any person contravening a relevant statutory provision, including you. Finally, the inspector can *seize*, *render harmless*, or *destroy* any substance or article that he or she considers to be the cause of imminent danger or personal injury.

Thus every employee, trainee or experienced worker must ensure he or she is a fit and trained person to carry out his or her assigned tasks. By law every employee must:

- Obey all safety rules of his or her place of employment.
- Understand and use, as instructed, the safety practices incorporated in particular activities or tasks.
- Not proceed with his or her task if any safety requirement is not thoroughly understood. Guidance must always be sought.
- Keep his or her working area tidy and maintain his or her tools in good condition.
- Draw the attention of the safety officer or his or her immediate superior to any potential hazard.
- Report all accidents or incidents to the responsible person (even if injury does not result from that incident).
- Understand emergency procedures in the event of an accident or an alarm.
- Understand how to give the alarm in the event of an accident or an incident, e.g. a fire.
- Cooperate promptly with the senior person in charge during an accident or incident, e.g. a fire.

Safety, health and welfare are very personal matters not only for the young worker just entering industry, but also for the experienced worker. This chapter sets out to identify the main hazards and discuss how they can be avoided. Factory life, and particularly engineering, is potentially dangerous and a positive approach must be taken towards safety, health and welfare at all times.

SELF-ASSESSMENT TASK 12.1

1. State who is responsible for ensuring safety in the workplace.

2. Describe the enforcement powers of HSE inspectors under the Health and Safety at Work Act.

12.5 Accidents

Accidents do not happen, they are caused. There is not a single accident that could not have been prevented by care and forethought on somebody's part. Accidents can and must be prevented. They cost millions of pounds every year in damage and loss of premises, plant and lost business. They cost millions of lost working hours each year, but these are of little importance compared with the immeasurable cost in human suffering.

Fig. 12.1 *Causes of industrial accidents*

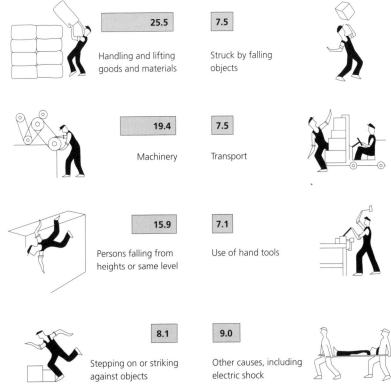

25.5 Handling and lifting goods and materials	**7.5** Struck by falling objects
19.4 Machinery	**7.5** Transport
15.9 Persons falling from heights or same level	**7.1** Use of hand tools
8.1 Stepping on or striking against objects	**9.0** Other causes, including electric shock

In every eight-hour shift nearly 100 workers are the victims of industrial accidents. Many of them will be blinded, maimed for life, or confined to a hospital bed for months. At least two of them will die. Figure 12.1 shows the main causes of accidents.

12.6 Protective clothing

For general workshop purposes the *boiler suit* is the most practical and the safest form of body protection. However, to be completely effective, certain precautions must be taken (Fig. 12.2).

12.6.1 *Sharp tools*

Sharp tools protruding from the breast pocket can cause severe wounds to the wrist. Since the motor nerves of the fingers are near the surface in the wrist, these wounds can paralyse the hand and fingers.

12.6.2 *Button missing*

Since the overall cannot be fastened properly, it becomes as dangerous as any other loose clothing and is liable to be caught in moving machinery.

Fig. 12.2 *Safe dress*

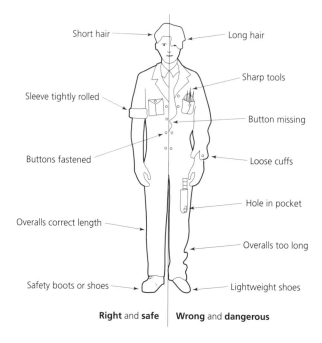

Short hair —► ◄— Long hair

◄— Sharp tools

Sleeve tightly rolled ◄— Button missing

◄— Loose cuffs

Buttons fastened —►

◄— Hole in pocket

Overalls correct length —► ◄— Overalls too long

Safety boots or shoes —► ◄— Lightweight shoes

Right and **safe** | **Wrong** and **dangerous**

12.6.3 *Loose cuffs*

Not only are loose cuffs liable to be caught up like any other loose clothing, they may also prevent the wearer from snatching his or her hand away from a dangerous situation.

12.6.4 *Hole in pocket*

Tools placed in the pocket can fall through onto the feet of the wearer. This may not seem potentially dangerous, as the feet should be protected by stout shoes, but it could cause an accident by distracting attention at a crucial moment.

12.6.5 *Overalls too long*

Excessively long overalls can cause falls, particularly when negotiating stairways.

12.6.6 *Lightweight shoes*

The possible injuries associated with lightweight and unsuitable shoes are:

- Severe puncture wounds caused by treading on sharp objects.
- Crushed toes caused by falling objects.
- Damage to the Achilles' tendon due to insufficient protection around the heel and ankle.

In addition to body protection, it is necessary to protect the head, eyes, hands and feet. Let's now consider suitable protective clothing.

12.7 Head protection

Long hair is a very real hazard in any workshop.

- Long hair is liable to be caught in moving machinery, particularly drilling machines and lathes. When the hair and scalp are torn away, the resulting wound is both extremely painful and dangerous. Brain damage may also occur.
- Long hair is also a health hazard, as it is almost impossible to keep it clean and free from infection in the workshop environment (Section 12.7).

If your hair becomes entangled in a machine (Fig. 12.3), you are very likely to be scalped – a very painful and serious injury. If a fitter or machinist persists in retaining a long hairstyle in the interests of fashion, then the hair must be contained within a close-fitting cap. This also helps to keep the hair and scalp clean and healthy.

When working on site, or in a heavy engineering erection shop involving the use of overhead cranes, all persons should wear a safety helmet complying with BS 2826, as even small objects such as nuts and bolts can cause very serious head injuries when dropped from heights. Figure 12.4(a) shows such a helmet. Safety helmets are made from moulded plastic or from fibreglass-reinforced polyester. They are colour-coded for personnel identification, and are light and comfortable to wear, yet despite their lightness they have a high resistance to impact and penetration. To eliminate the possibility of electric shock, safety helmets have no metal parts. The materials used to manufacture the outer shell have to be non-flammable and their electrical insulation must withstand 35,000 volts. Figure 12.4(b) shows the harness inside a safety helmet. This provides ventilation and a *fixed safety clearance* between the outer shell of the helmet and the wearer's skull. This clearance must always be maintained at 32 mm. The entire harness is removable for cleaning and sterilising. It can be adjusted for size, fit and angle to suit the individual wearer's head.

Fig. 12.3 *The hazards of long hair*

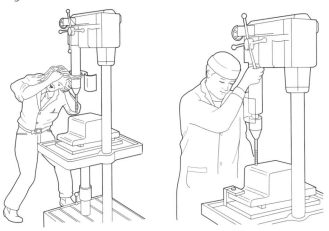

Fig. 12.4 *Safety helmets: (a) a typical fibreglass safety helmet made to BS 2826; (b) a safety helmet harness*

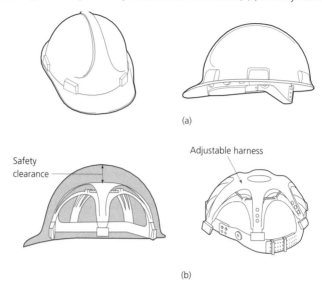

(a)

Safety clearance

Adjustable harness

(b)

12.8 Eye protection

Although it is possible to walk on a wooden leg, nobody has ever seen out of a glass eye. Therefore eye protection is possibly the most important safety precaution to take in the workshop. Eye protection is provided by wearing suitable goggles or visors (Fig. 12.5).

When welding, special goggles have to be worn with coloured lenses to filter out harmful rays (Sections 12.21 and 12.22). Gas welding goggles are not suitable when arc welding. Eye injuries fall into three main categories:

- Pain and inflammation due to abrasive grit and dust getting between the lid and the eye.
- Damage caused by exposure to ultraviolet radiation (arc welding) and to high-intensity visible radiation. Particular care is required when using laser equipment.
- Loss of sight due to the eyeball being punctured or the optic nerve being severed by flying splinters of metal (swarf, or by the blast from a jet of compressed air).

Fig. 12.5 *Safety goggles and visor: (a) transparent goggles suitable for machining operations; (b) complete plastic face visor for protection against chemical splashes*

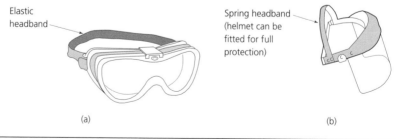

Elastic headband

Spring headband (helmet can be fitted for full protection)

(a)

(b)

12.9 Hand protection

An engineer's hands are in constant use and they run the risk of handling dirty, oily, greasy, rough, sharp, brittle, hot and maybe toxic and corrosive materials. Gloves and palms in a variety of styles and materials are available to protect the hands, whatever the nature of the work.

Gloves are sometimes inappropriate, e.g. for working precision machines, but hands still need to be protected from oil and grime, though not cuts and abrasions, by rubbing them in a barrier cream before starting to work. This is a mildly antiseptic, water-soluble cream which fills the pores of the skin and prevents the ingress of dirt and subsequent infection. The cream is easily removed by washing, which carries away the dirt and removes sources of infection.

12.10 Foot protection

Unsuitable footwear should always be discouraged. It is not only false economy, but extremely dangerous to wear lightweight casual or sports shoes in the workplace. They offer no protection from crushing or penetration. In safety footwear, protection is provided by a steel toecap (inside the boot or shoe) which conforms to a strength specification in accordance with BS 1870. Safety footwear is available in a wide range of styles and prices. It can be attractive in appearance and comfortable to wear. Figure 12.6 shows sections through safety footwear.

Fig. 12.6 *Footwear: (a) lightweight shoes offer no protection; (b) industrial safety shoe; (c) industrial safety boot.*

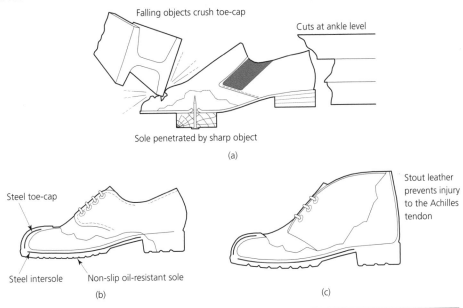

SELF-ASSESSMENT TASK 12.2

List the items of protective clothing and accessories that are normally available for use in workshops, and explain under what circumstances they should be worn or used.

12.11 Health hazards

12.11.1 *Noise*

Excessive noise can be a dangerous pollutant of the working environment. The effects of noise can be:

- Fatigue that leads to carelessness and accidents.
- Mistaken communications that lead to accidents.
- Ear damage that leads to deafness.
- Permanent nervous disorders.

The level at which a noise becomes dangerous varies with its frequency band (pitch) and the length of time the worker is exposed to it. Noise is energy and it represents waste since it is useless. Ideally it should be suppressed at source to avoid waste and to improve the working environment. If this is not possible, the operator should be insulated from it by sound-absorbent screens and/or earmuffs.

12.11.2 *Narcotic (anaesthetic) effects*

Exposure to small concentrations of narcotic substances causes drowsiness, giddiness and headaches. Under these conditions the worker is obviously prone to accidents, since his or her judgement and reactions are adversely affected. Injury can result from falls, and a worker who has become disorientated by the inhalation of narcotics is a hazard to other workers.

Prolonged or frequent exposure to narcotic substances can lead to permanent damage to the brain and other organs of the body, even in relatively small concentrations. Exposure to high concentrations can cause rapid loss of consciousness and end in fatality. Many industrial solvents are narcotic substances; they are used in paint adhesives, polishes and degreasing agents. Careful storage and use is essential and should be closely supervised. Fume extraction and adequate ventilation of the workplace must be provided when working with narcotic substances. Suitable respirators should be available for use in emergencies.

12.11.3 *Irritant effects*

Many substances cause irritation to the skin, both externally and internally. They may also sensitise the skin so it becomes irritated by substances not normally considered toxic.

External irritants can cause industrial dermatitis by coming into contact with the skin. The main irritants in the workshop are oils, particularly coolants, and adhesive solvents.

Internal irritants are the more dangerous as they may have deep-seated effects on the major organs of the body, e.g. inflammation, ulceration, internal bleeding, poisoning and the growth of cancerous tumours. Internal irritants are usually air pollutants in the form of dusts (asbestos fibres), fumes and vapours. They may also be carried into the body on food handled without washing (Section 12.12), or from storing noxious substances in soft-drink bottles without proper labelling. Many domestic tragedies happen this way.

12.11.4 *Systemic effects*

Substances known as *systemics* affect the fundamental organs and bodily functions. They affect the heart, the brain, the liver, the kidneys, the lungs, the central nervous system and the bone marrow. Their effects cannot be reversed and they lead to chronic illness and early death. These toxic substances may enter the body in various ways:

- Dust, vapour and gases can be breathed in through the nose. Observe the safety codes when working with them and wear the respirators provided.
- Liquids and powders which contaminate the hands can be transferred to the digestive system by handling food or cigarettes with dirty hands. Wash before eating.
- Liquids, powders, dusts and vapours may enter the body directly trough the pores in the skin.
- Liquids, powders, dusts and vapours may enter the body by destroying the outer horny layers of the skin and attacking the sensitive layers underneath.
- Liquids, powders, dusts and vapours may enter the body through undressed puncture wounds. Therefore, it is essential to maintain regular washing, use of a barrier cream, use of suitable protective (rubber or plastic) gloves and immediate dressing of cuts, however small.

12.12 Personal hygiene

Personal hygiene is most important. There need be no embarrassment about rubbing a barrier cream into your hands before work, about washing thoroughly with soap and hot water after work, about changing your overalls regularly so they can be cleaned. Personal hygiene can go a long way towards preventing skin diseases, both irritant and infectious. And where gloves would hinder manual dexterity, a barrier cream is sometimes the only protection available.

Skin disease due to continual contact with mineral oil forms the main health hazard in the engineering industry. The effects range from skin irritations and dermatitis to the formation of skin cancers, and will depend upon the type of oil, its temperature, and the degree and duration of exposure. They will also depend upon the condition of the skin, cuts and abrasions from handling rough or sharp components and swarf, irritation from additives, and infection.

The first effect is usually simple irritation accompanied by redness and pimples. If treatment is not sought, the condition deteriorates until cracking, scaling and skin growths appear. Even mild cases may cause the skin to become sensitised, then the operator may need to change to a job where oil and other irritants are not present.

Soluble oils (suds) are particularly difficult as the water content causes the skin to macerate (become soggy); this reduces its natural resistance. If excessive contact with oil cannot be reduced by modification to the plant or process, then additional water- and oil-proof protective clothing must be made available and worn. Dirty and oil-impregnated overalls are also a major source of skin infection. Overalls should be changed and cleaned regularly.

12.13 Behaviour in workshops

Horseplay is reckless, foolish and boisterous behaviour such as pushing, shouting, throwing things and practical joking by a person or group. Horseplay can distract a worker's attention, break concentration and lead to serious – even fatal – accidents. A workshop is no place for such foolish activity, yet it occurs daily in some form in every type of firm.

Horseplay observes no safety rules. It has no regard for safety equipment. It can defeat safe working procedures and undo the painstaking work of the safety officer by the sheer foolishness and thoughtlessness of the participants. The types of accidents due to horseplay depend largely on the work of the factory concerned and the circumstances leading to the accident.

Generally, accidents are caused when:

- A person's concentration is disturbed so that they incorrectly operate a machine or come into contact with moving machinery.
- Someone is pushed or knocked against moving machinery or factory transport.
- Someone is pushed against ladders and trestles upon which people are working at heights.
- Someone falls against and dislodges heavy stacked components.
- Electricity, compressed air and dangerous chemicals are involved.

12.14 Lifting and carrying

As shown by Fig. 12.1, the movement of materials is the biggest single cause of factory accidents. Manual handling accidents can be traced to one or more of the following:

- Incorrect lifting technique.
- Carrying too heavy a load.
- Incorrect gripping.
- Failure to wear protective clothing.

Figure 12.7(a) shows the wrong technique for lifting, which can lead to ruptures, strained backs, sprains, slipped discs and other painful and permanent injuries. The correct technique is shown in Fig. 12.7(b). The back is kept straight and the lift comes from the powerful leg and thigh muscles. Figure 12.7(c) is a reminder that, to avoid falls and injury, the load being carried must not obstruct forward vision.

Fig. 12.7 *Lifting and carrying: (a) an incorrect way to lift; (b) the correct way to lift; (c) the correct way to carry*

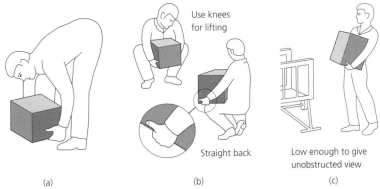

Use knees for lifting

Straight back

Low enough to give unobstructed view

(a) (b) (c)

Protective clothing should be worn when lifting and carrying. Crushed toes caused by dropped loads can be avoided by wearing safety shoes. Cuts and splinters can be avoided by wearing suitable gloves when handling rough and sharp materials. Burns can be prevented when handling caustic and corrosive fluids by wearing face shields and rubber or plastic suits as protection against spillage and splashes.

Team lifting should be employed when handling heavier loads, but remember that there can be only *one captain* to the team and only he or she gives the orders. All members of the team should lift together so as to spread the load evenly. All members of the team should be of similar physique and build.

Heavy loads should be lifted using mechanical lifting gear such as hoists, cranes and fork-lift trucks. These appliances must only be used by persons trained in their use or under the direct supervision of a person trained in their use.

SELF-ASSESSMENT TASK 12.4

1. Explain briefly why safe codes of behaviour in workshops should always be obeyed.

2. Briefly describe the main causes of injury you are likely to sustain when lifting and carrying loads.

12.15 Hazards associated with hand tools

The newcomer to industry often does not realise the potential danger in hand tools that are maintained badly or used incorrectly. Unfortunately the newcomer is often influenced by more experienced workers; experienced workers should know better, but they may still misuse hand tools, often through laziness. The time and effort taken in fetching the correct tool from the stores or in servicing a worn tool is considerably less than the time taken in convalescing from an injury. Figures 12.8 and 12.9 show some of the hazards arising from the incorrect use of hand tools.

Fig. 12.8 *Dangers in the use of hand tools – hand tools in a dangerous condition: (a) hammer faults; (b) hammer faults; (c) spanner faults (d) file faults.*

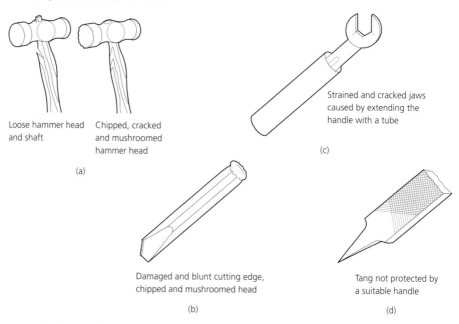

Loose hammer head and shaft

Chipped, cracked and mushroomed hammer head

(a)

Strained and cracked jaws caused by extending the handle with a tube

(c)

Damaged and blunt cutting edge, chipped and mushroomed head

(b)

Tang not protected by a suitable handle

(d)

Fig. 12.9 *Dangers in the use of hand tools – misuse of hand tools: (a) do not use an oversize spanner and packing – use the correct size of spanner for the nut or bolt head; (b) do not use a file as a lever*

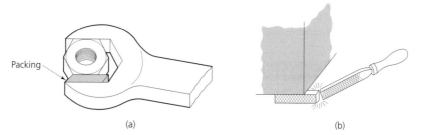

Packing

(a)

(b)

12.16 Hazards associated with portable power tools

Besides maintaining portable power tools in good mechanical order, it is imperative to keep them in good electrical order, especially the power leads, otherwise they could give the user a fatal electric shock. Make sure that every tool is properly earthed or double-insulated. Report the following unsafe conditions and do not use an unsafe tool until it has been repaired by a qualified electrician:

- Defective or broken insulation both to the tool and its flex.
- Improperly or badly made connections to terminals.
- Broken or defective plug.
- Loose or broken switch.
- Sparking brushes.

Do not overload the motor of a portable power tool, as the heat generated will damage the insulation. And do not use a portable electrical tool in the presence of flammable vapours and gases unless it is designed for such use; it is far better to use a tool powered by compressed air. A spark from the switch or motor of an electrically powered tool could cause a fire or an explosion. Under industrial conditions and when out on site, portable power tools should be operated from low-voltage isolating transformers at all times. Normally these transformers have an output potential of 110 volts, but on wet sites, out of doors, it may be as low as 50 volts. The transformer should be fed from a circuit breaker incorporating a residual current detector (RCD). This trips the supply immediately a fault current flows to earth. Since modern practice favours using double-insulated equipment without an earth wire, a residual current detector should be incorporated so that any imbalance in current flow between the live and neutral conductors causes the contactor to trip and isolate the equipment from the supply. An imbalance occurs if a fault current flows to earth by any path, including the body of the user.

12.17 Hazards associated with machine tools

Metal-cutting machines are potentially dangerous. Tools designed to cut through solid metal will not be stopped by fragile flesh and bone.

- Before operating machinery be sure that you have been properly taught how to control it and the dangers associated with it.
- Do not operate a machine unless all guards and safety devices are in position and working correctly.
- Make sure you understand any special rules applicable to the *particular machine* you are about to use, even if you have been trained on machines in general.
- Never clean or adjust a machine whilst it is still in motion.
- Report any dangerous aspect of the machine immediately and stop working it until it has been made safe again by a qualified person.
- A machine may have to be stopped in an emergency. Learn how to make an emergency stop without having to pause and think about it.

12.17.1 *Transmission guards*

By law no machine can be sold or hired out unless all gears, belts, shafts and couplings making up the transmission equipment are guarded so that they cannot be touched whilst in motion. Sometimes guards have to be removed to replace, adjust or service the components they are covering. Before removing guards or covers:

- Stop the machine.
- Isolate the machine from its energy supply.

- Lock the isolating switch so that it cannot be turned on again whilst you are working on the exposed equipment; keep the key in your pocket.
- If it is not possible to lock the isolating switch, remove the fuses and keep them in your pocket.

If an interlocked guard is removed, an electrical or mechanical trip will stop the machine from operating. This trip is only provided in case you forget to isolate the machine; it is no substitute for full isolation. Guards must only be removed or adjusted by a suitably qualified person.

12.17.2 *Cutter guards*

The machine manufacturer does not normally provide cutter guards, because of the wide range of work a machine may have to do.

- It is the responsibility of the owner or the hirer of the machine to supply their own cutter guards.
- It is the responsibility of the operator to make sure the guards are fitted and working correctly before operating a machine, and to use the guards as instructed.
- It is an offence in law for an operator to remove or tamper with the guards provided.
- A technician may have to fit and adjust a guard, or supervise a craftsperson fitting or adjusting a guard or other safety device for an unskilled operator. This is a great responsibility and the technician must thoroughly understand the function and correct fitting of the guard before performing such a duty.
- If ever you are doubtful about the adequacy of a guard, or the safety of a process, consult your safety officer immediately.
- Figure 12.10(a) and (b) shows typical drilling machine guards; Fig. 12.10(c) shows two typical milling cutter guards and Fig. 12.10(d) a typical lathe chuck guard.

12.17.3 *Use of grinding wheels*

Because of its apparent simplicity, the double-ended off-hand grinding machine comes in for more than its fair share of abuse. A grinding wheel does not just 'rub off' the metal; it is a precision multi-tooth cutting tool in which each grain has a definite cutting geometry (Section 2.7). Therefore the grinding wheel must be mounted, dressed and used correctly if it is to cut efficiently.

A grinding wheel that is damaged or incorrectly mounted, or is unsuitable for the machine on which it is mounted, can burst at speed and cause serious damage and injury. Thus the guard for a grinding wheel not only stops the operator from coming into accidental contact with the wheel, but also provides burst containment in case the wheel shatters in use; that is, the broken pieces of wheel are contained within the guard and are not thrown out from the machine at high speed.

The tool rest must also be properly adjusted so there is no chance of the work being dragged down between the tool rest and the wheel. The mounting of grinding wheels, except by a trained and registered person, is prohibited under the Abrasive Wheel Regulations 1970.

Fig. 12.10 *Typical machine guards: (a) simple drill guard; (b) telescopic drill guard; (c) milling cutter guards; (d) lathe chuck guard*

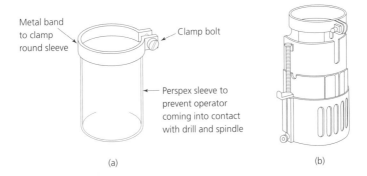

Metal band to clamp round sleeve

Clamp bolt

Perspex sleeve to prevent operator coming into contact with drill and spindle

(a)

(b)

(c)

(d)

SELF-ASSESSMENT TASK 12.5

List the main causes of accidents and state how they may be avoided when:

(a) using hand tools
(b) using machine tools and grinding machines

12.18 Electrical hazards

The installation and maintenance of electrical equipment must be left to qualified electricians. However, all engineers should have an understanding of the principles and practices concerned with the installation and maintenance of electrical equipment and protective devices, so as to avoid misuse of the equipment and to recognise potential dangers.

An electrical shock from a factory power supply can easily kill a man or woman. (Note that most machines operate at 415 volts RMS.) Even if the shock is not severe, the convulsion it causes can throw the victim from a ladder or against a machine, and this can cause serious injury or even death. Be very careful when pulling a victim clear of the fault that caused the shock, as it is possible to receive an equally severe shock from the victim. Always pull the victim by his or her clothing which, if dry, will act as an insulator. Never touch the flesh of the victim until he or she is clear of the fault, as flesh is a conductor.

Artificial respiration must be commenced immediately the victim is pulled clear of the fault or the live conductor. Figure 12.11 gives details of how to treat the victim of severe shock.

12.19 Firefighting

Firefighting is a highly skilled operation and most medium and large firms have properly trained teams to contain the fire locally until the professional service arrives.

The best way you can help is to learn the correct fire drill – how to give the alarm and how to leave the building. It only requires one person to panic and run in the wrong direction to cause a disaster.

In an emergency never lose your head and panic

The main cause of panic is smoke; it spreads quickly through a building, reducing visibility and increasing the risk of falling down stairways. It causes choking and even death by asphyxiation. Smoke is less dense near the floor – as a last resort, crawl. To reduce the spread of smoke and fire, keep fire doors closed at all times but *never locked*. The plastic materials used in the finishes and furnishings of modern buildings give off highly toxic fumes. Therefore it is best to vacate the building and leave the firefighting to the professionals, who have breathing apparatus. Saving human life is more important than saving property.

If you do help to fight a fire, remember some basic rules. A fire is the rapid oxidation (burning) of combustible (burnable) materials at relatively high temperatures. Remove the air or the fuel, or lower the temperature, and the fire goes out. Figure 12.12 shows that different fires need to be dealt with in different ways. The main classes of fire can be correlated to the common portable extinguishers.

Fig. 12.11 *Treatment for electric shock*

Order of action

1. Switch off current
Do this immediately. If not possible do not waste time searching for the switch.

2. Secure release from contact
Safeguard yourself when removing casualty from contact. Stand on non-conducting material (rubber mat, DRY wood, DRY linoleum). Use rubber gloves, DRY clothing, a length of DRY rope or a length of DRY wood to pull or push the casualty away from the contact.

3. Start artificial respiration
If the casualty is not breathing artificial respiration is of extreme urgency. A few seconds delay can mean the difference between success or failure. Continue until the casualty is breathing satisfactorily or until a doctor tells you to stop.

4. Send for doctor and ambulance
Tell someone to send for a doctor and ambulance immediately and say what has happened. Do not allow the casualty to exert himself by walking until he has been seen by a doctor. If burns are present, ask someone to cover them with a dry sterile dressing.

If you have difficulty in blowing your breath into the casualty's lungs, press his head further back and pull chin further up. If you still have difficulty, check that his lips are slightly open and that the mouth is not blocked, for example, by dentures. If you still have difficulty, try the alternative method, mouth-to-mouth, or mouth-to-nose, as the case may be.

Mouth-to-mouth

Lay casualty on back, if immediately possible, on a bench or table with a folded coat under shoulders to let head fall back. Kneel or stand by casualty's head. Press his head fully back with one hand and pull chin up with the other.

Breathe in deeply. Bend down, lips apart and cover casualty's mouth with your well open mouth. Pinch his nostrils with one hand. Breathe out steadily into casualty's lungs. Watch his chest rise.

Turn your own head away. Breathe in again.

Repeat 10 to 12 times per minute.

If the patient does not respond

 NORMAL PUPILS   DILATED PUPILS

Check carotid pulse, pupils of eyes and colour of skin.
Pulse present, pupils normal – continue inflations until recovery of normal breathing (Steps 1, 2 and 3.)

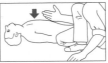

Pulse absent, pupils dilated, skin grey – strike smartly to the left part of breast bone with edge of hand.
Response of continued pulse, pupils contract – continue inflations until recovery of normal breathing.

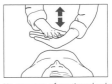

No response of continued pulse, pupils unaltered, skin grey – commence external heart compression.

When normal breathing commences, keep warm, place casualty in the recovery position.

External heart compression

1. Place yourself at the side of the casualty.

2. Feel for the lower half of the breastbone.

3. Place the heel of your hand on this part of the bone, keeping the palm and fingers off the chest.

4. Cover this hand with the heel of the other hand.

5. With arms straight, rock forwards pressing down on the lower half of the breastbone (in an unconscious adult it can be pressed towards the spine for about one and a half inches (4 cm)).

6. The action should be repeated about once a second

Continue as above until a continued pulse is felt and pupils contract.
Continue inflations until recovery of normal breathing

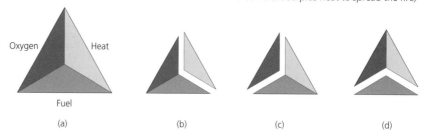

Oxygen   Heat

Fuel

(a)   (b)   (c)   (d)

12.19.1 *Water*

Used in large quanitites, water reduces the temperature and puts out the fire. The steam generated also helps to smother the flames; by displacing the air, it removes the oxygen that is needed for burning. Water should only be used on burning solids such as wood, paper and some plastics; there are various technical reasons for this.

12.19.2 *Foam extinguishers*

Foam is used for fighting oil and chemical fires; the foam smothers the flames, preventing oxygen in the air from feeding the fire. Water alone cannot be used as the burning fuel floats on its surface and spreads the fire.

Since both water and foam are electrically conductive, do not use them on fires associated with electrical equipment, otherwise the person wielding the hose or the extinguisher could be electrocuted.

12.19.3 *Carbon dioxide (CO_2) extinguishers*

Carbon dioxide is used on burning gases and vapours. They can also be used for oil and chemicals in confined places. The carbon dioxide gas replaces the air and smothers the fire. It can only be used in confined spaces where it is not dispersed by drafts.

If the fire cannot breathe then neither can you, so care must be taken to evacuate living creatures from the vicinity before operating the extinguisher. Back away from the bubble of CO_2 gas as you operate the extinguisher; do not advance into it.

12.19.4 *Vaporising liquid extinguishers*

Many vaporising liquids have acronyms, e.g. CTC, CBM and BCF. The heat from the fire causes rapid vaporisation of the liquid spray from the extinguisher, and this vapour smothers the fire. It will also smother living creatures unless precautions are taken. This type of extinguisher is suitable for oil, gas, vapour and some chemical fires. Like carbon dioxide, vaporising liquids are safe to use on fires associated with electrical equipment.

12.19.5 *Dry powder extinguishers*

Dry powders are suitable for small fires involving flammable liquids and small quantities of solids such as paper. They are also suitable for fires in electrical equipment. The main active ingredient is powdered sodium bicarbonate (baking powder), which gives off carbon dioxide when heated. They are suitable for offices and canteens as the powder leaves little mess and does not contaminate foodstuffs when it comes into contact. Residual powder can be removed with a vacuum cleaner.

12.19.6 *Containing a fire*

Since a fire spreads rapidly, it can only be contained by quick action. Sound the alarm and send for assistance before attempting to fight a fire. Remember:

- Extinguishers are only provided to fight small fires.
- Take up a position between the fire and the exit, so that your escape cannot be cut off.
- **Do not** continue to fight the fire if it is dangerous to do so.
- **Do not** continue to fight the fire if there is a possibility that the escape route may be cut off by fire, smoke or collapse of the building.
- **Do not** continue to fight the fire if the fire spreads despite your efforts.
- **Do not** continue to fight the fire if toxic fumes are being generated by the burning of plastic finishes or furnishings.
- **Do not** continue to fight the fire if there are gas cylinders or explosive substances in its vicinity.

If you have to withdraw, close windows and doors behind you wherever possible. Ensure that extinguishers are recharged immediately after use.

12.20 Fire prevention

Prevention is always better than cure, and fire prevention is better than firefighting. Tidiness is of paramount importance in reducing outbreaks of fire. Fires have small beginnings and many of them originate amongst rubbish left lying around. So make a practice of constantly removing rubbish, shavings, offcuts, cans and bottles, waste paper, oily rags and other unwanted materials to a safe place at regular intervals. Discarded foam plastic packing, now widely used, is not only flammable but gives off highly dangerous toxic fumes when burnt.

Highly flammable materials should be stored in specially designed and equipped compounds away from the main working areas. Only minimum quantities should be allowed in the workshop at a time, and then only into non-smoking zones. Seek advice from the local authority fire prevention officer.

It is good practice to provide metal containers for rubbish, preferably with airtight, hinged lids, and with proper labels to describe their contents; some types of rubbish will ignite spontaneously when mixed. The lids of the bins should be kept shut so that any fire inside will soon use up the limited air available and go out of its own accord without doing any damage.

Liquid petroleum gases (LPG) such as propane and butane are increasingly used for process heating and space heating in workshops and on site. Full and empty cylinders should be stored separately in isolated positions away from the working areas and shielded from the sun's rays. There should be plenty of air around them with free circulation and ventilation. They should be protected from frost. Bulk storage in fixed containers, e.g. large cylinders and spheres, requires secure fencing as an extra precaution and the containers should be defended against damage from passing vehicles. Pipe runs, joints and fittings for an LPG supply must be regularly inspected by a qualified person, and any flexible tubing connected to gas cylinders should be regularly inspected for cuts and abrasions and replaced as necessary.

Special procedures govern the storage and use of gas cylinders for welding equipment; they are dealt with in Section 12.21.

SELF-ASSESSMENT TASK 12.6

1. Describe the possible consequences of badly installed and badly maintained electrical equipment.

2. For any company or institution with which you are familiar, describe the arrangements for:

   (a) giving the alarm in the event of fire
   (b) fire prevention
   (c) firefighting
   (d) evacuation of the premises

12.21 Gas welding hazards

12.21.1 *Eye injuries*

Eye injuries are caused by glare from the incandescent (white-hot) weld pool and splatter from molten droplets of metal. They can be prevented by wearing proper welding goggles (Fig. 12.13). The goggles need to have the correct filter glasses for the welded materials and

Fig. 12.13 *Oxyacetylene welding goggles: goggles with lenses may be specified for gas welding or cutting but they are not suitable for arc welding*

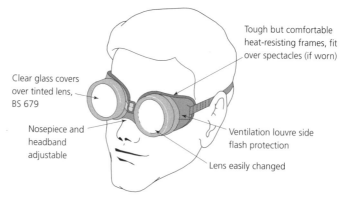

Tough but comfortable heat-resisting frames, fit over spectacles (if worn)

Clear glass covers over tinted lens, BS 679

Nosepiece and headband adjustable

Ventilation louvre side flash protection

Lens easily changed

the flux because both metal and flux affect the radiation. Note that gas welding goggles offer no protection during arc welding; the two methods produce different kinds of radiation.

12.21.2 *Burns*

Careless handling of the torch and the hot metal can lead to burns and other fire hazards. Sparks may also occur. Ensure all rubbish and flammable materials are removed from the working area. Wear fire-resistant overalls but avoid cuffs, which present traps for sparks and globules of hot metal. For the same reason, fasten overalls at the neck. Wear a leather apron and leather spats, and if the conditions are very hot, wear gloves.

12.21.3 *Explosions*

Explosions are caused by the improper storage and use of compressed gases and gas mixtures. They can occur when acetylene gas is present in the air at any proportion between 2 and 82 per cent. Acetylene explodes spontaneously at high pressure even without air or oxygen. The working pressure for acetylene should not exceed 620 millibars. Explosions in the equipment itself may result from *flashbacks* due to improper use or lighting-up procedures, incorrect setting of the equipment or faulty equipment. Flashback arrestors must therefore be fitted and regularly inspected. An exploding acetylene cylinder is equivalent to a large bomb. It can demolish a building and kill the occupants. Great care must be taken in their use.

Gas cylinders are not themselves dangerous as they are regularly inspected to government standards. However, the following safety precautions should be observed:

- Cylinders must be protected from mechanical damage during storage, transportation and use. *Acetylene cylinders must always be kept upright when in use.*
- Cylinders must be kept cool. On no account should the welding flame be allowed to play on the cylinder or any other part of the equipment. They must also be protected from extremes of hot sunlight and frost.
- Cylinders must be sited in well-ventilated surroundings to prevent the build-up of explosive mixtures of gases. Even slight oxygen enrichment can cause spontaneous ignition of clothing and other flammable materials, leading to fatal burns. Explosions can occur with acetylene concentrations as low as 2 per cent in air.
- Correct automatic pressure regulators must be fitted to all cylinders before their use. The cylinder valve must always be kept closed when the equipment is not in use, and whilst changing cylinders or equipment.
- Keep cylinders free from contamination. Oils and greases can explode spontaneously in the presence of pure oxygen. Similarly, oil or dirty cloths can ignite spontaneously in an oxygen-enriched atmosphere.

These notes are the barest outlines of the safety precautions for gas welding. Not only must you understand the safe handling of cylinders and gases, but also the correct procedures for lighting up and shutting down welding equipment. BOC Gases issues a number of booklets on welding safety; they should be studied and fully understood before handling gas welding equipment.

12.22 Arc welding hazards

The hazards of mains-operated welding equipment are set out in Table 12.1.

Table 12.1 *Electrical hazards of arc welding*

Fault	Hazard
High-voltage primary	
Damaged insulation	Fire: loss of life and damage to property
	Shock: severe burns and loss of life
Oversize fuses	Overheating: damage to equipment and fire
Lack of adequate earthing	Shock: if fault develops, severe burns and loss of life
Low-voltage primary*	
Lack of welding earth	Shock: if a fault develops, severe burns and loss of life
Welding cable: damaged insulation	Local arcing between cable and any adjacent metalwork at earth potential, causing fire
Welding cable: inadequate capacity	Overheating: damage to insulation and fire
Inadequate connections	Overheating: severe burns and fire
Inadequate return path	Current leakage through surrounding metalwork may cause overheating and fire

*Very high current.

To eliminate these hazards as far as possible, the following precautions should be taken. These are only the basic precautions, so check whether the equipment and working conditions require any extra precautions.

- Make sure the equipment is fed from the mains via an isolating switch-fuse incorporating a residual current detector (RCD) as described in Section 12.12. Easy access to this switch must be provided at all times.
- Make sure the trailing high-voltage primary cable is armoured against mechanical damage and heavily insulated against a supply potential of 415 volts.
- Make sure that all cable insulation and armouring is undamaged and in good condition and that all terminations and connecting plugs and sockets are also secure and undamaged. If in doubt, do not operate the equipment until it has been checked and made safe by a skilled electrician.
- Make sure that all the equipment and the work is adequately earthed with conductors capable of carrying the high currents used in welding.
- Make sure the welding current regulator has an 'off' position so that, in the event of an accident, the welding current can be stopped without having to trace the primary cable back to the isolating switch.
- Make sure the external welding circuit is adequate for the high currents it has to carry.

For all arc welding operations it is essential to protect the welder's head, face and eyes from radiation, spatter and hot slag, using either a helmet or a hand shield (Fig. 12.14). The injurious effects of the radiation emitted by an electric arc are similar for AC and DC welding currents. Exposing the head, face and eyes to infrared (heat) rays would lead to the welder becoming uncomfortably hot and would lead to serious eye troubles. If too much ultraviolet radiation is received by the welder, or anyone else in the vicinity, it can cause effects similar to sunburn, and a painful condition known as arc-eye. Too much visible light from the incandescent weld pool will dazzle the operator, but too little can cause eyestrain and headaches.

Fig. 12.14 *Eye and face protection for arc welding: (a) helmet; (b) hand shield*

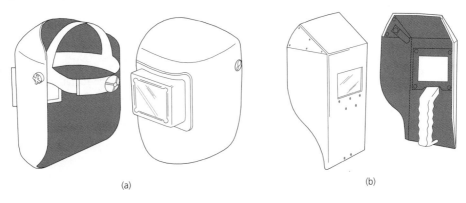

(a)　　　　　　　　　　　　　　　　　(b)

The obvious precaution is to prevent harmful radiation from the welding arc and the molten weld pool from reaching unprotected skin and eyes. Eye protection is by special glass filters of suitable colour and density. Fitted in the face mask or helmet, they not only absorb harmful radiation but also reduce the intensity of the visible light. The expensive filter glass is protected by a cover glass on the outside. This cover glass is clear and toughened.

The slag left by the flux used when arc welding has to be chipped away when the weld has cooled down. Clear goggles should be worn whilst chipping, as the slag breaks away in glass-like splinters. Protective screens should be provided so that adjacent workers are not put at risk.

The welder's body and clothing must be protected from radiation and burns. The minimum protection should be a leather apron. Wherever possible each arc welding station should be screened in such a way as to keep stray radiation to a minimum, either by the use of individual cubicles or portable screens.

Although the electrodes are flux coated, the welder is not likely to suffer any ill effects from welding fumes, provided there is reasonable ventilation. Localised ventilation can be provided by a suction fume extractor; this not only dilutes and removes fumes but assists in keeping down the temperature and adds to the comfort and efficiency of the welder. Extraction should be at a low level so that the fumes are not drawn up past the face of the welder.

1. With reference to appropriate manufacturers' literature concerning oxyacetylene welding equipment:
   (a) list the safety equipment that should be used
   (b) briefly describe the correct way to set up oxyacetylene welding equipment ready for use
   (c) briefly describe the correct way to light up and shut down oxyacetylene welding equipment.

2. Summarise the precautions that should be taken when using manual metal-arc welding equipment.

12.23 Workshop layout

Much can be done to prevent accidents by the layout of the machines and equipment in a workshop. Figure 12.15 shows a well laid-out workshop. Notice there are ample gangways, clearly marked and free of obstructions. The machines are arranged so that bar stock does

Fig. 12.15 *Layout of a machine shop*

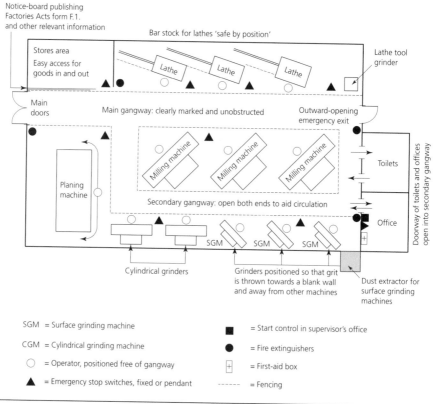

SGM = Surface grinding machine
CGM = Cylindrical grinding machine
○ = Operator, positioned free of gangway
▲ = Emergency stop switches, fixed or pendant

■ = Start control in supervisor's office
● = Fire extinguishers
⊞ = First-aid box
------ = Fencing

not protrude into the gangways. They are also arranged so that operators cannot be distracted by other workers constantly passing close by.

Grinding machines are arranged so that grit is *not* thrown towards other machines and so that dust extraction out of the working area can easily be provided. There is easy access to emergency stop switches for the whole shop, and fire extinguishers are strategically placed.

Scale models of the more common types and makes of machine tools are available, and these can be used in conjunction with scale plans of the workshop floor for experimenting with different layouts in order to achieve an efficient and safe workshop.

EXERCISES

12.1 Briefly explain why goggles should be worn:
  (a) when machining
  (b) when oxyacetylene welding

12.2 Describe the essential differences between the type of goggles worn when machining, the type of goggles worn for gas welding and the visor used when arc welding.

12.3 Is a machine manufacturer legally required to supply transmission guards, cutter guards or both?

12.4 Describe with the aid of sketches what is meant by an interlocked guard and how a typical guard works.

12.5 Explain briefly why motorised machine tools should be fitted with:
  (a) an isolating switch
  (b) no-volt protection.

12.6 Describe the correct way to lift a load manually without suffering back injury.

12.7 Describe in detail six checks that should be made to ensure a portable power tool is electrically safe to use.

12.8 Describe the precautions that should be taken to ensure a reasonable level of personal hygiene when working in an engineering workshop. State three harmful effects that may result from not taking reasonable precautions to maintain personal cleanliness.

12.9 (a) Name the type of fire extinguisher that should be used in the following circumstances, giving reasons for your choice:
    (i) burning oil
    (ii) a fire in electrical equipment
    (iii) a fire involving general rubbish (waste paper, wood, etc.)
  (b) Describe the action you should take if you discover a fire has broken out in a building in which you are working.

12.10 A workmate has been rendered unconscious by electric shock. Describe how you would:
  (a) separate the victim from the source of the shock
  (b) render artificial respiration

Index